Entomology

Current Status and Future Strategies

The Editors

Dr. Arijit Ganguly is presently working as an Assistant Professor of Zoology in the Achhruram Memorial College, Jhalda, Purulia in West Bengal, India. He has completed his master and PhD degrees from Visva-Bharati University, Santiniketan, West Bengal, India. He was further trained as a postdoctoral fellow in Rajiv Gandhi University, Arunachal Pradesh, India. Since last ten years, Dr. Ganguly is working in the field of utilisation of insects as food for livestock. He has published more than 20 research articles in the journals of international repute, and authored one book. He has visited Akdeniz University in Turkey for academic purpose.

Khokan Naskar is an Assistant Professor of Zoology, Achhruram Memorial College, Jhalda, Purulia, West Bengal, India. For several years he has been investigating different behavioural aspects of ants occurring in and around Kolkata, West Bengal. His publications on ants are impressive. He obtained his M.Sc. degree from University of Calcutta and taught in a higher secondary school as Biology teacher for approximately 8 years before joining as an Assistant Professor. He has a keen interest regarding foreign languages and learnt German language for three years from Ramakrishna Mission Institute of Culture, Kolkata. He is also a licensed amateur radio operator. His call sign is VU3KHO.

Entomology

Current Status and Future Strategies

— *Chief-Editor* —

Dr. Arijit Ganguly

— *Associate Editor* —

Khokan Naskar

2018

Daya Publishing House®

A Division of

Astral International Pvt. Ltd.

New Delhi – 110 002

ISBN: 9789387057593 (International Edition)

Publisher's Note:

Every possible effort has been made to ensure that the information contained in this book is accurate at the time of going to press, and the publisher and author cannot accept responsibility for any errors or omissions, however caused. No responsibility for loss or damage occasioned to any person acting, or refraining from action, as a result of the material in this publication can be accepted by the editor, the publisher or the author. The Publisher is not associated with any product or vendor mentioned in the book. The contents of this work are intended to further general scientific research, understanding and discussion only. Readers should consult with a specialist where appropriate.

Every effort has been made to trace the owners of copyright material used in this book, if any. The author and the publisher will be grateful for any omission brought to their notice for acknowledgement in the future editions of the book.

Published by : **Daya Publishing House®**
A Division of
Astral International Pvt. Ltd.
– ISO 9001:2015 Certified Company –
4736/23, Ansari Road, Darya Ganj
New Delhi-110 002
Ph. 011-43549197, 23278134
E-mail: info@astralint.com
Website: www.astralint.com

Dedication

This book is dedicated to Dr. Rene Arzuffi, one of the contributors of this book, whose sudden demise took us by storm while this volume was in press. May his soul rest in peace.

Dr. Arijit Ganguly
Khokan Naskar

Foreword

As insects are the numerically dominant animal group of the planet, they represent an important biomass in the ecosystems; this is the reason why it is important to gather knowledge, among other aspects, on their taxonomy, biodiversity, behavior, conservation and the impact they have in the health of various ecosystems.

Moreover, these animals have lived in close association with man since time immemorial, as they have been used as food for animals, for example, poultry and human societies. On one hand they are beneficial with medicinal or nutraceutical values, and on the other hand they are vectors of varied diseases, as well as devastating pests, as in the case of the locusts whose integral biological or chemical control is necessary in order to minimize their effects on ecosystems and crops.

Because of the aforementioned reasons, insects are the object of basic and applied research in the universities, institutes and industries of the whole world. In this context, the objective of this book is to encourage the interest in the study and research on insects in the fields that have been little addressed so far, and ultimately to appraise the significance of these animals, which are amply distributed throughout the world. Finally, we are sure that readers will enjoy the contents of this book whose information can be used by experts worldwide to enrich their projects.

Dr. Jose manuel Pino Moreno

Faculty of Science

National Autonomous University of Mexico

Preface

We are glad to present our edited volume entitled *"Entomology: Current status and Future Strategies"* to the Etomology enthusiasts. This book has an international flavour in true sense because 23 contributors from 3 continents (*i.e.* Asia, Africa, and the Latin America) came in a single platform to share their works. We have a firm belief that the information provided here would be useful to the students, researchers, scholars, faculty members and all who are interested in insect science.

Dr. Arijit Ganguly

Khokan Naskar

Contents

Section 7: Insect Pest Management

The Contributors

Abhijit Mazumdar

Dr. Abhijit Mazumdar is a Professor in Department of Zoology, University of Burdwan, India. He is working on Bionomics, Taxonomy and Systematics of Dipteran flies (Chironomidae and Ceratopogonidae) and Strepsiptera. At present, his central research focus is on bionomics of *Culicoides* sp. and their life history strategy. Dr. Mazumdar has a glorious academic and research carrier. He has produced 9 Ph.D. students and has published more than 100 articles in various International journals of repute. Dr. Mazumdar obtained his Ph.D. in Zoology from The University of Burdwan, India. He has authored numerous scientific publications, book, book chapters and has conducted researches in collaboration with international institutions.

Alejandro García Flores

Dr. García earned his PhD on Agricultural Sciences and Rural Development and he is a Professor of Ethnozoology in the Faculty of Biological Sciences and is the chairperson of Theory of Social Management at the Research institute in the Humanities and Social Sciences of the Autonomous University of the State of Morelos, Mexico. He also holds the position of Associate Research Professor "C" in Ecology Laboratory of Biological Research Center. Currently he is engaged with the Academic Management Body of Traditional Production Units. He is a member of the state system of researchers. His research is concerned with social development and use of bio-cultural diversity.

Ana Isabel Arias Barrera

She is a food chemist from the Faculty of Chemistry at the National Autonomous University of Mexico having a keen interest on the role of various chemicals as nutraceuticals found in insects. She has worked on the development of new food products named "palanqueta" from grasshoppers. She has also investigated the difference between consumption in mice when fed on decaffeinated and normal coffee. Currently she is looking for viable and safe food alternatives in order to improve the diet of Mexican people, as well as extending the shelf life and food safety.

Arijit Ganguly

Dr. Arijit Ganguly is presently working as an Assistant Professor of Zoology in the Achhruram Memorial College, Jhalda, Purulia in West Bengal, India. He has completed his masters and PhD degrees from Visva-Bharati University, Santiniketan, West Bengal, India. He was further trained as a postdoctoral fellow in Rajiv Gandhi University, Arunachal Pradesh, India. Since last ten years Dr. Ganguly is working in the field of utilisation of insects as food for livestock. He has published 20 research articles in the journals of international repute, and authored one book. He has visited Akdeniz University in Turkey for academic purpose.

Atanu Seni

Dr. Atanu Seni is working as an Assistant Professor in Orissa University of Agriculture and Technology, Odisha. He has completed his Graduation from Visva Bharati, Santiniketan and Post Graduation from Punjab Agricultural University, Ludhiana. He joined Ph.D. programme in Bidhan Chandra Krishi Viswavidyalaya, Mohanpur, West Bengal. He has more than 4 years of experience in research, teaching and extension in the field of entomology and pest managment. He worked on some important insect pests such as *Helicoverpa armigera, Paracoccus marginatus* and different rice insect pests. Till now, he has published 18 research papers and 1 book chapter in various national and international journals. He is life member of several scientific societies in the field of entomology and plant protection and also associated with various journals as referee. Presently, he is working as Scientist under All India Coordinated Rice Improvement Project.

Chandrik Malakar

Dr. Chandrik Malakar is an Assistant Professor in the Department of Zoology in Suri Vidyasagar College. He had completed his M. Sc with specialization in Environmental Biology and accomplished his PhD from Visva Bharati University. His PhD topic was "Effects of heavy metals on the development and reproductive traits of *Oxya fuscovittata* (Marschall) (Acrididae:Orthoptera)." He has published many national and international research articles in the field of insect biology and toxicology.

Dipanwita Das

Dr. Dipanwita Das is presently working as Assistant Professor in Bagnan College, Howrah, West Bengal, India. Previously she has worked as Assistant Professor in the Department of Zoology, Jagannath Kishor College, Purulia, West Bengal, India. Dr. Das has completed her Ph.D. work on ecology and microbial control of the dengue vector *Aedes aegypti* from the University of Burdwan, West Bengal, India. Presently she is working in the field of vector biology, ecology and biodiversity of different parts of West Bengal.

Hiroj Kumar Saha

Dr. Hiroj Kumar Saha is an Assistant Professor in the Post Graduate Department of Zoology, Bethune College, Kolkata. He obtained his PhD degree in acridid diversity and ecosystem health from Visva-Bharati Central University, West Bengal, Bengal, India. He has published his papers in six national and five international journals.

Humberto Reyes Prado

Dr. Humberto Reyes Prado is an accomplished biologist, currently working at the Autonomous University of Morelos, Mexico. Dr. Prado has completed his PhD studies from the National Polytechnic Institute (IPN), Mexico. His research interest includes insect chemical ecology, with special emphasis in the study of antennal activity through electro-antennography, and gas chromatography coupled to electro-antennography as a tool in the characterization of pheromones and plant volatiles. He is an experienced worker in the field of sexual behavior and chemical ecology of moths and fruit flies, as well as host-seeking behavior and oviposition of herbivorous insects of agricultural importance. He is a member of the National System of Researchers, Mexico and Mexican Entomological Society.

Jaydeep Halder

Dr. Jaydeep Halder is currently working as a Scientist (Agricultural Entomology) at ICAR-Indian Institute of Vegetable Research, Varanasi, Uttar Pradesh. He has studied B.Sc (Ag) from Bidhan Chandra Krishi Viswavidyalaya (B.C.K.V.), Mohanpur, Nadia, West Bengal; M.Sc (Ag) Entomology from Acharya N.G. Ranga Agricultural University (A.N.G.R.A.U.), Hyderabad; and Ph.D in Entomology from Indian Agricultural Research Institute (I.A.R.I.), New Delhi. He was awarded several scholarships (Merit Certificates under National Scholarship Scheme for secondary and higher secondary examinations; BCKV Merit Scholarship) and fellowships (ICAR-JRF and IARI Senior Scholarship) during his study period. He has qualified ICAR- NET. His areas of interest include insect toxicology and biological control of major vegetable insect pests. He has so far published more than 45 research papers in many national and international journals of repute, 5 book chapters, 3 review papers, 3 lead/status papers, 3 training manuals, 40 research abstracts, 30 popular articles, 27 training chapters, 6 extension folders and 7 technical bulletins. He is the recipient of *"Young Scientist Award"* by the Society of Scientific Development in Agriculture and Technology (SSDAT), Uttar Pradesh. He is also a member of editorial board of an international journal and reviewer of many national and international journals. He is a fellow of the Society for Biocontrol Advancement, Bangalore and Society of Plant Protection Sciences, New Delhi.

Jose Manuel Pino Moreno

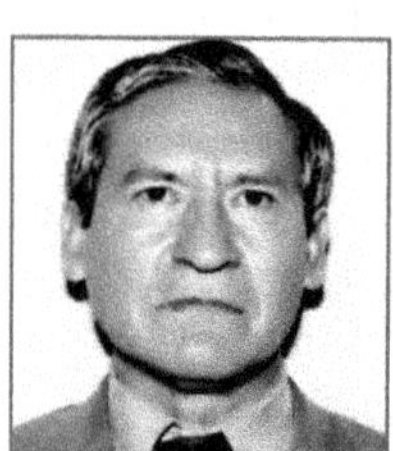

Dr. Pino completed his Masters of Science from the Faculty of Sciences of the National Autonomous University of Mexico. Later he earned his PhD from the Pacific Western University, USA. At present he is engaged as a Technical Academician at the Institute of Biology, and Professor of the Faculty of Science at the National Autonomous University of Mexico. His research is popularly known as— Insects as a protein source for the future. He has published 75 articles in national and international journals, 17 book chapters and co-authored 3 books and 38 broadcast articles. He has participated in 117 national and 86 international congresses.

Khokan Naskar

Khokan Naskar is an Assistant Professor of Zoology, Achhruram Memorial College, Jhalda, Purulia, West Bengal, India. For several years he has been investigating different behavioural aspects of ants occurring in and around Kolkata, West Bengal. His publications on ants are impressive. He obtained his M.Sc. degree from University of Calcutta and taught in a higher secondary school as Biology teacher for approximately 8 years before joining as an Assistant Professor. He cleared several national level competitive examinations such as NET, SET, GATE, ICMR-JRF. He has a keen interest regarding foreign languages and learnt German language for three years from Ramakrishna Mission Institute of Culture, Kolkata. His hobbies are reading story books and listening to music. He has an off-bit hobby which is dxing. Here "d" means distance, " x"- unknown station, listening to unknown long distance radio stations and send them reception report for verification. He is also a licensed amateur radio operator. His call sign is VU3KHO.

Nilotpol Paul

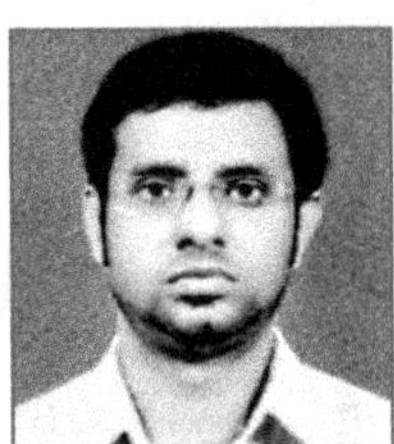

Dr. Nilotpol Paul is a district entomologist, Department of Health and Family Welfare, Rampurhat HD. He also acted as a teaching faculty (Contractual) of Animal Science, Department in Kazi Nazrul University, Asansol, India. He is working on Taxonomy and Systematics of Dipteran flies (Chironomidae and Ceratopogonidae). At present, his central research focus is on immature systematics and life history strategy of Dipteran flies. Dr. Paul obtained his Ph.D. in Zoology from The University of Burdwan, India. He has authored numerous scientific publications and has conducted researches in collaboration with international institutions.

Norma Robledo

Dr. Norma Robledo is a professor of National Polytechnic Institute (IPN), Mexico. Dr. Robeldo graduated as a Biochemist Engineer from the Autonomous Metropolitan University, Iztapalapa. Later, she completed her M.Sc. from the Autonomous University of Morelos and Ph.D. from IPN. She teaches courses of Insect Chemical Ecology, Phytochemistry, and Agro-ecological Management of Pests and Diseases.Her area of research interest includes study of semiochemicals for agricultural insect pest management, plant secondary metabolites, functional foods and environmental bioremediation.

Parimalendu Haldar

Dr. Parimalendu Haldar is an accomplished entomologist and was a Professor of Zoology in Visva-Bharati University, India. He has a broad area of interest in the fields of Entomology, Ecology, and Environmental Biology. Prof. Haldar has successfully completed several Govt. of India funded research projects and came out with 12 PhD, 1 edited book and 82 research articles. For his research in the field of grasshopper mass production as an alternative food source, he was awarded by the Orthopterists' Society, USA. Later, he was also invited as an expert member of FAO (UNO) at Rome. Zoological Society, Calcutta, and West Bengal Academy of Science and Technology conferred fellowship to him. His academic visits to abroad include: Thailand, China, France, Turkey, and Italy.

Rafael Monroy Martínez

Rafael Monroy Martínez is currently the head of the Academic Management Traditional Production Units. He earned his Master of Science (Biology) from the National Autonomous University of Mexico and is engaged as a Research Professor of Ecology in the Biological Research Center of the University of the State of Morelos, Mexico. He is a specialist in environmental impact on ecological lands. His research interests include – social development and use of bio-cultural diversity through the application of ethnobiological techniques, applied ecology and sustainable community development. He is also a member of the state system of researchers.

Ranita Chakravorty

Ranita Chakravorty is currently employed as an Assistant Grade-III (Technical), Food Corporation of India. She completed her M.Sc. in the year 2009 from Visva Bharati University with Fish Biology as special paper. Later she joined as a project fellow in the Zoology Department with the topic Acridid Biodiversity in the Grasslands in and around Santiniketan. Ranita Chakravorty has attended two international seminars and published three articles in international journals.

René Arzuffi

Dr. René Arzuffi is a professor of CEPROBI, National Polytechnic Institute (IPN), Mexico since 1997, and presently is the coordinator of the Master's course of "Agro-ecological Pest and Disease Management", also he is a member of National System of Researchers and of the Entomological Society of America. Dr. Arzuffi graduated as a biologist from the Autonomous University of Mexico, and later received his M.Sc. and PhD from the National Polytechnic Institute (IPN), Mexico. Dr. Arzuffi teaches courses on "Experimental design and statistical analysis", "Insect Chemical Ecology" and "Management of agro-based pest insects and diseases". His area of interest includes the study of physiology and ethology of arthropods, particularly insects. His major research focus is the use of semiochemicals for insect management of agricultural importance.

Sandra D. Barreto Sánchez

She has earned her Masters' in Natural Resource Management, and is a Collaborator of the Laboratory of Ecology of the Center for Biological Research of the Autonomous University of the State of Morelos, Mexico.

Santanu Ray

Dr. Santanu Ray is presently working as a professor of Zoology in Visva-Bharati University, India. Dr. Ray had been trained by some stalwarts of ecology like Prof. Sven Erik Jorgensen, Prof. Bernard Christopher Patten and Prof. Eugene P. Odum, and himself is a world renowned ecologist with special interest in integrated ecosystem theories and their application, ecological phenomenology and ecological modeling. Prof. Ray has an excellent research profile with publication over 100 research articles in prestigious international journals, 2 books and 11 book chapters. He has completed 3 research projects and supervised 8 PhD students and 3 postdoctoral scholars till date. He is the Associate Editor of the journal - Ecological Modelling - An International Journal on Ecological Modelling and system ecology, published by ELSEVIER, Netherlands, and Energy Ecology and Environment, published by SPRINGER. He is also the editorial board member of various prestigious journals published by Taylor and Francis, ELSEVIER, *etc.* For conducting research, teaching and delivering academic lectures he has visited countries like Czech Republic, Germany, Denmark, Sweden, USA, Japan, Italy, Norway, China, Sri Lanka, France *etc.*

Sory Cisse

Dr. Sory Cisse is a Locust Information Officer in Centre National de Lutte contre le criquet pelèrin, in Mali, Africa. He is also working on geographic information system (GIS) focused on pest dynamics monitoring, remote sensing analysis, satellite images management and forecasting. At present his central research focus is on desert locust management, because since old times desert locust has threatened agricultural production in Africa (particularly in Mali), the Middle East, and Asia. Desert locusts are well known as swarming locust which is characterized by density dependent continuous phase polyphenism and most dangerous of all pests from Africa to South west Asia via the Middle East.

Dr. Cisse obtained his PhD in plant pest management from Institut Agronomique et Vétérinaire Hassan II (IAV) in Marocco. He has authored numerous scientific publications and has conducted several researches in collaboration with international institutions such as CIRAD in France. He's still working with CIRAD team on little-known aspects of the dynamics of the desert locust populations, the preventive control strategy and land monitoring with satellite eyes.

Srimanta Kumar Raut

Dr. Srimanta Kumar Raut is a senior faculty member at the Department of Zoology, University of Calcutta. He is engaged in research on different aspects of ecology and ethology of animals occurring in and around West Bengal, India since 1974. He has published many papers in National and International Journals with good impact factor. Moreover, he has contributed "Chapters" in books of National and International Standard, and published Monographs. He is the editor of Proceedings of Zoological Society, Kolkata, being published by Springer.

Sudipta Mandal

Dr. Sudipta Mandal is a young Zoologist presently working as Assistant Professor, Department of Zoology, Bangabasi College, Kolkata, West Bengal, India. Previously he has worked as Assistant Professor and Head, Department of Zoology, Dasaratha Deb Memorial College, Khowai, Tripura for more than six years. He graduated from Maulana Azad College, Kolkata and did his masters from Presidency University, Kolkata. Dr. Mandal received his PhD degree from The University of Burdwan, West Bengal in 2013. Dr. Mandal is a nature lover and passionate about wildlife photography. Presently he is working on the documentation of biodiversity of North East India and West Bengal.

Editorial

Chapter 1

From the Chief Editor's Desk

☆ *Arijit Ganguly*

The ancestors of modern day insects came into being about 400 million years ago, much before the dinosaurs' reign on the earth. They were also the foremost animals to fly, conquering the sky for the first time. Since then these creatures thrived in the planet and are obviously the most successful group of animals which is exhibited by their super diversity and adaptive styles. Scientists have described more than one million insect species till date, and many are waiting to be discovered.

Insects have a very interesting relationship with human. We have been fascinated by the colourful wings of a butterfly, we are amazed by the twinkling lights of a group of fireflies flashing in the dark, we love the sugary honey in our breakfast, we are disgusted by the monotonous but persistent buzzing of the housefly spoiling our sweet slumber in a lazy summer afternoon, we are scared of the swarming locust speculating the loss of our crop and possible impact on economy, we have been frightened by the sight of a bullet ant fearing its painful sting, some of our ancestors even have worshiped insects for the good that it bestows upon us. Yes, there are so many species of insects that only a few of them create menace, but there are innumerable numbers of them living in our surroundings that are silently working behind the scene to make this world beautiful. It is because of these interesting facts man tried to understand these marvellous creations of nature for long.

Initial goal of Entomology *i.e.* study of insects was "pest control". Scientists studied taxonomy, internal and external anatomy of pest insects, their physiology and all only to understand and design a control measure for them. Many chemical pesticides were designed that later proved to be harmful for the environment and human as well. Another group of scientists were engaged with some beneficial insects that offer us some products to exploit such as, honey, silk, lac *etc.* which lead to overexploitation of these insects. Then we started rearing of these species in mass scale. Gradually people began to understand the importance of the presence of insects and their key role in the ecosystem. Since then the concept of pest control changed into "pest management" which means maintaining insect population under a threshold level. Instead of using chemical pesticides people were suggested to utilize bio-pesticides that could be obtained from various plants in the form of extracts or semiochemicals. From this point of time study of the documentation of various species from various habitats, diversity, reproductive biology, behaviour, ecology, conservation *etc.* were conducted. Later on various ethnic groups around the world were found to consume insects as food and medicine. Scientific studies revealed the components that can cure many diseases. Some others were proved to be extremely rich in nutrients.

Modern day entomology has gone far from only pest control and exploitation to estimation of biodiversity and conservation. On one hand agricultural and medical entomologists are trying to enlist various pest and disease causing insects and attempts are being made to curb their impacts on crop and human health in an eco-friendly manner, so that non-target species are not affected. On the other hand environmental entomologists are trying their best to evaluate the impacts of the changing world on the insect populations and how to mitigate them. In this context we felt a need to come out with a complete work comprising both these aspects of entomology that could offer a wide range of works for the scientists, scholars and students of insect biology. Thanks to the scientists and research scholars from around the world who have contributed chapters in this book and we could finally present our book *"Entomology: Current Status and Future Strategies"* that is divided into seven segments: Insects diversity and conservation, Environmental toxicology of insects, Insect taxonomy, Insect behaviour, medicinal value of insects, Insects as food for human and livestock, and insect pest management.

In the first segment you will find a general discussion on the biodiversity and conservation of insects with emphasis on India, next will be chapter on grasshopper diversity and yet another one that describes how acridid grasshoppers plays an important role in conserving forest ecosystem health. In the second segment you will discover how heavy metal pollution is impacting grasshopper population. The third segment will be dealing with a wide taxonomical review on the order Diptera that have medical importance. In the fourth segment the foraging behaviour of two ant species has been revealed. In the fifth segment you will gain knowledge about nutraceuticals and medicinal insects that is yet a very less explored field. In the sixth segment we have tried to put some lights on the edible insects of Latin American countries, which will be followed by a discussion on a very special group of Mexican edible hemipteran: The Jumiles. The last chapter of this segment will disclose nutrients and anti-nutrients of a potential food insect from India. In the concluding seventh segment ideas of pest control using semiochemicals will be revealed that are most recent eco-friendly way for pest management. This segment will be concluded with a discussion on the management of a very special and devastating group of insects: The Locusts.

With this note I hope that the readers will enjoy reading this book which is written in a very lucid way and the information in this book will be utilized by various workers on insect science across the globe.

– Segment 1 –

Insect Diversity and Conservation

Chapter 2

Insect Diversity and Needs for its Conservation with Special Emphasis on the Indian Subcontinent

☆ *Dipanwita Das and Sudipta Mandal*

ABSTRACT

Insects are the most diverse group of organisms that comprises 66 per cent of all the animals in the world. Insects are considered as one of the most important group of organisms in the world that dominate the food webs of the terrestrial as well as aquatic ecosystems and help to maintain the stability and diversity of most of the ecosystems. They may influence the life and welfare of humans in various ways – while some insects cause harm as pests and carrier of disease causing organisms, others are beneficial to humans. However, as a result of the economic and population growth and development, world insect diversity is facing serious problems like habitat degradation, global climate change, pollution, invasive alien organisms and over exploitation causing species extinction. Considering huge diversity of insects and their worldwide distribution, a major portion of the insect species of the world are yet to be described taxonomically. Proper knowledge base about the huge diversity of insects and their role in ecosystem processes are essential to design conservation plans for insects which will help the world towards sustainable development.

1. Introduction

Insects, defined by the possession of three pairs of walking legs and two pairs of functional and/or non-functional wings, conquered almost each habitat of the world except the benthic zone of the oceans (Grimaldi and Engel, 2005). Insects dominate the food webs of the terrestrial as well as aquatic ecosystems and help to maintain the stability and diversity of the species and the ecosystem wholly. They play important functional roles in their respective niches and are significant contributors to a variety of ecosystem processes (Obrist and Duelli, 2010; Chapman, 2012; Santorufo *et al.*, 2012). Entomological studies provide better understanding about genetics, ecology, evolution, developmental biology, physiology, ethology, biomechanics and climate change. Insects are ideal for studying natural sciences (Wigglesworth, 1976), and are directly related with human since evolution began. Culture of *Bombyx mori* was started before 4700 BC (Konishi and Ito, 1973). Insect products widely used by human include lac from lac insects (*Laccifer lacca* Kerr), honey and wax from honeybees (*Apis indica, Apis mellifera etc.*), and the dye- cochineal from *Dactylopius coccus* Costa *etc.* Insects, especially honeybees play a major role in plant pollination (Scudder, 2009). However these ecologically and economically important creatures may act as vectors by transmitting diseases in many organisms including human as well. A group of insects, known as "pests" are responsible for huge economic loss for human society.

Insects account for a large proportion of the animal species found on the earth and even the exact number of species of insects still not known. The class "Insecta" encompasses 66 per cent of all the animals in the world (Chandra, 2013). The number of identified insect species is about 900,000- 10,20,007 till date (Scudder, 2009; Chandra, 2013; The Columbia Encyclopedia, 2015) which is supposedly increasing each year.

Considering the origin and evolution of life on the earth, insects may be crowned as the most successful animal group, representing around two to ten million species on earth existing at the present time. The success of this animal group lies behind their ability to survive many geological events during the last few millions of years. But the scenario is changing dramatically in recent times due to huge anthropogenic disturbances, impact of which is estimated to be threatening the existence of more than 25 per cent of all insect species (Samways, 2005). Less than 10 per cent of the insects present on the earth have been taxonomically described and named so far. There is threat of extinction for many insect species before being discovered. There is an urgent need to protect and conserve this enormous diversity of life which is so vital to many ecosystem functions.

1.1 Diversity Among Different Insect Orders

According to different schemes of classification, there are 27-32 orders under the Class Insecta (Table 1.1). Coleoptera is the largest order among class Insecta (38 per cent of the insect species of the world) (165 families) and is followed by Hemiptera (92 families), Diptera (87 families), Lepidoptera (121 families) and Hymenoptera (65 families).

1.1.1 Coleoptera

Coleoptera, which consists of beetles, are the most diverse orders among the described fauna of the world. About 3,00,000 to 4,50,000 coleopterans are illustrated (Zhang, 2011) and 165 families are included in this order. Evolution of elytra replacing the forewing is the key factor for their immense diversity as this sclerotized hard elytra protect the group from harsh environment (Scudder, 2009). Beetles occur in both terrestrial and aquatic habitats and even marine environments (Doyen, 1976; Lawrence and Britton, 1994).

Table 1.1: Present Scenario of Insect Diversity (Chandra, 2013)

Insect Orders	*Number of Species Described in World (Zhang, 2011)*	*Number of Species in India (ZSI, 2012)*	*Number of Species in West Bengal (Chandra, 2013)*
Archaeognatha	513	10	-
Zygentoma	561	28	9
Ephemeroptera	3240	124	17
Odonata	5899	463	185
Dermaptera	1978	298	79
Plecoptera	3788	116	21
Embioptera	464	31	8
Phasmida	3029	144	28
Orthoptera	24,276	1033	278
Mantodea	2400	174	35
Blattaria	7314	186	23
Isoptera	2864	271	55
Psocoptera	5720	105	39
Phthiraptera	5102	400	35
Thysanoptera	6019	686	124
Hemiptera	1,03,590	6479	966
Coleoptera	3,87,100	17,455	1570
Raphidioptera	254	5	-
Megaloptera	354	25	-
Neuroptera	5868	312	79
Hymenoptera	1,16,861	12,605	130
Mecoptera	757	23	3
Siphonaptera	2075	46	13
Strepsiptera	609	21	14
Diptera	1,59,294	6337	681
Trichoptera	14999	1046	79
Lepidoptera	1,57,424	15000	1020
Total	**10,20,007**	**64,423**	**5818**

1.1.2 Lepidoptera

Lepidoptera, which include butterflies and moths, are highly diverse orders. Being larger they are easy to identify and sampling. Lepidopterans occur in different types of habitat and can indicate endemism in landscape (Scudder, 2009). The estimated number of butterfly species is more than 1,55,208 because of the large number of newly discovered species in the tropics (Scudder, 2009). There are 121 families under the order Lepidoptera.

1.1.3 Hymenoptera

Among the order Hymenoptera, bees, ants and wasps are familiar with the common people but there are more. They are highly diversified morphologically. Pagliano (2006) estimated about 1,40,470 species among this order, though according to Zhang (2011), the exact number is 1,16, 861 (Table 1.1).

1.1.4 Diptera

Possession of single pair of wings in the dipteran adults makes the difference from other insect fauna. Mosquitoes, midges, fruit flies, horse flies, house flies and black flies are included in the Diptera. About 1,59,294 species are described among Diptera (Zhang, 2011).

1.2 Aquatic Insects

Like terrestrial ecosystems, aquatic ecosystems are also dominated by the class Insecta. They invade almost every types of aquatic habitat like lakes, streams, coastal waters, estuaries, ground water and even hot springs (Yule and Young, 2004). About 45,000 insect species have been reported from the freshwater ecosystems (Balaram, 2005). Insect orders which mainly dominate the aquatic ecosystems are Coleoptera, Hemiptera, Odonata, Diptera, Trichoptera, Ephemeroptera and Heteroptera. They form an important node of aquatic food web and are very good indicators of water quality as some of them are extremely sensitive to pollutants and some are resistant (Arimoro and Ikomi, 2008; Hepp *et al.*, 2013; Prommi and Payakka, 2015), however, some of them act as pathogen vectors (Chae *et al.*, 2000).

1.3 Diversity of Insects around the World

Biogeographical realms get their own identity through the differences in community structure, species richness and species abundance in the local assemblages among them (Davis *et al.*, 2002a; 2004). Tropical areas are the most diversified regions of the world and this stands true for insect species. About 1,00,000 insect species are described from Afrotropical region (Miller and Rogo, 2001). Amazonian rainforests harbour almost 85 per cent of insects (May, 1998) among the 1,00,000 described arthropod species, of which 25 per cent are beetles (Erwin, 2004). The highest insect diversity is found in Brazil (Rafael *et al.*, 2009). The β diversity value for the insects living in the canopy layer of the tropical rain forest is quite high which alters according to the change in latitude and altitude and also with climatic gradients (Novotny *et al.*, 2006). Surprisingly, the tropical lowland rainforest of Papua New Guinea harbours low β diversity for the herbivorous insects and this might be true for all the tropical lowland forests (Novotny *et al.*, 2007). Although the numbers of the identified insects are vast but the most surprising fact is that almost 80-95 per cent of insect species are still waiting for detection (Stork, 2007). According to Scudder (2009), among the 1,44,000 insect species of the Nearctic region only 65 per cent are described in details. Arctic regions are poorer than the Southern areas (Mexico, Arizona, Florida and Texas). Insect biodiversity of this region is low compared to other biogeographical realms for example, Palearctic region. Insect fauna of the Australian Region are quite well characterised. Yeates *et al.* (2003) estimated 2,05,000 species of insects which include 12,000 Hemiptera,

20,000 Lepidoptera, 30,000 Diptera and 40,000 Hymenoptera in Australian Region. Coleopterans are the largest order and they occupy 40-50 per cent of the total estimated insects. Though about 1,00,000 species are identified worldwide but the detail record about their biology are still unknown which is very essential for understanding their role in global ecosystem (Stork, 2007).

1.4 Indian Scenario

Insect community of India is enormous which is expected for a mega-diversity country. Insect fauna of India comprises 7.06 per cent of the total insect fauna of the world (Chandra, 2011). In the year 1909, Maxwell and Howlett reported 25,700 insect species in the book entitled 'Indian Insect Life'. Beeson (1941) and Menon (1965) enlisted the number of insect species in India as 40,000 and 50,000 respectively. Then in 1998 Varshney enumerated 59,353 species in India. Chandra (2011) reported 61,181 number of insect species from India in the book entitled 'Biodiversity of Sikkim'. Presently, 63,423 species of Class Insecta comprising 631 families under 27 orders are reported from India (Chandra, 2013). Among the 27 insect orders present in India, 17,455 Coleoptera, 15,000 Butterfiles and Moths, 12,605 Hymenoptera, 6,337 Diptera, and 6479 species of Hemiptera are described (Varshney, 1998; Alfred, 2003; Kazmi and Ramamurthy, 2004; Chandra, 2013) (Table 1.1). Till now new insect species are being reported in each year from India; for example, 31 Hemiptera-, 13 Hymenoptera-, 19 Coleoptera-, 12 Orthoptera- and 5 Lepidopteran species have been first time reported in January 2015 (ZSI e-news, Jan. 2015). Around 5,818 insect species, which represents 6.83 per cent of the recorded insect fauna of India, are reported from West Bengal, India (Sarkar *et al.*, 2015). Almost 80 per cent of insects in India are endemic (Varshney, 1998; Murugan, 2006). Varshney (1998) estimated 20,717 insect species as endemic in India among which Hymenoptera comprises for 9600 species and is followed by Coleoptera (3100), Hemiptera (2421), Diptera (2135), Lepidoptera (1500) and others (Vershney, 1998). The greatest insect diversity in India is found in Sikkim (5941), followed by West Bengal (5818), Meghalaya (5118) and Uttarakhand (4160) (Chandra, 2013). These four States are quite well explored but other States such as Assam, Kerala, Tamil Nadu, Arunachal Pradesh, Odisha, Rajasthan, Chandigarh, Goa, Lakshadweep, Puducherry *etc.* are either under-explored or unexplored (Chandra, 2013). Though large numbers of Insects are found to inhabit Indian forests and agricultural field, the ratio of insects to flowering plants is low due to higher plant biodiversity in this country.

1.5 Factors Related to the Diversity of Insects

The incredible diversity of insects is the result of multiple aspects such as ability to metamorphose, possession of wings, evolutionary interactions with plants and animals *etc.* Presence of wings is one of the major factors for long term existence of these creatures. Evolution of flight mechanism helps them easily avoid the predators and to search for suitable habitat.

As insects are r-selected species, their average size is very small and generation time is short. The small size of insects provides more niches than the large sized animals and this leads to more diversification. The shorter generation time induce to propagate in large number which in turn increases the chances for mutation.

Almost half of the whole insect world is dependent on plants for foraging. Diversity of plants in the tropics provides diverse habitats which create the opportunity for niche shift and this leads to speciation or in other word, diversification (Simpson, 1949; 1953). The highly diverse coleopterans are mostly herbaceous and feed on angiosperms. Due to repeated origins of angiosperm-feeding, beetles get opportunity to diversify widely (Futuyma and Scheffer, 1993; Farrell, 1998). Both the angiosperms and beetles diversify in a parallel path. Novotny *et al.* (2006) suggested that the latitudinal gradient in insect species richness might be directly related with plant diversity and the β diversity value increased almost seven times from temperate to tropical areas. The higher β diversity for the tropical insects results from the ecological specialization (Dyer, 2007; Janson, 2008).

About 75 per cent of the Hymenopterans are parasitoid and parasitism play crucial role in Hymenopteran diversity (Whitfield, 2003).

2. Conserving Insect Diversity: An Urgent Need

Insects on the earth are facing threats of extinction for a variety of reasons, with habitat destruction and fragmentation the worst threat, especially in the tropics where most insect diversity lives (Samways, 2009). Other major threats include use of pesticides and certain biological control practices, invasive alien organisms, genetically modified crops and global climate change. Synergistic effects as a result of interaction among many of these threats may aggravate the situation and expedite insect extinction. Special efforts are necessary to counteract these threats. Application of landscape planning along with special plans for certain species may be helpful.

2.1 Why Should we Conserve Insect Diversity? Importance of Insects

McKelvey (1973) rightly stated that "Insects nurture and protect us, sicken us, kill us. They bring both joy and sorrow." Insects are important from various aspects. They play important ecological roles; they have profound impact on crop yield and agriculture as well as human and animal health. They are also important as natural resources because of their huge diversity. Insects have been used in revolutionary studies in different field of science like biomechanics, climate change, developmental biology, ecology, evolution, genetics, paleo-limnology, and physiology (Scudder, 2009).

2.1.1 Insects Acting as Keystone Organisms

Insects may be considered as the pillars for biological foundation of all terrestrial ecosystems. They are the natural pollinators and disperse seeds, cycle nutrients, help in maintaining soil fertility, acts as regulators to control populations of other organisms, and provide a major food source for other taxa (Carpenter, 1928; Majer, 1987). They are virtually key components in any food web of a terrestrial or freshwater ecosystem (Shurin *et al.*, 2005). Considering multifaceted ecological roles played by insects they are well designated as Keystone organisms. Their impact on some terrestrial ecosystems is so vital that those ecosystems may completely collapse within few weeks in their absence.

2.1.2 Ecosystem Engineers

Insects act as Ecosystem Engineers by influencing many ecological processes. Insects also have huge impact on the physical and chemical properties of soils, especially in arid areas. For example, the Australian funnel ant *Aphaenogaster longiceps* is reported to be able to move about 80 per cent of the soil to the surface that is moved by all soil fauna combined (Samways, 2009). Termites are other important engineers, and their nests are reported to cover almost 10 per cent of the land surface in West Africa. They influence the global carbon cycle as they have been known to produce 1.5 per cent of carbon dioxide and 15 per cent of methane produced from all sources in tropical forests. The action of termites can be so extensive that they can influence plant communities (Samways, 2009).

2.1.3 Natural Pollinators

Insects are important natural pollinators and survival of many plant species are dependent on particular insect species because of the mutualism between specialist insect and specialist flower (Samways, 2009). Loss of insect diversity may negatively affect crop yield for the same reason.

2.1.4 Population Regulators

Insects have great influence on many aspects of population dynamics; they regulate population by acting as predators, parasites, or parasitoids (Tamarin, 1978).

Apart from the above mentioned roles played by insects they also possess esthetical value that beautify our nature like butterflies, dragonflies, damselflies, moths and many other insects.

2.2 Threats to Insects

Many insect species are facing the threat of extinction all over the globe. Though the current extinction rate of insects is not clearly known, estimates based on extrapolation suggest that 11200 species of insects have gone extinct since 1600, and there is a probability that possibly half a million insect species may become extinct in the next three centuries (Samways, 2009). Other estimates also suggest very high extinction rates, with 25 per cent of all insect species present on the earth today on the verge of extinction (McKinney, 1999).

Major threats to the insect includes habitat loss, degradation, and fragmentation, invasive alien species and diseases, pollution, pesticides and global climate change. Most of these are common in the tropical rain forests, which are home to at least half of all insect species.

2.2.1 Habitat Loss, Degradation and Fragmentation

Habitat loss, degradation and fragmentation are serious threats to insects. Major reasons which attributes for this kind of threats include agriculture, resource extraction, and urban and suburban development. Most of the species are habitat-specific, and the loss of habitat that provides sites for overwintering, foraging for pollen and nectar, or nesting can be detrimental to these species (New, 2009).

Deterioration of habitat quality known as habitat degradation is another major concern. For example, ground-nesting insects require loose, sandy soil for making their nests which may be crushed by heavy foot movement or the use of off-road vehicles. In metropolises, these insect species may be particularly limited due to the large amount of landscape that has been covered with concrete, building materials or other water-resistant surface materials (New, 2009).

Large, intact areas of habitat may be broken up into smaller, isolated patches by anthropogenic activities like conversion of natural forest or grasslands into new agricultural fields, other developmental activities like construction of dams or industry and road construction. All of these lead to habitat fragmentation which adversely affects many insect species.

2.2.2 Invasive Alien Species

Plants or animals which are not native and brought to a place from elsewhere are called invasive or alien species. They can decrease the quality of the habitat. If non-native plant species replaces native plant species that might result into scarcity of feeding and/or nesting plants needed by certain insect species like pollen, nectar, or larval food. For example, if the food plant of butterfly larvae is replaced by some alien plant species the young caterpillars fail to thrive. In other cases, alien species may compete with indigenous plants or animals for resources. European honey bees have been shown to compete with native bees for pollen and nectar (New, 2009)

2.2.3 Diseases

Diseases caused by virus, bacteria, protozoa or other parasites are another serious threat to many insects. The effects of these parasites are usually species-specific. Parasites introduced by human for biological control may severely affect other non-target species.

2.2.4 Pollution and Pesticides

Air pollution may cause serious problem for those insects that rely on scent trails to find food sources like nectar containing flowers. Bees and other insects are reported to be unable to locate their food source in areas with high levels of air pollution (New, 2009). Nocturnal insects like moths are affected by light pollution. When they are attracted to artificial lights at night they become more susceptible to predation by bats or birds.

Indiscriminate use of pesticides is another serious threat to insects, use of pesticides containing non-biodegradable or persistent chemicals are particularly detrimental for insect diversity as these poisonous chemicals remain in the environment for a long time causing problems for many non-target insect species. Systemic insecticides applied to seeds can contaminate the pollen grains that are an essential source of food for pollinators and their larvae (New, 2009). Pollution of water bodies from agricultural runoff or other sources severely affects aquatic insects and insects whose life cycles are dependent on water bodies. Herbicides used for weed control may destroy important adult and larval food plants adding to the problem.

2.2.5 Global Climate Change

As for most of the species on the earth insects are also likely to be affected by global climate change. Climate change may cause shifting of many plant species from tropical to temperate regions or from low altitude to high altitude areas. If insects fail to synchronise with this migration they may be deprived of their nectaring or larval food plants while plants may also face problem due to lack of their pollinators. Increase temperature and other climatic changes may negatively affect the life cycle of many insects.

2.2.6 Taxonomic Challenges

There is huge taxonomic void as far insect taxonomy is concerned as less than 10 per cent of all insects have scientific names till date. Even majority of them have not been biologically described. In short, the vast majority of insects remain unknown, and getting them described before they go extinct, is the taxonomic challenge (Samways, 2009). So there is need to name and study insects so that they can be conserved.

2.2.7 Lack of Awareness

Not all insects are looked upon in a similar way. While many of us are very fond of colourful butterflies few if any have a similar affinity for cockroaches, the word 'cockroach' even being used in disrespect. The underlying problem is that humans are by nature inclined to the idea that insects at large are carriers of disease or at least irritants. Samways (2009) have rightly pointed out that all organisms have the right to live, and this includes the vast majority of insects that neither impact on humans nor are even for a moment in the collective conscience of humankind.

2.3 Strategies for Conserving Insect Diversity

Conservation of biodiversity is a complex and comprehensive issue. Though insects are enormously diverse taxonomic group distributed worldwide, conservation measures taken till date to protect insect diversity is limited to certain geographical areas and have been emphasized on groups of insects with large body size. According to Samways (1993) insect conservation has to be in a broad scale which should cover all possible types of landscapes and biotopes like high mountains, caves, deserts, forests, grasslands, islands and river systems which are liable to be affected by anthropogenic disturbances and natural calamities.

2.3.1 Habitat Conservation at Global and Regional Levels

The key to insect conservation is to protect and enhance or create habitats (if and when necessary) throughout the landscape, so that even the least mobile species can find somewhere to live when conditions become locally unsuitable.

At least 25 areas that are hotspots of world bio-diversity and that are also threatened (Myers *et al.*, 2000) have been identified at the global scale. 17 countries have been designated as mega diversity countries including India. These are very likely to be major areas for insect diversity. Various conservation practices have been initiated in most of these regions of the world which should be strengthened further.

At the regional scale, insects have a role in systematic conservation planning. This planning aims to identify locations and landscapes with high insect diversity in a particular region that is selected as a priority for conservation action (Pressey *et al.*, 1993). The identification and protection of keystone species are vital to the conservation of an ecosystem.

Managing and maintaining the legally protected areas like Sanctuaries and National parks are not enough; preservation of all the insect habitats should be given priority for a successful conservation strategy. Maintaining heterogeneity of land-scape and undisturbed or minimally disturbed habitats will definitely help in maintaining insect diversity.

2.3.2 Reducing Pollution and Pesticide Use

Even if their habitats are conserved, insects may be harmed by pesticides and other forms of pollution of water, soil and air. Only 1 per cent of the insects is harmful or causes problem for humans. Unfortunately pesticides, the only economically practicable means of controlling pests often harms non-target species, many of which are beneficial to farmers and growers. Harm from pesticides can be reduced by using methods such as integrated pest control, in which different methods are combined so as to minimise or eliminate the dosage. If pesticides have to be applied, they should be used cautiously to minimise the harm to non-target organisms and possible contamination of soil and water bodies.

2.3.3 Legal Framework for Conservation of Biodiversity in India

As insects are only a part of the diverse groups of living organisms in any given ecosystem, it is impossible to protect insects as an independent group without taking the entire ecosystem into account (You *et al.*, 2005). India has an extensive body of constitutional provisions, laws and policies to promote conservation and sustainable use of biodiversity and natural resources. The Indian Constitution clearly assigns responsibilities between the Union and State Governments (Part XI and article 246) on various subjects. India is a signatory to various international conventions and treaties related to environmental protection and have also taken numerous initiatives towards their implementation. The most relevant national policies and legislation are the Biological Diversity Act of 2002, National Policy and Macro Level Action Strategy on Biodiversity of 1999, National Forest Policy of 1988, National Water Policy of 2002, National Environmental Policy (NEP) of 2006, Indian Forest Act of 1927 (and related state legislation), Forest (Conservation) Act of 1980, Wildlife (Protection) Act of 1972, Environmental (Protection) Act of 1986, Schedule Tribes and other Traditional Forest Dwellers (Recognition of Forest Rights) Act of 2006, Environmental Impact Assessment Notification of 2006, Factories Act of 1948, Mines and Minerals (Development and Regulation) Act of 1957, Energy Conservation Act of 2001, Air (Prevention and Control of pollution) Act of 1981,Water (Prevention and Control of pollution) Act of 1974. One of the most significant recent legislative steps taken by the Government of India has been the setting up of the National Green Tribunal (NGT). In addition, the Supreme Court of India has also played a significant role in the conservation of biodiversity. Under Article 32 and Article 226, the Supreme Court and the High Court have played a proactive role in the conservation of biodiversity. In 2013, the Supreme Court of India set up a 'Green Bench' to deal with environmental issues replacing the existing Forest Bench (Anonymous, 2014). All these Acts and laws should be applied properly to protect habitats, reduce pollution and preserve biodiversity which will definitely help in the conservation of insect diversity.

Conclusions

Healthy biodiversity of insects can provide mankind with food, fibres, medicines and other valued materials. Values of many insects not evident at the present time may be found to have important use in the future (Vasseur *et al.*, 2002). For example, the genetic information contained in many insect species may provide important clues for pest management, utilization of beneficial species and conservation of rare species (You *et al.*, 2005). Increased knowledge may even be of use in the research of molecular biology and genetic engineering. Biodiversity of insects, therefore, is one of the most important components of a life supporting system (You *et al.*, 2005). Therefore it is obvious that loss of insect species and their genetic diversity will cause the degradation and collapse of ecosystems that will threaten our existence on earth. The conservation of insect diversity has, therefore, become an urgent task for entomologists and all citizens (You *et al.*, 2005).

The main challenge for conservation of habitat and biodiversity is over exploitation by local people which could significantly alter these ecosystems. Laws should be implemented properly and awareness among common people and students should be prioritized for a long term solution. Farmers should be educated about the problems of pesticide use and encouraged for biological pest control and integrated pest management.

An effective participation and initiative from all the stake holders for conservation of insect diversity will definitely help India as well as the world to move a step forward towards sustainable development.

References

Alfred, J.R.B., 2003. Diversity, dimension and significance of insects: an overview in the Indian context. India: Proceedings of the national Symposium on Frontier Areas of Entomological Research, Nov. 5-7. IARI, New Delhi.

Anonymous, 2014. India's fifth National Report to the Convention on Biological Diversity. Ministry of Environment and Forests, Government of India.

Arimoro, F.O., Ikomi, R.B., 2008. Ecological integrity of upper Warri River, Niger Delta using aquatic insects as bioindicators. Ecol. Indicat. 395, 1-7.

Balaram, P., 2005. Insect of tropical streams. Curr. Sci. 89, 914.

Carpenter, G. H., 1928. The Biology of Insects. Sidgwick and Jackson, London, pp. 473.

Chae, S.J., Purstela, N., Johnson, E., Derock, E.S., Lawler, P. and Madigan, J.E., 2000. Infection of aquatic insects with trematode metacercariae carrying *Ehrilichia risticii*, the case of the Potomac house fever. J. Med. Entomol. 37, 619-625.

Chandra, K., 2011. Insect diversity of Sikkim, India. Biodiversity of Sikkim- Exploring and Conserving a Global Hotspot. Arrawatia ML, Tambe S. Information and Public Relations Department, Government of Sikkim.

Chandra, K., 2013. Insect species diversity in Indian states and union territories: An Introduction. e- Newsletter West Bengal Biodiversity Board 4(2),11–18.

Chapman, P.M., 2012. Adaptive monitoring based on ecosystem services. Sci. Total Environ. 415, 56-60.

Davis, A.L.V., van Aarde, R.J., Scholtz, C.H., and Delport, J.H., 2002a. Increasing representation of localised dung beetles across a chronosequence of regenerating vegetation and natural dune forest in South Africa. Global Ecol. Biogeogr. 11, 191–209.

Doyen, J.T., 1976. Marine beetles (Coleoptera excluding Staphylinidae) in: L. Cheng (ed.), Marine Insects. North-Holland Publishing, Amsterdam, pp. 497–519.

Dyer, L.A., Singer, M.S., Lill, J.T., Stireman, J.O., Gentry, G.L., Marquis, R.J., Ricklefs, R.E., Greeney, H.F., Wagner, D.L., Morais, H.C., Diniz, I.R., Kusar, T.A., Coley, P.D., 2007. Host specificity of Lepidoptera in tropical and temperate forests. Nature. 448, 696-699.

Evenhuis, N.L., Pape, T., Pont, A.C., and Thompson, F.C. (eds.), 2007. Bio Systematics Database of World Diptera, Version 9.5. http://www.diptera.org/biosys.htm [Accessed 20 January 2008.

Farrell, B.D., 1998. "Inordinate Fondness" Explained: Why are There so Many Beetles? Science. 281(5376), 555-559.

Futuyma, D.J.M.C.K., Scheffer, S.J., 1993. Genetic constraints and the phylogeny of insect-plant associations: responses of Ophraella communa (Coleoptera: Chrysomelidae) to host-plants of its congeners. Evolution. 47, 888–905.

Grimaldi, D., and Engel, M.S., 2005. Evolution of the Insects, Cambridge University Press, New York, pp. 755.

Hepp, L.U., Restello, R.M. and Milesi, S.V., 2013. Distribution of aquatic insects in urban headwater streams. Acta Limnologica Brasiliensia 25(1), 1-9.

Janson, E.M., Stireman, J.O., Singer, M.S., Abbot, P. and Mauricio, R., 2008. Phytophagous insect-microbe mutualisms and adaptive evolutionary diversification. Evolution. 62, 997–1012.

Kazmi, S.I. and Ramamurthy, V.V., 2004. Coleoptera (Insecta) fauna from the Indian Thar Desert, Rajasthan. Zoo's Print J. 19, 1447-1448.

Konishi, M. and Ito, Y., 1973. Early entomology in East Asia, in: History of Entomology. Annual Reviews Inc. Palo Alto, California. pp. 517.

Lawrence, J.F., and Britton, E.B., 1994. Australian Beetles, Melbourne University Press, Australia. pp. 184.

Majer, J.D., 1987. The conservation and study of inverte-brates in remnants of native vegetation, in: D. A. Saunders, G. W. Arnold, A. A. Burbridge, and A.

J. M. Hopkins (eds.), Nature Conservation: The Role of Remnants of Native Vegetation, Surrey Beatty and Sons, Sydney, pp. 333–335.

Maxwell, L.H., and Howlett, F.M., 1909. Indian Insect Life, Thacker, Spink and Co., Calcutta. pp. 786

May, R.M., 1988. How many species are there on earth? Science 241, 1441–1449.

McKelvey, J.J.Jr., 1973. Man against Tsetse. Struggle for Africa, Cornell University Press, Ithaca, New York. pp. 306

McKinney, M.L., 1999. High rates of extinction and threat in poorly studied taxa. Conserv. Biol. 13,1273–81

Menon, M.G.R., 1965. Systematics of Indian insects, in: Entomology in India (supplement), Entomological Society of India, Delhi, pp. 70-87.

Miller, S.E. and Rogo, L.M., 2001. Challenges and opportunities in understanding and utilization of African insect diversity. Cimbebasia. 17, 197–218.

Murugan, K., 2006. Biodiversity of Insects- Meeting Report. Curr. Sci. 91(12), 1602-1603.

Myers, N., Mittermeier, R.A., Mittermeier, C.G., da Fonseca, G.A.B., Kent, J., 2000. Biodiversity hotspots for conservation priorities. Nature. 403, 853–58.

New, T.R., 2009. Insect species conservation, Cambridge University Press.

Novotny, V., Drozd, P., Miller, S.E., Kulfan, M., Janda, M., Basset, Y., Weiblen, G.D., 2006. Why are there so many species of herbivorous insects in tropical rainforests? Science 313(5790), 1115-1158.

Novotny, V., Miller, S.E., Hulcr, J., Drew, R.A.I., Basset, Y., Janda, M., Setliff, G.P., Darrow, K., Stewart, A.J.A., Auga, J., Isua, B., Molem, K., Manumbor, M., Tamtiai, E., Mogia, M., Weiblen, G.D., 2007. Low beta diversity of herbivorous insects in tropical forests. Nature. 448, 692-695.

Obrist, M.K., and Duelli, P., 2010. Rapid biodiversity assessment of arthropods for monitoring average local species richness and related ecosystem services. Biodiv. Conserv. 19, 2201-2220.

Pagliano, G., 2006. Gli Imenotteri del Piemonte. www.storianaturale.org/anp/pagliano.htm.

Pressey, R.L., Humphries, C.J., Margules, C.R., Vane-Wright, R.I., Williams, P.H., 1993. Beyond opportunism: key principles for systematic reserve selection. Trends. Ecol. Evol. 8,124–128.

Prommi, T., Payakka, A., 2015. Aquatic Insect Biodiversity and Water Quality Parameters of Streams in Northern Thailand. Sains Malaysiana 44(5), 707–717

Rafael, J.A., Aguiar, A.P., Amorim, D.deS., 2009. Knowledge of insect diversity in Brazil: challenges and advances. Neotrop. Entomol. 38(5), 565-70.

Samways, M.J., 1993. Insects in biodiversity conservation: some perspectives and directives. Biodiv. Conserv. 2, 258-282.

Samways, M.J., 2005. Insect Diversity Conservation. Cambridge University Press, New York, pp. 342.

Samways, M.J., 2009. Insect Conservation, in: Kleber Del Claro, Paulo S. Oliveira, Victor Rico-Gray (Eds.), Tropical Biology and Conservation Management - Volume 7, Eolss Publishers Co. Ltd. Oxford, United Kingdom.

Santorufo, L., VanGestel, C.A.M., Rocco, A., Maisto, G., 2012. Soil invertebrates as bioindicators of urban soil quality. Environ. Poll. 161, 55-63.

Sarkar, S., Saha, S., Raychaudhuri, D., 2015. Click beetle diversity of Buxa Tiger Reserve, West Bengal, India. World Sci. News. 19, 120-132.

Scudder, G.G.E., 2009. Insect Biodiversity- Science and Society. Foottit and Adler, Wiley Blackwell Publishing Ltd., pp. 20.

Shurin, J.B., Gruner, D.S., and Hillebrand H., 2005. All wet or dried up? Real differences between aquatic and terrestrial food webs. Proc. Biol. Sci. 273, 1–9.

Simpson, G.G., 1949. Tempo and Mode in Evolution. Columbia University Press, New York.

Simpson, G.G., 1953. The Major Features of Evolution. Columbia University Press, New York.

Stork, N.E., 2007. World of insects. Nature. 448(9), 657-658.

The Columbia Encyclopedia, 6th Ed. 2015. http://www.encyclopedia.com

Varshney, R.K., 1998. Faunal diversityin India: Insecta, in: Alfred, J.R.B., Das, A.K. and Sanyol, A.K (Eds.), Faunal Diversity in India. ENVIS Centre, Zoological Survey of India, Calcutta, pp. 145-157.

Vasseur, L., Rapport, D., and Hounsell, J., 2002. Linking ecosystem health to human health: a challenge for this new century, in: Costanza B. and Jorgensen S. (eds.), Integrating Science to Policy – Toward the 21st Century. Elsevier, Cambridge, UK, pp. 167–190.

Whitfield, J.B., 2003. Phylogenetic insights into the evolution of parasitism in hymenoptera. Adv. Parasitol. 54, 69-100.

Wigglesworth, V.B., 1976. Insects and the life of Man. Collected Essays, in: Pure Science and Applied Biology. Chapman and Hall, London. pp 217.

Yeates, D.K., Harvey, M.S., and Austin, A.D., 2003. New estimates for terrestrial arthropod species-richness in Australia. Records of the South Australian Museum, Monograph Series.7, 231–241.

You, M., Xu, D., Cai, H. and Vasseur, L., 2005. Practical importance for conservation of insect diversity in China. Biodiv. Conserv. 14, 723–737

Yule, C.M., and Yong, H.S., 2004. Freshwater Invertebrates of the Malaysian Region, Academy of Sciences Malaysia, Kuala Lumpur Malaysia.

Zhang, Z.Q., 2011. Animal biodiversity: an introduction to higher- level classification and taxonomic richness. Zootaxa 3184, 7-12.

ZSI e-news. January 2015, 7(1).

ZSI. 2012. COP XI publications. www.zsi.gov.in/Cop-11/cop-11.html/

Chapter 3

Diversity of Grasshopper Communities with the Cropping Patterns of a Rice Agro-ecosystem in West Bengal, India

☆ ***Ranita Chakravorty, Parimalendu Haldar and Santanu Ray***

ABSTRACT

Being primary consumer grasshoppers hold a key point in the food chain. Their abundance and response to climate change make them a good indicator. Thus these insects have become an object of special interest to the ecologists. In the present scenario of global climate change and altered agricultural practices it has become a necessity to understand their ecology and diversity. The present study attempts to find out how grasshopper diversity is related to plant diversity in a rice agro-ecosystem of West Bengal, India. For this study the selected area is divided into twelve plots, each measuring 1.5mX1.5m, and sampling has been conducted for two consecutive years, 2011 and 2012. Grasshoppers are collected from each plot by using an insect aerial net. Vegetation composition is examined in the same plots after completing grasshopper collection. The collected grasshoppers and plant samples are taken to the laboratory for identification. Statistical analyses show a quite high diversity for both the plant and animal species in the study area. The results of regression analyses depict that grasshopper richness is positively related to the plant richness and diversity. Hence it is concluded that vegetation cover of an area has great influence on the distribution of grasshopper species. The results may be utilized by the policy makers and forecasting modellers worldwide to comment on the pest status of the grasshoppers and their management, to speculate the wellbeing of the ecosystem, and to take preventive measures to save the ecosystem from destruction.

Introduction

Ecological indicators are a used to demonstrate the effect of environmental changes such as habitat alteration, fragmentation and climate change on biotic systems (McGeoch 2007, Bazelet and Samways, 2011). To be a good ecological indicator some of the criteria such as quantifiable sensitivity to changes in the environmental conditions, abundance in the habitat, ease of capture and taxonomic stability are necessary (Fleishman and Murphy 2009). Choice of a specific microhabitat by a particular insect depends on number of different factors used to monitor habitat suitability (Kramer *et al.*, 2012). Selection of their microhabitat is regulated by a set of very definite factors which include microclimatic variables (temperature, humidity, light intensity, *etc.*), availability of food, structural qualities, oviposition sites, suitable hiding places, and the presence of predators (Joren, 2005). Thus insects could be widely used as ecological indicators for many habitats and regions (Fleishman and Murphy 2009). Besides being sensitive to environmental changes and their abundance they also act as primary consumer in a food chain and thus hold a key position. Variation in their number may disrupt the food chain thus disturbing the entire ecological balance.

Among insects, the order Orthoptera represents one of the most abundant ground dwelling groups. Being primary consumers vegetation structure plays an important role for habitat selection by grasshoppers (Branson, 2010), as vegetation determines the availability and distribution of all resources required (Joern *et al.*, 2009). Therefore change in vegetation cover of an area affects their composition. Anthropogenic activities, livestock grazing and fire shape the plant species composition of an area (Joren, 2005; Branson *et al.*, 2006). As per reports of several studies changes in grasshoppers composition is observed to grazing intensity (Jauregui *et al.*, 2008), fire frequency (Hochkirch and Adorf 2007), and mowing regime (Gardiner and Hassall 2009). Besides, both plant and grasshopper species composition is reported to be affected due to changing environmental gradients (Kemp *et al., 1990*).Thus in regions of grasshoppers abundance, they may be successfully used as ecological indicators (Steck *et al.*, 2007).

Besides, being a voracious primary feeder grasshoppers also waste a lot of food plant on which they feed. Such a habit causes both loss and benefit to human and the ecosystem. This helps in plant decomposition and re-growth and plays a vital role in preventing plant overgrowth (Belovsky and Slade, 2000, 2002). Grasshoppers' droppings, known as frass along with plant waste, help to fertilize soil and facilitate plant growth. They also act as food for birds, small mammals, spiders, reptiles (Pitt, 1999) and even human beings of some countries. They can be used to control weeds like snakeweed (*e.g.* snakeweed grasshopper *Hesperotettix viridis*), ragweed (eg. ragweed grasshopper *Campylancantha olivacea*), rabbit-brush and locoweeds (Fielding and Brusven, 1993).

Although ecologically grasshoppers have an important role to play, throughout the world they are considered to be one of the most devastating pests of agriculture that may cause widespread and severe damage to grasslands, forage, cereal, vegetable, and orchard crops during outbreaks (Weiland *et al.*, 2002; Lockwood *et al.*, 2002;Usmani *et al.*, 2012). Crop damage occurs early in the growing season when newly hatched grasshoppers invade the field (Begna and Fielding, 2003). According to their host plant choice and feeding habit, Mulkern (1967) classified grasshoppers as graminivores, forbivores and ambivores or mixed feeders. Hence it is clear that along with crop plants, forest plantations and grasslands also remain under constant threat.

Despite being economically and ecologically so important, the relationship of grasshopper abundance with plant diversity is yet to be clearly defined. Various workers have proposed hypotheses in order to explain the relationships between the plant diversity of an ecosystem and the species richness of herbivores present there (Hawkins and Porter, 2003; Haddad, 2009). According to Branson (2010) grasshopper species diversity is often positively influenced by plant diversity. In spite of these kind of studies being conducted worldwide, challenge remains regarding grasshopper ecology because the patterns of their species diversity is still very much uncertain (Joren, 2005; Koch *et al.*, 2013). Unless and until more and more authentic statistical data is acquired from all over the world the actual relationship of the diversity of plant species with grasshopper species will remain unknown (Haddad *et al.*, 2001; Joren, 2005).

In this context the present study aims to determine if variation in grasshopper abundance and diversity in rice agro-ecosystem could be explained by its plant species richness and diversity. The study has been carried out in Santiniketan, in the state of West Bengal which is one of the most important agro-based states of India. The study area has a vast array of plots with different cropping patterns where rice is cultivated as the principal crop. There are two main seasons for growing rice in this region summer and winter crop. The summer crop is sown during January to February and harvested in April-May; while the winter crop is sown during the months of June to August and harvested in November to January (Ganguly *et al.*, 2013).To accomplish the present study grasshopper and plant samples are collected from the month of August to November for two consecutive years 2011 and 2012. During these months the winter rice remains in the field, and grasshoppers are abundantly available; so, there could be a possible relation between these insects and plants. Defining this relation will help to predict their occurrence with the availability of different plants. This information in future may be used to speculate their pest status and help in their management.

Materials and Methods

Study Site and Procedure

For the present study, a rice agro-ecosystem in Santiniketan (23°39´N, 87°42´E), Birbhum, West Bengal, India is chosen. The area of study consists of cultivable lands surrounded by grasslands.

The entire sampling area is divided into twelve plots (P1-P12) each measuring 1.5m × 1.5m. Each day 4 plots are selected for the field study, and 30 minutes are allotted per plot for sampling. Gap of the sampling is such that we return to a particular plot fortnightly. Grasshoppers are collected from each plot by sweeping method (Haldar *et al.*, 1999) using an insect aerial net while walking slowly through the vegetation. The collected grasshoppers are taken to the laboratory for species identification. Vegetation composition is examined in the plots the same day after completing grasshopper collection. The intact plants with all parts are handpicked and stored in plastic bags to be taken to the laboratory for identification. This is to be noted that sowing of crops are not done in all the plots at a time, hence when there is crop in some plots there are fallow plants present in some other adjacent plot. Hence different species of plants are collected during sampling period.

Statistical Analyses

Biodiversity is one of the primary interests of ecologists, but quantifying the species diversity of ecological communities is complicated. Species diversity itself has two separate components: 1) the number of species present (species richness), and 2) their relative abundances (dominance or evenness). To find out whether there is any relationship between the number of plant and grasshopper species, graphs are plotted for both the years. Further, various indices are used to verify the relationship. Diversity and richness is measured following Aslam (2009) and evenness is measured following Ludwig and Reynolds (1988). The indices are as follows:

1) Margalef's index (R1) (Margalef, 1958)(Clifford and Stephenson, 1975)–It is a measure of species diversity. The calculation is done using the given formula:

   $R1 = (S - 1)/ln\ N$

   Where, S = total number of species

   N = total number of individuals in the sample

   ln= natural logarithm

2) Shannon-Wiener Index (H2) (Shannon and Wiener,1949)–According to this index the diversity of a community is similar to the amount of information available. The formula used for calculation is:

   $$H' = \sum_{i=1}^{S^*}(p_i \ln p_i)$$

   Where, Pi = S/N

   S*= total number of species in the community

   S = number of individuals of one species

   N = total number of all individuals in the sample

   ln= natural logarithm

3) Simpson's index (λ) (Simpson,1949) – It measures the strength of dominance. The formula is as follows:

   $$\lambda = \sum_{i=1}^{S} p_i^2$$

Initially regression analysis of number of plant and grasshopper species are carried out to find whether there is any relation between plant species and grasshopper species. Later to verify the relationship, regression analysis is carried out using the results of statistical data of plant and animal species diversity indices. All the results of plant and grasshopper species are subjected to "Paired t test" to find out if any significant variation is there between the results of two years. All the calculations are done using PAST, version 3.02 and Microsoft excel 2010 software.

Results

Rice is the major crop grown in the study area; grasses that dominate the vegetation are *Cynodon* sp., *Brachiaria* sp., *etc.* Out of the total twenty grasshopper species found, eighteen are observed during the year 2011 and fourteen species during 2012. Among these *Oedaleus abruptus, Acrida exaltata, Spathosternum prasiniferum, Atractomorpha crenulata* are dominant in both the years (Table 2.1). *Phlaeoba infumata* is quite high in number in the first year but is nearly absent in the samples of second year (Table 2.1). *Truxalis* sp., *Oxya hyla, Gesonula punctiforns* and *Hieroglyphus banian* are totally absent in the samples of second year, whereas *Trilophidia annulata* is present in the second year but not in the first year (Table 2.1).

Table 2.1: Number of Nymph, Male and Female of different Grasshopper Species Collected during the Year 2011 and 2012

Species	*2011*			*2012*		
	Nymph	*Male*	*Female*	*Nymph*	*Male*	*Female*
Oedaleus abruptus (Thunberg)	87	38	30	237	38	55
Acrida exaltata (Walker)	99	19	5	100	10	5
Atractomorpha crenulata (Fabricius)	31	8	21	49	12	12
Spathosternum prasiniferum (Walker)	92	66	86	9	7	6
Chrotogonus trachypterus (Blanchard)	1	1	1	6	4	1
Truxalis sp. (Fabricius)	79	17	3	0	0	0
Leva sp.	5	7	8	5	17	12
Aulacobothrus lutipes (Walker)	9	2	1	27	0	0
Phlaeoba infumata (Brunner)	41	31	15	5	0	0
Oxyafus covittata (Marschall)	13	2	11	0	0	0
Trilophidia annulata (Thunberg)	0	0	0	2	1	0
Aiolopus thalassinus (Fibricius)	28	2	1	7	2	0
Oxya hyla (Serville)	0	0	3	0	0	0
Gesonula punctifrons (Stal)	0	7	1	0	0	0
Gastrimargus marmoratus (Thunberg)	2	0	0	7	1	0
Phlaeoba pantelli (Bolivar)	19	3	9	0	1	0
Dittopternes venusta	16	6	3	12	4	5

Species	2011			2012		
	Nymph	Male	Female	Nymph	Male	Female
Epis taurus (Bolívar)	0	0	0	0	1	1
Hierogylphusbanian (Fabricius)	8	20	38	0	0	0
Tylotropidius varicornis (Walker)	0	0	1	0	0	0

Figures 2.1a and 2.1b show the results of grasshopper variation along with plant species variation in 2011 and 2012 respectively. The regression analysis of the data shows that there is a prominent relationship between the grasshopper and plant species (R^2= 0.69).

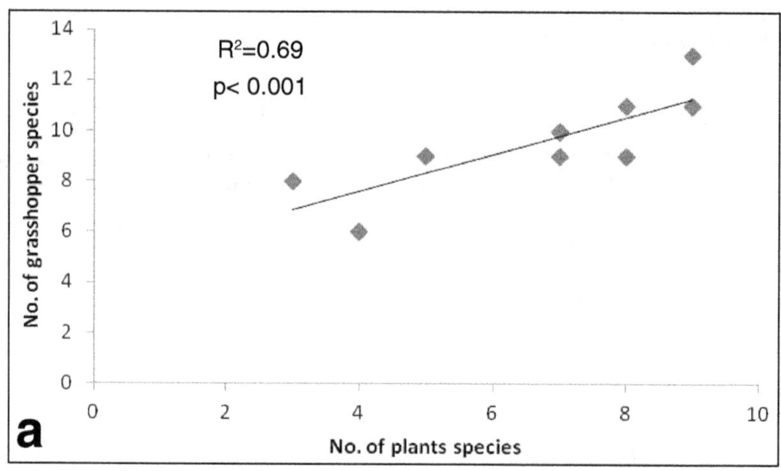

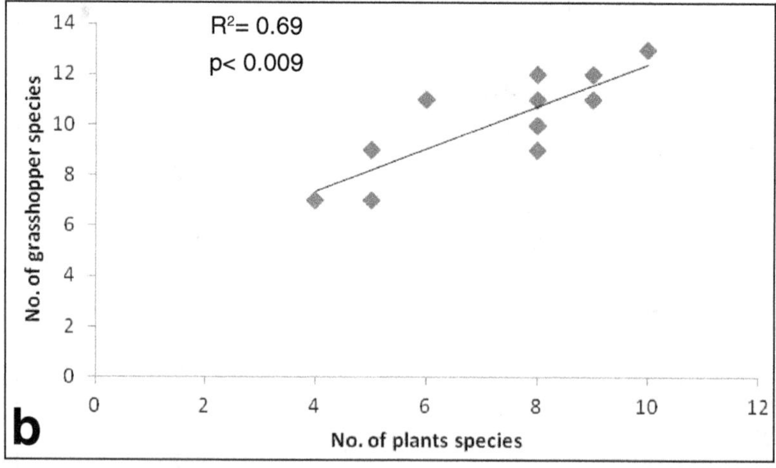

Figure 2.1a-b: Year-wise Variation of Grasshopper Diversity and Plant Diversity. a: Shows the result of 2011 and b: Shows the Result of 2012.

Table 2.2 show the calculated results of the diversity indices for plant samples collected in 2011 and 2012. Plant diversity results for the first year (2011) shows that Margalef's index is highest in P10 (1.197) and lowest in P11 (0.405). Simpson index is lowest in P6 (0.113) and highest in P11 (0.332). Shannon index is highest in P6 (2.184) and lowest in P 11 (1.093). The result of plant data for the second year (2012) shows Margalef's index is highest in P11 (1.192) and lowest in P1 (0.422). Simpson index is highest in P1 (0.259) and lowest in P11 (0.109). Shannon index is highest in P11 (2.250) and lowest in P1 (1.366).

Table 2.2: Results of Various Statistical Indices Considered for Plant Samples Collected in the Year 2011 and 2012

Plot No.	*Margalef's index*		*Simpson*		*Shannon*	
	2011	*2012*	*2011*	*2012*	*2011*	*2012*
P1	0.956	0.422	0.135	0.259	2.038	1.366
P2	0.903	1.090	0.154	0.116	1.899	2.171
P3	0.951	0.990	0.159	0.132	1.878	2.048
P4	1.046	0.597	0.115	0.211	2.176	1.580
P5	0.588	0.716	0.210	0.183	1.580	1.744
P6	1.058	0.960	0.113	0.137	2.184	2.027
P7	0.820	0.982	0.146	0.146	1.932	1.987
P8	0.852	0.954	0.149	0.134	1.922	2.042
P9	1.055	1.081	0.131	0.121	2.046	2.151
P10	1.197	0.592	0.128	0.213	2.101	1.576
P11	0.405	1.192	0.332	0.109	1.093	2.250
P12	0.638	1.041	0.283	0.131	1.295	2.051

Table 2.3 shows the results for the grasshopper species collected in the year 2011 and 2012 respectively. The result of diversity indices of grasshopper species for first year (Table 2.4) shows that Margalef's index is highest in P10 (2.339) and lowest in P12 (1.369). Simpson index is highest in P8 (0.291) and lowest in P1 (0.049). Shannon index is highest in P5 (1.93) and lowest in P1 (0.609). The result of grasshopper species diversity indices for the second year (Table 2.5) shows that Margalef's index is highest in P11 (2.617) and lowest in P1 (1.280). Simpson index is highest in P4 (0.442) and lowest in P9 (0.180). Shannon index is highest in P11 (1.986) and lowest in P4 (1.219).

The above statistical data (richness and diversity indices) are used for regression analysis to study the relation between plant and grasshopper diversity. The results of the regression analysis with significant R^2 values only are shown in Figures 2.2–2.4. Significant R^2 values are obtained between grasshopper richness and richness and diversity indices of plants. Rest of the indices show insignificant R^2 values which indicate that there exist no relationship. Thus the regression analysis shows that there is a strong relation between grasshopper richness and plant richness and diversity for both the years with respect to Margalef's index, Simpson index and Shannon Index.

Table 2.3: Results of Various Statistical Indices Considered for Grasshopper Species Collected in the Year 2011 and 2012

Plot No	*Margalef's index*		*Simpson Index*		*Shannon Index*	
	2011	*2012*	*2011*	*2012*	*2011*	*2012*
P1	1.963	1.280	0.049	0.326	0.609	1.370
P2	2.048	2.130	0.148	0.280	0.758	1.596
P3	1.897	1.852	0.178	0.327	1.880	1.466
P4	2.018	1.297	0.189	0.442	1.922	1.219
P5	1.643	2.153	0.161	0.257	1.930	1.700
P6	2.103	1.735	0.214	0.363	1.800	1.438
P7	1.765	1.926	0.221	0.256	1.695	1.711
P8	2.048	1.757	0.291	0.220	1.627	1.713
P9	1.643	2.463	0.276	0.180	1.405	1.905
P10	2.339	1.847	0.213	0.184	1.685	1.850
P11	1.808	2.617	0.164	0.202	1.587	1.986
P12	1.369	2.383	0.118	0.260	1.205	1.786

Discussion

Variation in species richness with change in location is one of the most observable patterns in biodiversity (Lennon *et al.*, 2004). Grasshoppers being polyphagous herbivores, naturally there could be a close ecological relationship between grasshopper richness and variation in vegetation cover of the area. Understanding the relationships between different species is a major task of conservation biologists and ecologists (Xu *et al.*, 2008).The most important factor for them is to determine whether there is any relation between animal and plant diversity. Besides being producers, plants provide food, along with shelter and hiding place from predators. This may act as a big cause that plant diversity in general leads to diversities of different groups of animals (Perner *et al.*, 2005).

Margalef's index and Shannon index are two well-known indices used for measuring richness and diversity. The present study shows that variation in the results of these two indices for plants and grasshopper species are correlated. Higher values for both these indices indicate higher diversity (Margalef, 1956; Ludwig and Reynolds, 1988). Study conducted by Pereira and De Luca (2003) in Rio Grande do Sul River, and Bhandarkar and Bhandarkar (2013) in Maharashtra for benthic macro invertebrates shows similar results. A study by Kemp *et al.* (1990) in different types of habitat also supports the fact that presence of grasshopper and plant species richness is positively correlated. The positive effect of plant species richness on grasshopper diversity may be due to two possible reasons: First, diversity of food plant results in a better diet (Specht *et al.*, 2008; Ibanez *et al.*, 2013), and second, increase in plant species richness results a decrease in top-down control of grasshoppers by natural enemies (Hoekman, 2010).

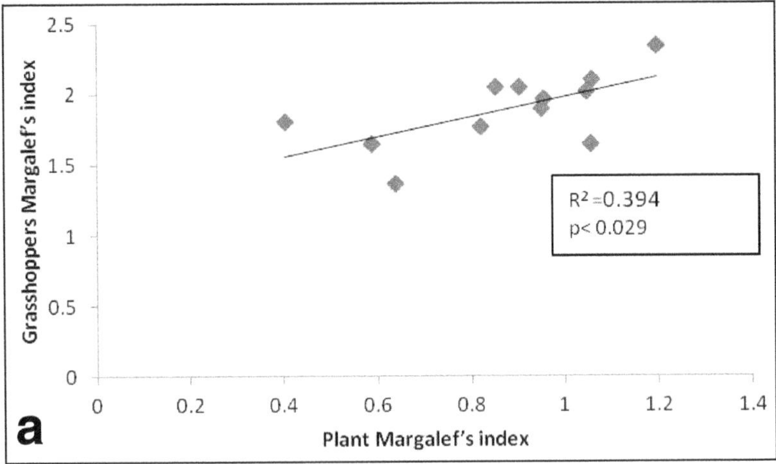

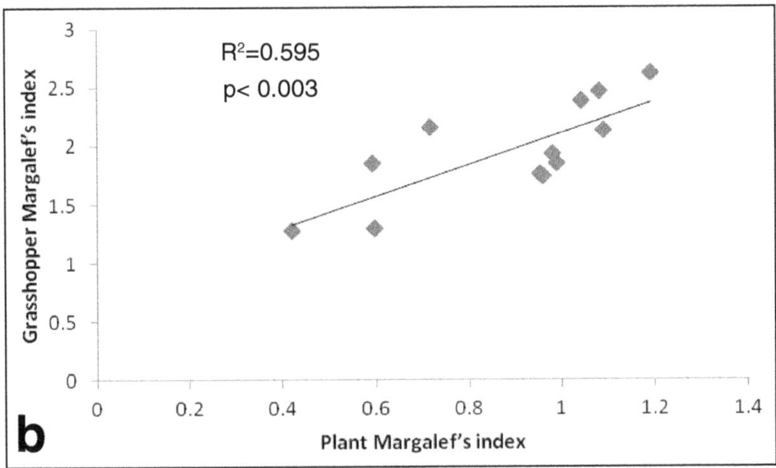

Figure 2.2a-b: R^2 and their Respective P Values for the Regressions of Plant Margalef's index vs. Grasshopper Margalef's Index. a: Shows the results of first year (2011) and b: Shows the Result of Second Year (2012).

Higher species diversity is important in maintaining ecosystem functioning (Chesson *et al.*, 2002). Besides Shannon index, Simpson index is another commonly used measure of diversity. The value of Simpson's Index varies from 0 to 1 (Sagar and Sharma, 2012); more the data tends towards 1 less diverse will be the community (Smith and Wilson, 1996). In the present study the value of this index in all the plots for both plants and grasshoppers are less than 0.5 which indicates that both the plant and grasshopper community are highly diverse. A study conducted by Guo (2006) shows that diversity in vegetation community cause variation in grasshopper diversity. Regression analysis of the richness and diversity indices data also shows that the presence of acridid species is positively related to the vegetation cover.

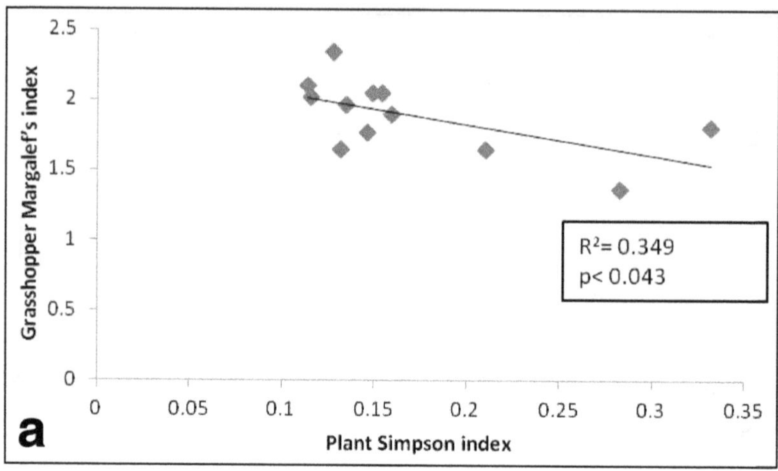

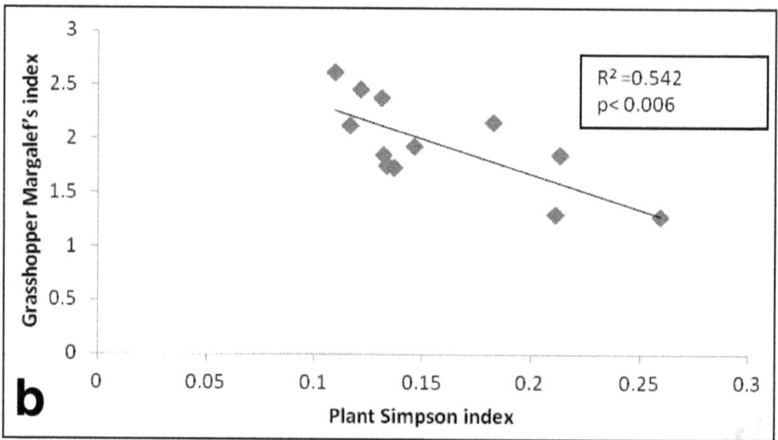

Figure 2.3a-b: R^2 and their Respective P Values for the Regressions of Plant Simpson index vs. Grasshopper Margalef's Index.
a: Shows the Results of First Year (2011)
b: Shows the Result of Second Year (2012).

Grasshoppers form an important component of ecosystems in terms of their abundance, species richness and functional contributions (Hamer *et al.*, 2006; Whiles and Charlton, 2006). Besides, being the primary consumer they also form an important trophic level in the food chain. Grasshopper assemblages are sensitive to change in vegetation cover (Barimo and Young, 2002) environmental parameters and anthropogenic activities (Saha *et al.*, 2011). Change in their number or their absence will disrupt the food chain leading to destruction of the ecosystem. A study on another ground dwelling insect, soil nematodes by Eisenhauer *et al.* (2011; 2013) also supports our findings that animal diversity and plant richness are related to each other.

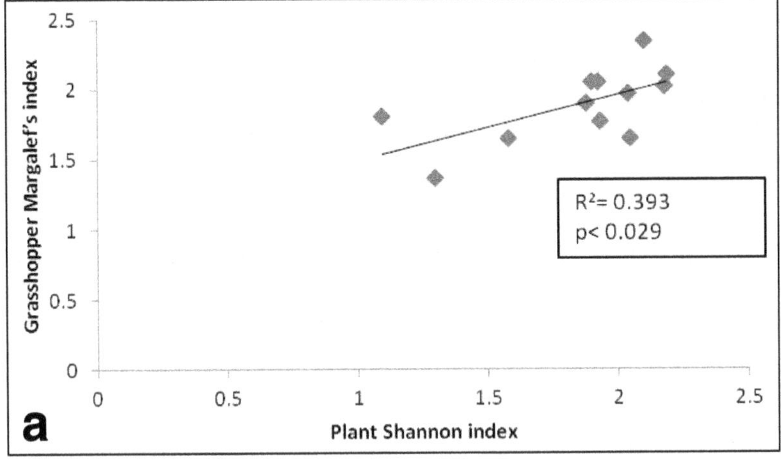

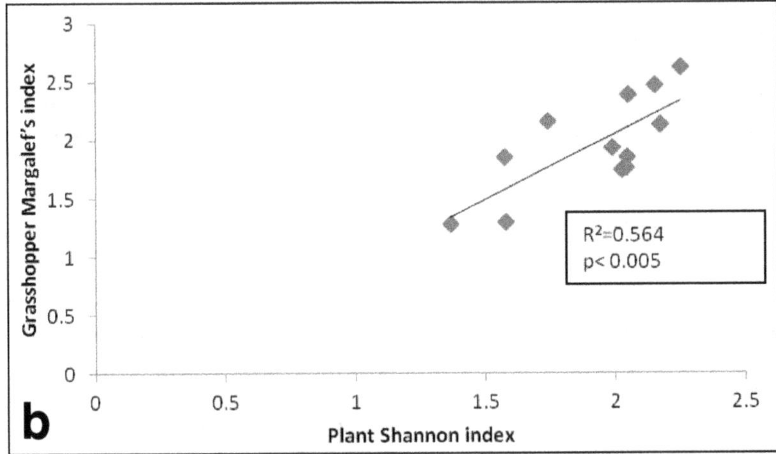

Figures 2.4a-b: R^2 and their Respective P Values for the Regressions of Plant Shannon Index vs. Grasshopper Margalef's Index. a: Shows the Results of First Year (2011) b: Shows the Result of Second Year (2012).

Thus this study contributes to a better understanding of grasshopper diversity, indicating that they are not randomly distributed in the different environments. They have distribution patterns associated primarily with feeding on the vegetation types. Most of the grasshoppers are multivoltine and their life cycle coincides with winter rice and Kharif crops. Grasshoppers usually attack the seedlings during the early stage of life and the adults usually feed on the leaves of the crop plant. This results to a huge economic loss for the farmers if proper measures are not taken. Knowing their relation with plant composition may help the agriculturists to predict their occurrence relating to crops growing in the region. This will help them to take proper measures and prevent crop loss due to attack of these pests.

Conclusion

Agricultural ecosystem is a man-made ecosystem. Depending on the types of crops grown, agricultural ecosystem has a specific composition of grasshoppers. Excessive use of chemical fertilizers and pesticides for agricultural purpose casts a negative impact on the health of the ecosystem. Grasshoppers are directly dependent on plants for their food and shelter. Accumulation of hazardous chemicals in the environment due to anthropogenic activities affects the plant composition; this in turn will result in a difference in the composition of grasshoppers. Change in their composition can be used as an indication of the health of the ecosystem. According to Steck *et al.* (2007) the number of grasshopper species in a particular area is directly related to the cultivated land available and to the frequency of open land in the surroundings. Several studies have suggested that they could be used as ecological indicators (Steck *et al.*, 2007; Saha and Haldar, 2009) or as biodiversity indicators (Sauberer *et al.*, 2004). As plant and grasshopper composition are positively correlated their presence or absence will help to speculate the well being of the ecosystem. This may in turn help the agriculturists to take preventive measures to save the ecosystem from destruction. Moreover gaining detail knowledge of the food habit of these insects may help to find a particular species to be very sensitive to anthropogenic changes. This may in future be selected as a prominent ecosystem health indicator.

Acknowledgements

The authors are thankful to the Head, Department of Zoology, Visva Bharati University for providing necessary laboratory facilities. University Grants Commission (UGC), Government of India is acknowledged for financial support.

References

Aslam, M., 2009. Diversity, species richness and evenness of moth fauna of Peshawar. Pak. J. Entomol. 31, 99-102.

Barimo, J. F., Young, D. R., 2002. Grasshopper (Orthoptera: Acrididae)—Plant-environmental interactions in relation to zonation on an Atlantic Coast Barrier Island. Environ. Entomol. 31, 1158-1167.

Bazelet, C.S., Samways, M.J., 2011. Identifying grasshopper bioindicators for habitat quality assessment of ecological networks. Eclo. Indicators. 11(5), 1259-1269.

Begna, S. H., Fielding, D. J., 2003. Damage potential of grasshoppers (Orthoptera: Acrididae) on early growth stages of small-grains and canola under subarctic conditions. J. Econ. Entomol. 96, 1193-20.

Belovsky, G. E., Slade, J. B., 2000. Insect herbivory accelerates nutrient cycling and increases plant production. Proc. Natl. Acad. Sci. 97, 14412–14417.

Belovsky, G. E., Slade, J. B., 2002. An ecosystem perspective on grasshopper control: possible advantages to no treatment. J. Orth. Res. 11, 29–35.

Bhandarkar, S. V., Bhandarkar, W. R., 2013. A study on species diversity of benthic macro invertebrates in freshwater lotic ecosystems in Gadchiroli district Maharashtra. Int. J. Lif. Sci. 1, 22-31

Branson, D. H., Joern, A., Sword, G. A., 2006. Sustainable management of insect herbivores in grassland ecosystems: new perspectives in grasshopper control. BioScience 56, 743–755.

Branson, D. H., 2010. Relationships between plant diversity and grasshopper diversity and abundance in the Little Missouri National Grassland. Psyche 2011, 1-7.

Chesson, P., Pacala, S., Neuhauser, C., 2002. Environmental niches and ecosystem functioning, in: Kinzig, A. P., Pacala, S. W., Tilman, D. (Eds), The Functional Consequences of Biodiversity: Empirical progress and Theoretical Extensions. Princeton University Press, Princeton, pp. 213-245.

Clifford, H. T., Stephenson, W., 1975. An introduction to numerical classification. Academic Press, New York.

Eisenhauer, N., Dobies, T., Cesarz, S., Hobbie, S. E., Meyer, R. J., Worm, K., Reich, P. B., 2013. Plant diversity effects on soil food webs are stronger than those of elevated CO_2 and N deposition in a long-term grassland experiment PNAS 110, 6889–6894.

Eisenhauer, N., Migunova, V. D., Ackermann, M., Ruess, L., Scheu, S., 2011. Changes in plant species richness induce functional shifts in soil nematode communities in experimental grassland. PLoS ONE. 6, e24087.

Ferrant, S., Caballero, Y., Perrin, J., Gascoin, S., Dewandel, B., Aulong, S., Dazin, F., Ahmed, S., Marechal, J., 2014. Projected impacts of climate change on farmers' extraction of groundwater from crystalline aquifers in South India. Sci. Rep.4, Article number: 3697

Fielding, D. J., Brusven, M. A., 1993. Grasshopper (Orthoptera: Acrididae) community structure and ecological disturbance on southern Idaho rangeland. Environ. Entomol. 22, 71-81.

Fleishman, E., Murphy, D.D., 2009. A realistic assessment of the indicator potential of butteries and other charismatic taxonomic groups. Conserv. Biol. 23, 1109–1116.

Ganguly, A., Chakravorty, R., Haldar, P., 2013.Assessment of consumption, utilisation and growth of *Oedaleus abruptus* (Thunberg) and *Spathosternum prasiniferum prasiniferum* (Walker) (Orthoptera:Acrididae) fed with various food plants in laboratory condition. Ann. de la Soc. Ent de Fr. 49, 160-171.

Gardiner, T., Hassall, M., 2009. Does microclimate affect grasshopper populations after cutting of hay in improved grassland? J. Insect Conserv. 13, 97–102.

Guo, Z. W., Li, H. C., Gan, Y. L., 2006. Grasshopper (Orthoptera: Acrididae) biodiversity and grassland ecosystems. Insect Sci. 3, 221-227.

Haddad, N. M., Tilman, D., Haarstad, J., Ritchie, M., Knops, J. M. H., 2001. "Contrasting effects of plant richness and composition on insect communities: a field experiment. Amer. Nat.158, 17–35.

Haddad, N. M., Crutsinger, G. M., Gross, K., Haarstad, J., Knops, J. M. H., Tilman, D., 2009. Plant species loss decreases arthropod diversity and shifts trophic structure. Ecol. Lett. 12, 1029–1039.

Haldar, P., Das, A., Gupta, R. K., 1999. A laboratory based study on farming of an Indian grasshopper *Oxya fuscovittata* (Marschall) (Orthoptera:Acrididae). J. Orth. Res. 8, 93-97.

Hamer, T. L., Flather, C. H., Noon, B. R., 2006. Factors associated with grassland bird species richness: the relative roles of grassland area, landscape structure, and prey. Springer 21, 569–583.

Hawkins, B. A., Porter, E. E., 2003. Does herbivore diversity depend on plant diversity? The case of California butterflies. Am. Naturalist.161, 40–49.

Heip, C., 1974. A new index measuring evenness. J. Mar. Biol. Assoc. 54, 555-557.

Hill, M, O., 1973. Diversity and evenness: A unifying notion and its consequences. Ecology 54, 427-432.

Hochkirch, A., Adorf, F., 2007. Effects of prescribed burning and wildres on Orthoptera in Central European peat bogs. Environ. Conserv. 34, 225–235

Hoekman, D., 2010. Relative importance of top-down and bottom-up forces in food webs of Sarracenia pitcher communities at a northern and a southern site. Oecologia 165, 1073–1082.

Ibanez, S., Manneville, O., Miquel, C., Taberlet, P., Valentini, A., Aubert, S., Coissac, P., Colace, M. P., Duparc, Q., Lavorel, S., Moretti, M., 2013. Plant functional traits reveal the relative contribution of habitat and food preferences to the diet of grasshoppers. Oecologia 173, 1459–1470.

Jauregui, B.M., Rosa-Garcia, R., Gracia, U., WallisDe Vries, M.F., Osoro, K., Celaya, R., 2008. Effects of stocking density and breed of goats on vegetation and grasshopper occurrence in heathland. Agri. Ecosyst. Environ. 123, 219–224

Joern, A., 2005. Disturbance by fire frequency and bison grazing modulate grasshopper assemblages in tallgrass prairie. Ecology 86, 861–873.

Joern, A., Kemp, W. P., Belovsky, G. E., O'Neill, K., 2009. Grasshoppers and vegetation communities. Southwest Nat. 63, 136-140.

Kemp, W. P., Harvey, S. J., O'Neill, K. M., 1990. Patterns of vegetation and grasshopper community composition. Oecologia 83, 299-308.

Koch, B., Buholzer, S., Edwards, P. J., Walter, T., Blanckenhorn, W. U., Wuest, R. O., Hofer, G., 2013. Vascular plants as surrogates of butterfly and grasshopper diversity on two Swiss sub-alpine summer pastures. Biodivers. Conserv. 22, 1451–1465.

Kramer, B., Kampf, I., Enderle, J., Poniatowski, D., Fartmann, T., 2012. Microhabitat selection in a grassland butterfly: a trade-offbetween microclimate and food availability. J. Insect Conserv. 16, 857–865.

Lennon, J. J., Koleff, P., Greenwood, J. J. D., Gaston, K. J., 2004. Contribution of rarity and commonness to patterns of species richness. Ecol. Letters. 7, 81–87.

Lockwood, A. J., Anderson-Sprecher, R., Schell, S. P., 2002. When less is more: Optimization of reduced agentarea treatments (RAATs) for management of rangeland grasshoppers. Crop. Prot.21, 551-562.

Ludwig, J. A., Reynolds, J. F., 1988. Statistical Ecology. New York: Wiley.

Margalef, R., 1956. Dos nuevos Gammarus de las aguas dulces espaiiolas. Publ. Inst. Biol. Apl. 23, 31-36.

Margalef, R., 1958. Temporal succession and spatial heterogeneity in phytoplankton, in: Buzzati-Traverso, (Eds.), Perspectives in Marine biology. University California Press, Berkeley, pp. 323-347.

McGeoch, M.A., 2007. Insects and bioindication: theory and progress. in: Stewart, A.J.A., New, T.R., Lewis, O.T. (Eds.), Insect Conservation Biology. CABI, Walling-ford, UK, pp. 144–174.

Mulkern, G. B., 1967. Food selection by grasshoppers. Annu. Rev. Entomol.12:59-78.

Pereira, D., De Luca, S. J., 2003. Benthic macro invertebrates and the quality of the hydric resources in Maratá Creek basin of Rio Grande do Sul State, Brazil. Acta. Limnol. Bras.15, 57-68.

Perner, J., Wytrykush, C., Kahmen, A., Buchmann, N., Egerer, I., Creutzburg, S., Odat, N., Audorff, V., Weisser, W. W., 2005. Effects of plant diversity, plant productivity and habitat parameters on arthropod abundance in Montane European grasslands. Ecography. 28, 429–442.

Pielou, E. C., 1977. Mathematical ecology. Wiley, New York.

Pielou, E. C., 1975. Ecological diversity. Wiley, New York.

Pitt, W. C., 1999. Effect of multiple vertebrate predators on grasshopper habitat selection: trade-offs due to predation risks, foraging and thermoregulation. Evol. Ecol.13, 499-515

Sagar, R., Sharma, G. P., 2012. Measurement of alpha diversity using Simpson index (1/λ): the jeopardy. Environ. Skep. Crit. 1, 23-24.

Saha, H. K., Haldar, P., 2009. Acridids as indicators of disturbance in dry deciduous forest of West Bengal in India. Biodivers. Conserv.18, 2343-2350.

Saha, H. K., Sarkar, A., Haldar, P., 2011. Effects of Anthropogenic Disturbances on the Diversity and Composition of the Acridid Fauna of Sites in the Dry Deciduous Forest of West Bengal, India. J. Biodivers. Ecol. Sci. 1, 313-320

Sauberer, N., Zulka, K. P., Abensperg-Traun, M., Berg, H., Bieringer, G., Milasowszky, N., Moser, D., Plutzar, C., Pollheimer, M., Storch, C., Trostl, R., Zechmeister, H., Grabherr, G., 2004. Surrogate taxa for biodiversity in agricultural landscapes of eastern. Austria Biol. Conserv.117, 181–190.

Shannon, C. E., Wiener, W., 1949.The mathematical theory of communication. University Illinos press, Urbana IL.

Sheldon, A. L., 1969. Equitability indices: dependence on species count. Ecology. 50, 466-467.

Simpson, E. H., 1949. Measurement of diversity. Nature. 163, 688.

Smith, B., Wilson, J. B., 1996. A consumer's guide to evenness measures. Oikos. 76, 70-82.

Specht, J., Scherber, C., Unsicker, S. B., Kohler, G., Weisser, W. W., 2008. Diversity and beyond: plant functional identity determines herbivore performance. J. Anim. Ecol. 77, 1047–1055

Steck, C. E., Burgi, M., Bolliger, J., Kienast, F., Lehmann, A., Gonseth, Y., 2007. Conservation of grasshopper diversity in a changing environment. Biol. Conserv. 138, 360– 370.

Usmani, M. K., Nayeem, M. R., Akhtar, M. H., 2012. Field observations on the incidence of grasshopper fauna (Orthoptera) as a pest of Paddy and pulses. Euro. J. Exp. Bio. 2, 1912-1917.

Weiland, R. T., Judge, F. D., Pels, T., Grosscurt, A. C., 2002. A literature review and new observations on the use of diflubenzuron for control of locusts and grasshoppers throughout the world. J. Orth. Res.11, 43-54.

Whiles, M. R., Charlton, R. E., 2006. The ecological significance of tall grass prairie arthropods. Annu. Rev. Entomol. 51, 387-412.

Xu, H., Wu, J., Liu, Y., Ding, H., Zhang, M., Wu, Y., Xi, Q., Wang, L., 2008. Biodiversity Congruence and Conservation Strategies: A National Test. Bioscience. 58, 632-639.

Chapter 4

Role of Acridids in Conservation and Management of Forest Ecosystem

☆ *Hiroj Kumar Saha and Parimalendu Haldar*

ABSTRACT

Biodiversity is very commonly used as a synonym for species diversity, particularly species richness, which is the number of species in a site or in a taxon. Biodiversity protects ecosystem's damage and degradation. Using species richness as a working measure of biodiversity, the area with the most species appear important as it offers the potential of conserving the largest 'amount' of diversity. Short-horned grasshopper (Orthoptera: Acrididae) plays a critical role in the forest ecosystem, and impacts on the structure, function and health of these forest ecosystems. The abundance, diversity, species richness and equitability index of acridids are lowest in disturbed habitats.

In forest ecosystems, acridid population explosions can result in a change in the stability of the acridid community, and thus can decrease the stability and finally damage the health of forest ecosystems. They regulate and maintain the biocoenotic balance and ensure in this way a proper functioning of the natural environment.

Introduction

According to Harper and Hawksworth (1995) the term 'biodiversity' was coined by Walter G Rosen for "The National Forum on Biodiversity", a conference held in Washington DC in 1986. Biodiversity is very commonly used as a synonym for species diversity, particularly species richness, which is the number of species in a site or in a taxon. Andersen *et al.* (2001) suggested terrestrial invertebrates to be used as bioindicator, because of their dominant biomass and diversity and their fundamental importance in ecosystem function. Biological diversity is now recognized increasingly as a vital parameter to assess global and local environmental changes and sustainability of developmental activities by Lovejoy (1995).

Grasshoppers are one of the critical groups to study biodiversity due to their abundance, species richness and diversity. Joern and Gaines (1990) and Lockwood (1997) observed that grasshoppers, especially acridids often emerge as one of the dominant groups of arthropods in terms of their contributions to diversity, abundance and biomass.

The importance of acrididoids to the insect community of deciduous forests in India was reported by Vats and Mittal (1991) who opined that these insects are of immediate relevance to conservation biology and ecological management. Conservation and restoration of biodiversity has become an important objective of forest management in recent years (Lindenmayer *et al.*, 1998 and Lachat *et al.*, 2006). Dobson *et al.* (2006) showed that losses of trophic diversity cause trophic collapse in nature, which ultimately cause losses of biodiversity and ecosystem services. Grasshopper (Orthoptera: Acrididae) play a critical role in the forest ecosystems, and impacts on the structure, function and health of these highly valued ecosystems.

Ecological Role of Acridids

Grasshoppers have a high reproductive potential and high nutritional value (Uvarov 1966, Julieta *et al.*, 1997) and may be important in energy and nutrient cycling. They play an important role in the functioning of forest ecosystems. Results from a variety of studies reveal that grasshoppers typically consume at least 10 percent of available plant biomass. They often harvest more plant biomass than they consume, influencing the availability and distribution of litter in the environment. This consumption and harvesting could be deemed negative from the perspective of available plant biomass for livestock production. But such harvesting process can serve important functions for the cycling of nutrients (Joern and Gaines 1990).

Microbes can break down the feces produced by grasshoppers more easily than those produced by larger herbivores, such as cattle or sheep. Grasshopper generated fecal nutrients are therefore more available for plant production. Also grasshoppers have a shorter lifespan and generally decompose where they die. The nutrients in their bodies return more rapidly to the soil for plant use than do nutrients found in the bodies of livestock. Even when grasshoppers create litter, they are enhancing plant production because increased litter increases the water retention of soils and reduces summer soil temperatures. Grasshoppers selectively feed on different plant species and, consequently, influence the plant species composition of the ecosystem (Capinera *et al.*, 1997).

According to Van Hook (1971) grasshoppers' are ecologically important and play a major role as food for wildlife, stimulating plant growth, creating plant litter for the soil, and cycling elements and nutrients was developed as a functional part of the whole ecosystem. They regulate and maintain the biocoenotic balance and ensure in this way a proper functioning of the natural environment. Mitchel and Pfadt (1974) showed that some of the species of grasshoppers have great role in weed control. They are much more important in the processing of plants to litter than in energy storage and trophic conversion. Grasshopper species also has the highest biomass, the highest productivity than other arthropods and best conserves the nutrient within the ecosystems (Andersen *et al.*, 2001). However, in an ecosystem point of view they may have net beneficial worth during most years as an important food source for at least a part of the life cycle of many species of mammals, birds, reptiles, *etc.*

Acridids Used as Bioindicator

The following criteria in insect fauna were suggested for indicators of temporal changes (Noss 1990, Pearson 1994): the indicator should be (1) sensitive to changes, (2) widely distributed, (3) easily and cost-effectively measurable, collectable and identifiable (stable taxonomy), (4) able to differentiate between natural and anthropogenic variation, (5) relevant to ecological phenomena and (6) economically important. These criteria are valid not only for monitoring programs but also for many programs addressing environmental questions.

According to Baldi and Kisbenedek (1997) the role of orthopterans as indicators is important for nature conservation, because they are good indicators of grassland disturbance. According to McGeoch (1998) "bioindicator" and the variety of terms used in relation to the concept can be apportioned into three categories corresponding to the three main applications of bioindicator. These are i) environmental indicators, ii) ecological indicators and iii) biodiversity indicator. The functions of bioindicators in each category are shown in **Figure 3.1**.

The applications of bioindicators can be expected to help not only in improving the environment but also in augmenting awareness of the living creatures around so that a better appreciation of the crucial role in sustaining life on the planet is obtained. Studies with bioindicators apply biodiversity as a principal tool to evaluate habitat characteristics or quality, structure and function of forest ecosystem.

It is argued that acridid assemblages are good indicators because, first, the structure of communities is sensitive to environmental changes. Second, the acridids

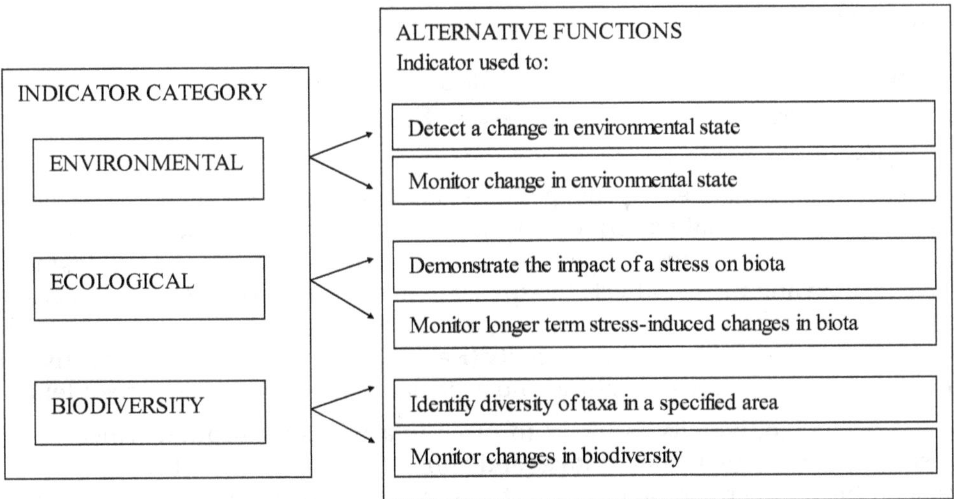

Figure 3.1: The Functions of Bioindicators in each Category of Bioindication (McGeoch 1998).

are habitat specific. Third, there is a simple sweep net sampling method, which can be standardized to obtain relative abundance data. Fourth, the identification is relatively easy, partly due to low number of species and their taxonomic stability. Fifth, they are one of the most dominant phytophagus insects, which play an important role in the functioning of forest ecosystem and potentially useful bioindicators for land disturbance. Sixth, the relationship between biodiversity and the stability of the acridid community, particularly species numbers and diversity for acridids of forest ecosystem interact on each other. It is not surprising that acridids are so important as food for wildlife because they (i) have high energy value and contain 50-70 per cent crude protein by Ueckert *et al.* (1972) (ii) are widely distributed and available in all habitats and (iii) are enough to easy exceed the energy cost of capture by foraging birds and wildlife.

Some acridid species are habitat specific. The population of acridids in different habitats indicates that their presence varies with habitat as well as seasonal variations. Saha and Haldar (2013) showed that the abundance, diversity, species richness and equitability index of acridids are lowest in disturbed habitats. Some species are only restricted in undisturbed habitat not found in the disturbed ones (Joshi *et al.*, 1999, Saha *et al.*, 2009).

Thus the population of acridid species are maintained a unique balance through their reproduction in the ecosystems, and so it is established that the number of multivoltine, bivoltine and univoltine acridid species maintain the population structure of forest ecosystem. They form the basis of many food webs and ecological interactions, promote soil fertility and structure and provide mechanism of biological control. Thus, they maintain the biocoenotic balance and ensure in this way a proper functioning of the forest ecosystem.

Being a voracious feeder and good recycler, acridids are an ecologically important community of insects. Moreover its predominance in the forest ecosystems may lead to indicate the health of respective ecosystems.

Relationship between Acridid Biodiversity and the Natural and Human Factors that affect Forest Ecosystems

Being a part of the global ecosystem, forest ecosystems surely make a response to global climate change. The changes in vegetation community diversity can cause variation in the special pattern of acridid biodiversity. Human activities, such as fire, grazing, clearing and farming in forests, can evidently change the character of vegetation and further affects the spatial pattern and the composition of acridid communities. The different ways of grazing can result in different changes in vegetation and grasshopper biodiversity (Kang 1995, Kang 1997). Human activities and the different actions of natural and artificial factors affect forest ecosystems.

Acridid biodiversity is likely to provide a window through which it can observe the effects of habitat changes and human activities on forest ecosystem. For relationship between acridid biodiversity and natural and human factors affect forest ecosystems. The causality between global climate change and dynamics in acridid biodiversities and the coupling relationship between human activities and acridid biodiversity are perhaps essential to the future evolution of forest ecosystems.

Acridid Biodiversity and the Health of Forest Ecosystems

India has a very rich flora and fauna, much of which is present in forest area. Each forest has a distinct faunal composition which varies in its species richness and the abundance of different species. Forests and Wildlife are essential for ecological balance of an area. It is in response to increasing evidence of human-dominated ecosystems becoming functionally disordered, that the concept of "health" to describe the state of regional ecosystems has been introduced (Vitousek *et al.*, 1997). A healthy ecosystem is defined as "steady and sustainable", and being able to maintain self-organization, self-government and resilience against pressures (Costanza, 1992). According to Rapport *et al.* (1995) ecosystem health has been described as an environment that maintains its biodiversity, is stable over time, and is resilient to change. In grassland ecosystems, some grasshopper population explosions can result in a change in the stability of the grasshopper community, and thus can decrease the stability and sustainability of the vegetation community and finally damage the health of grassland ecosystems (Guo *et al.*, 2006).

Biological indicators are used that reflect community derived environmental values to infer overall ecosystem health. A healthy forest maintains its unique species and processes while maintaining its basic structure, composition and function. Recent studies suggest that biodiversity contributes to ecosystem stability, structure and productivity (Naeem *et al.*, 1994, Larsen 1995 and Tilman *et al.*, 1996); which in turn contribute to sustainability. Ecological diversity may contribute importantly to various aspects of ecosystem stability (Walker 1995, Hobbs *et al.*, 1995, Peterson *et al.*, 1998). Higher species diversity may be important in maintaining ecosystem functioning (Yachi and Loreaue 1999 and Chesson *et al.*, 2002). Biologists have identified India as one of the top twelve mega diversity countries of the world.

The relationship between biodiversity and ecosystem functioning has emerged as a major scientific issue today. Here, we propose to examine the relationship between grasshopper biodiversity and the stability of the grasshopper community, particularly the relationship between the stability of species numbers and diversity for grasshoppers. Biodiversity provides the natural resources such as food, clothing, shelter and a number of useful products. In an ecosystem all the components are related to one another and it occurs in a state of dynamic equilibrium. This system of checks and balances is of a fundamental importance in an ecosystem which is maintained in a fundamental state by the activity of a large number of organisms.

Higher species richness and diversity may be important in maintaining ecosystem structure and functioning. In forest ecosystems, acridid population explosions can result in a change in the stability of the acridid community, and thus can decrease the stability and finally damage the health of forest ecosystems.

The relationship between biodiversity and the stability of the acridid community, particularly species numbers and diversity for acridids of forest ecosystem interact on each other. Moreover, based on the interaction between acridid biodiversity and health of forest ecosystem, expect to consider acridid biodiversity as an index for appraising the health of forest ecosystems. Healthy forest ecosystems provide greater opportunities for the production of economic and ecological goods and services valued by society than those ecosystems that are impaired.

There are many reasons for conservation; the major ones include (1) Protecting biodiversity keeps options open for future generations. (2) It has economic values in everything from medicine to pest control. (3) It has aesthetic values- enjoying life for its own sake. (4) A species is gone forever, once it has become extinct. We must save the blue print of the pieces of ecosystems. (5) Finally we need "indicator species", meaning those species that are more sensitive than ourselves to environmental changes that could warn us of environmental degradation. It is concluded that acridids can be used as indicator for habitat characteristics (disturbance, climatic conditions, edaphic factors, topography, *etc.*) of a forest ecosystem and ultimately it helps in conservation planning and management. There is a need therefore to conserve and protect our forests and that would, besides other advantages, also preserve the rich grasshopper fauna. Conservation therefore makes important contributions to social and economic development.

Conclusion

Changes in the amount and distribution of acridid populations can impact on the whole of a forest ecosystem and play main roles in forming and maintaining the biodiversity and the stability of various kinds of forest ecosystems. In forest ecosystems, acridid population explosions can result in a change in the stability of the acridid community, and thus can decrease the stability and finally damage the health of forest ecosystems.

They regulate and maintain the biocoenotic balance and ensure in this way a proper functioning of the natural environment. Acridid biodiversity is likely to provide a window through which it can observe the effects of habitat changes and human activities on forest ecosystem. They can be used as indicator for habitat characteristics of a forest ecosystem and ultimately it helps in conservation planning and management.

The community structure of acridids may be a viable diagnostic tool in assessing ecological conditions to ecosystem stability and finally develop healthy ecosystem which represents a desired endpoint of environmental management.

Acknowledgments

Authors are thankful to the Ministry of Environment and Forest, New Delhi for the financial assistance. Our thanks are also due to Prof. P. Nath, Head, Dept. of Zoology, Visva- Bharati University for providing laboratory facilities.

References

Andersen, A. N., Ludwig, J. A., Lowe, L. M., Rentz, D. C. F., 2001. Grasshopper biodiversity and bioindicators in Australian tropical Savannas: responses to disturbance in kakudu National Park. Austral Ecology 26, 213-222.

Baldi, A., Kisbenedek, T., 1997. Orthopteran assemblages as indicators of grassland naturalness in Hungary. Agriculture Ecosystems and Environment 66(2), 121-129.

Capinera, J. L., Scherer, C. W., Simkins, J. B,. 1997. Habitat associations of grasshoppers at the MacArthur Agro-Ecology Research Center, Lake Placid, Florida. *Florida Entomologist* 80(2), 253-261.

Chesson, P., Pacala, S., Neuhauser, C., 2002. Environmental niches and ecosystem functioning. In: The Functional Consequences of Biodiversity: Empirical progress and Theoretical Extensions. (eds kinzig AP, Pacala SW, Tilman D.). Princeton University Press, Princeton, Pp. 213-245.

Costanza, R., 1992. Ecosystem Health: New Goals for Environmental Management. Island Press, Washington D. C. pp. 239-256.

Dobson A, Lodge D, Alder J, Cumming GS, Keymer J, McGlade J, Mooney H, Rusak JA, Sala O, Wolters V, Wall D, Winfree R, Xenopoulos MA., 2006. Habitat loss, trophic collapse, and the decline of ecosystem services. Ecology 87, 1915–1924.

Guo, Z., Hong, Li., Gan, Y., 2006. Grasshopper (Orthoptera: Acrididae) biodiversity and grassland ecosystems. *Insect Science* 13, 221-227.

Herper, L. J., Hawksworth, L. D., 1995. Preface. Biodiversity Measurement and Ecosystem, pp 6-7. The Royal Society, Chapman and Hall.

Hobbs, R. J., Groves, R., Hopper, S. D, Lambeck, R. J, Lamont, B. B., Lavorel, S., Main, A. R., Majer, J. D., Saunders, D. A., 1995. Function of biodiversity in Mediterranean ecosystems in Australia. The function of biodiversity in Mediterranean ecosystems (ed. By G.W. Daris and D.M. Richardson). Springer Verlag, Heidelberg. Ecological Studies 109, 233-284.

Joern, A., Gaines, S. B., 1990. Population dynamics and regulation in grasshoppers, pp. 415-482. IN: Chapman RF. Joern A (eds). *Biology of grasshoppers*. Wiley; New York.

Joshi, P. C., Lockwood, J. A., Vashishth, N., Singh, A., 1999. Grasshopper (Orthoptera: Acrididae) community Dynamics in a Moist Deciduous Forest in India. Journal of Orthoptera Research 8, 17-23.

Julieta, R. E., Jose, M. P. N., Esteban, E. P., Manuel, A. P., Jaime, L. O., Oralia, L.de G., 1997. Nutritional value of edible insects from the state of Oaxaca, Mexico, Journal of Food Consumption and Analysis 10, 142-157.

Kang, L., 1995. Grasshopper-plant interactions under different grazing intensities in Inner Mongolia. Acta Ecologica Sinica 15, 1-11.

Kang, L., 1997. Changes of grasshopper communities in response to livestock grazing in grasslands. Grassland Ecosystem Research V, Science Press, Beijing, Pp 43-61.

Lachat, T., Attignon, S., Djego, J., Goergen, G., Nagel, P., Sinsin, B., Peveling, R., 2006. Arthropod diversity in Lama forest reserve (South Benin), a mosaic of natural, degraded and plantation forests. Biodiversity and Conservations 15, 3–23.

Larsen, J. B. 1995. Ecological stability of forests and sustainable silviculture. Forest ecology and management 73, 85-96.

Lindenmayer, D.B., Margules, C.R., Botkin, D.B., 1998. Indicators of biodiversity for ecologically sustainable forest management. Conserv Biol 14, 941–950.

Lockwood, J. A., 1997. Grasshopper population dynamics: a prairie perspective. Pp. 103-127. KinS.K. Gangwere, M.C. Muralirangan M. Muralirangan (Eds). The bionomics of grasshoppers, katydids and their kin. CAB International.

Lovejoy, T. E., 1995. The quantification of biodiversity: an esoteric quest or a vital component of sustainable development. In: Biodiversity Management and Estimation, Hawksworth DL. (ed), pp 81-87. Chapman and Hall, Londan.

McGeoch, M. A., 1998. The selection, testing and application of terrestrial insects as bioindicators. Biological Reviews 73, 181-201.

Mitchell, J. E., Pfadt, R. E., 1974. A role of grasshoppers in a short grass prairie ecosystem. Environmental Entomology 3, 358-360.

Naeem, S., Thompson, L. J., Lawler, S. P., Lawton, J. H., Woodfin, R. M., 1994. Declining biodiversity can alter the performance of ecosystems. Nature 368, 734-736.

Noss, R. F., 1990. Indicators for monitoring biodiversity: A hierarchical approach. Conservation Biology 4, 355-364.

Pearson, D. L., 1994. Selecting indicator taxa for the quantitative assessment of biodiversity. *Philos. Trans. R. Soc.* London, Ser.B. 345, 75-79.

Peterson, G., Allen, C. R., Holling, C. S., 1998. Biological resilience, biodiversity and Scale. Ecosystems I, 6-18.

Rapport, D. J., Gaudet, C., Calow, P., 1995. Evaluating and Monitoring the Health of Large-Scale Ecosystems. Springer-Verlag. Heidelberg.

Saha, H. K., Haldar, P., 2009. Acridids as indicators of disturbance in dry deciduous forest of West Bengal in India. Biodiversity and Conservation. Biodiversity and Conservation 18, 2343-2350.

Saha, H. K., Haldar, P., 2013. Response of acridid diversity (Orthoptera: Acrididae) in different disturbed habitat across seasons on dry deciduous forest. Caspian Journal of Applied Sciences Research. 2(7), 26-35.

Tilman, D., Wedin, D., Knops, J., 1996. Productivity and sustainability influenced by biodiversity in grassland ecosystem. Nature 379, 718-720.

Ueckert, D. N., Hansen, R. M., 1972. Egestion time of grasshoppers determined by a microscopic technique. J. Eco. Entomol. 65(5), 1263-1265.

Uvarov, B., 1966. Grasshoppers and locusts: a hand-book of general acridology, I. Cambridge University of press, Cambridge.

Van Hook, R. I., 1971. Energy and Nutrient Dynamics of Spider and Orthopteran Populations in a Grassland Ecosystem. Ecological Monographs 41(1), 1-26.

Vats, L. k., Mittal, K., 1991. Population density, biomass and secondary productivity of Orthoptera of Forest floor vegetation. Indian Journal of Forestry 14, 61-64.

Vitousek, P. M., Mooney, H. A., Lubchence, J., Melillo, J. M., 1997. Human domination of earth's ecosystems. Science 277, 494-499.

Walker, B., 1995. Conserving biological diversity through ecosystem resilience. Conservation Biology 9, 747-752.

Yachi, S., Loreau, M., 1999. Biodiversity and ecosystem productivity in a fluctuating environment: the insurance hypothesis. Proc. Nat. Acad. Sci. (USA). 96, 57-64.

– Segment 2 –

Environmental Toxicology of Insects

Chapter 5

Influence of Heavy Metals on the Hatching Abilities and Embryonic Duration of Two Common Indian Short-Horn Grasshoppers

☆ *Chandrik Malakar*

ABSTRACT

Effects of $HgCl_2$, $CdCl_2$ and $ZnCl_2$ on hatching success and embryonic duration of eggs of Oxya fuscovittata and Spathosternum prasiniferum (Acrididae) were studied under laboratory conditions. Same aged healthy eggs from laboratory cultures were incubated in sand treated with different concentrations of the metals. Compared with the control, egg hatching significantly decreased with the increase of metal concentrations in both the experimental species. Similar trends were also observed in case of embryonic duration.

Introduction

Heavy metals are one of the many environmental hazards that organisms can face. Various anthropogenic sources such as mining (Navarro *et al.*, 2008) smelting procedures (Brumelis *et al.*, 1999) and agriculture (Vaalgamaa *et al.*, 2008) are the chief sources of heavy metal pollution in recent past. Along with these, natural activities as well as chemical and metallurgical industries are the most important pathways of heavy metals in the environment (Cortes *et al.*, 2003). Transport of heavy metals from the atmosphere to the soil takes place by dust fall, bulk precipitation and gas or aerosol adsorption (Andersen *et al.*, 1978).

Eggs are the first life stage of a new generation and the embryonic development is very important in insect life history. In grasshoppers, eggs are immobile and remain in contact with soil; they cannot surpass environmental pollution. Like adult ones their resistance to soil pollutants might be relatively weak (Devkota and Schmidt, 1999). Hatching of eggs is an important event in insect reproduction. Hatching success depends in part on the quality or health of the embryo in an egg. Because pores and tiny channels in the egg hull allow egg respiration and water uptake from the environment, heavy metals in soil may enter eggs along with water that are toxic to embryos and consequently reduce hatching success (Xu *et al.*, 2009). These noxious substances in the soil influence the vitality and reproduction of grasshoppers, and eggs laid in heavy metal loaded soil exhibit a marked reduction in hatching rate (Schmidt *et al.*, 1991; Devkota and Scmidt 1999).

One approach to study the effects of heavy metals on eggs is to expose healthy eggs to different concentrations of metals in soil and measure effects on hatching success. In the present study, healthy eggs of *Oxya fuscovittata* and *Spathosternum prasiniferum* (Orthoptera: Acrididae) were exposed to Hg, Cd and Zn at different concentrations in soil under laboratory conditions, and their hatching success and embryonic duration was noted.

Materials and Methods

Insects and Eggs

Oxya fuscovittata and *Spathosternum prasiniferum* were obtained from the insectaria of Entomology Research Unit of the Zoology Department in Visva-Bharati University, Santiniketan, India. In order to obtain a colony of the acridid species of interest, mass culture strategies proposed by Hinks and Erlandson (1994) was adopted with slight modifications according to Haldar *et al.* (1998). Same aged individuals were separated and allowed to oviposit in sterilized sand trays and collected with special care within an hour after oviposition.

Preparation of Sterilized Sand

Sand (approximately 0.05 mm diameter) was collected from the bank of Kopai river, near the Department of Zoology, Visva-Bharati University. The samples were thoroughly washed with running tap water and rinsed with de-ionized water to remove any soil or metal particles attached to it. After washing the sand was oven dried and kept for the experiment.

Experimental Set Up

The experimental set up was followed by the method of Xu *et al.* (2009) with slight modifications. Grasshoppers lay eggs in moist soil. To recreate such an environment for the experiment, sand was provided as the incubation medium. Sterilized dried sand was taken in plastic cups (10cm deep and 6 cm diameter) with 100 gm of oven dried sand for each species. For every concentration of metal chlorides and controls, five replications were taken keeping single egg pod per cup for either grasshopper species. Thus a total of 180 plastic cups were used for the experiment. After that egg pods laid on the same day were put in separate cups for both the species.

Sand was contaminated by adding solutions of $HgCl_2$, $CdCl_2$ and $ZnCl_2$. Solutions were made by dissolving salts into double distilled water and further diluting them. The diluted solutions were added into plastic cups containing same amount of sand to obtain a range of metal concentrations of 0 mg/kg (control), 5 mg/kg, 10 mg/kg, 20 mg/kg, 40 mg/kg and 80 mg/kg dry sand.10 ml solutions of different concentrations of these heavy metals were given in each plastic cup for moistening the substrate (sand) to achieve 70-80 per cent humidity. Any loss of moisture due to evaporation was compensated by adding proper amount of demineralized water.

All the plastic cups were placed in an environmental chamber at 35±2°C for 7 weeks. When the nymphs started hatching, their numbers were recorded twice daily. The period between start of treatment and hatching was regarded as the embryonic duration. Developmental periods of each nymph from each set were considered to calculate the mean duration of development. After completion of the experiment (7 weeks) the number of undeveloped eggs was recorded from each set. This could provide information about the effects the heavy metals have on egg mortality.

Analysis of Data

All the data for successful hatching and embryonic duration for each metal treatment were subjected to one way analysis of variance (ANOVA) and Tukey's posthoc test were carried out to compare the values between different doses using PAST version 3.02.

Results and Discussion

Effects of Heavy Metals on Hatching

Experiment shows that number of hatchlings reduces with increased doses of metal concentration. In *O. fuscovittata* Hg, and Zn shows decrease in hatching success significantly ($F=65.85778$, $P<0.01$ for Hg; $F=49.92821$, $P<0.01$ for Zn), when metal concentrations reached 20 mg/kg, 40 mg/kg and 80 mg/kg dry sand. But for Cd significant decrease ($F=47.96735$, $P<0.001$) was observed only in the two highest treatments. Zn exerted less effect on *S. prasiniferum* where in the highest dose significant decrease (39.67273, $P<0.001$) was found but in Hg and Cd treatments hatching success decreased significantly from second dose onwards ($F=65.31282$, $P<0.001$for Hg; $F=73.54667$, $P<0.001$for Cd) (Table 5.1).

Results of the present study are in concert with several other workers. While working on *Aiolopus thalassinus* Devkota and Schmidt (1999) observed that egg hatching rate decreases with the increase of Hg, Cd and Pb concentrations. In *Oncopeltus fasciatus* reduction in the percentage of hatchlings throughout the oviposition period was observed with increasing Cd concentrations in drinking water (Cervera *et al.*, 2004). Xu *et al.* (2009) observed that increased Zn concentration results in the decrease of egg hatching in *Folsomia candida*. Some authors have also attributed the reduction in hatchability to a possible harmful effect Hg and Cd have on embryonic development (Schmidt *et al.*, 1991 and Cervera *et al.*, 2004). The present study supports these facts.

Table 5.1: Effects of Hg, Cd, and Zn on Number of Hatched Eggs per Egg Pod in *O. fuscovittata* and *S. prasiniferum* after 7 Weeks Exposure in Soil

Metal Concentration (mg/kg dry sand)	Eggs hatched per Pod					
	O. fuscovittata			*S. prasiniferum prasiniferum*		
	Hg	Cd	Zn	Hg	Cd	Zn
Cont	10.60±0.89a	10.60±1.14a	10.80±1.0a	9.20±0.45a	9.80±0.45a	9.40±0.89a
5	9.20±0.84a	9.80±0.84a	10.20±1.48a	8.80±0.84a	9.20±0.45a	9.60±0.55a
10	7.60±0.89a,b	8.80±0.84a	9.20±0.45a	7.40±0.55b	8.00±0.70b	8.60±0.55a
20	6.20±0.45b	7.20±0.84a	8.20±0.45b	6.60±1.14b	7.80±0.84b	7.80±0.84a
40	3.80±0.84c	4.80±0.84b	5.40±0.55c	3.80±0.84c	4.80±0.84c	6.80±0.84a,b
80	2.40±1.14c	3.40±0.89b	4.60±0.55c	1.80±0.84d	2.80±0.84d	4.00±0.70b

Notes: Means within a column bearing the same letter were not significantly different (P<0.05)

Effects of Heavy Metals on Embryonic Duration

In case of both *O. fuscovittata* and *S. prasiniferum* a trend of prolonged embryonic duration was observed along with increase of doses of all the three metals. But significant increase was observed in the two higher doses of Cd and in the highest dose of Hg in case of *O. fuscovittata*. However, Zn showed insignificant results in those mentioned doses (Figure 5.1). On the other hand in *S. prasiniferum* significant

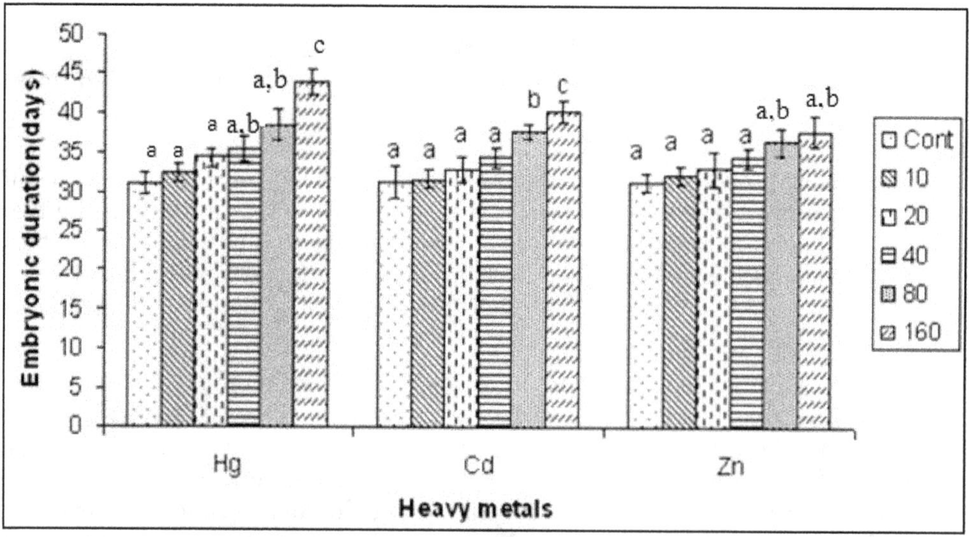

Figure 5.1: Effects of Hg, Cd, and Zn on Embryonic Duration of *O. fuscovittata*. Each bar represents mean ± SD. Different superscripts denote significant differences between the mean values (P<0.01).

increase was observed only in the highest dose of Zn but in case of Hg and Cd insignificant results were observed (Figure 5.2).

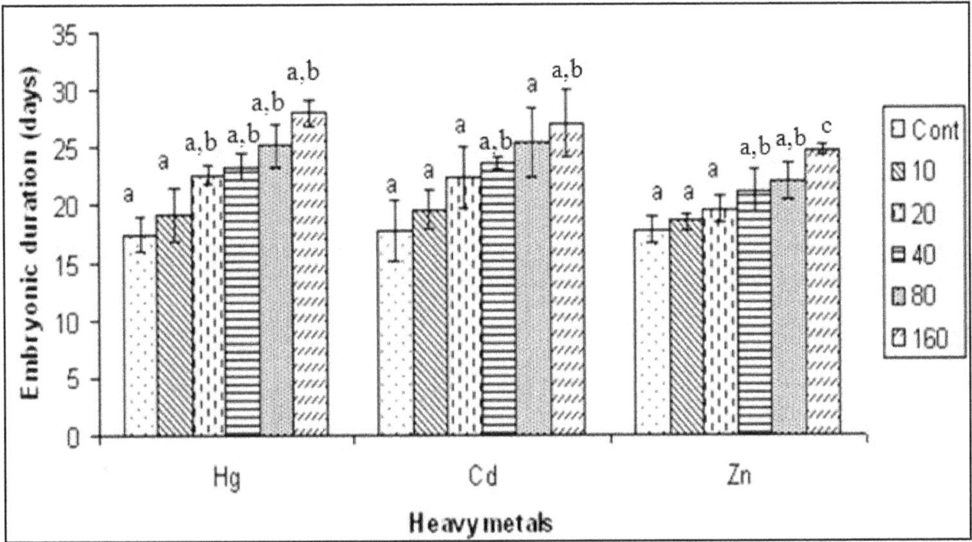

Figure 5.2: Effects of Hg, Cd, and Zn on Embryonic Duration of *S. prasiniferum*. Each bar represents mean ± SD. Different superscripts denote significant differences between the mean values (P<0.01).

Devkota and Schmidt (1999) observed that differences in the concentrations of Hg and Cd were found to have no influence on the mean duration of *A. thalassinus*. But in the present experiment the case was somewhat different. I conclude that heavy metals reduce the hatching rate and increases the embryonic duration in acridid eggs.

Conclusion

In the present study I have determined the toxicity of selected heavy metals on hatching abilities and embryonic duration of two short horned Indian grasshoppers, *Oxya fuscovittata* and *Spathosternum prasiniferum*. The results revealed that in *O. fuscovittata* the range of severity of Hg and Zn is more than Cd. On the other hand in case of *S. prasiniferum* Zn exerts less effect than Hg and Cd. This reveals that selective permeation of various metal ions is in part responsible for determining the embryonic development in non shelled eggs. In nature gradual increase of heavy metal lode both from anthropogenic and geogenic sources may pose a serious problem among the insects who lays their eggs in soil. The present experiment also revealed that heavy metals produces prolonged embryonic duration particularly in Hg and Cd treatment groups of all the experimental acridid species indicating both these metals delay the embryonic pathways during development. Whether effects (low hatching rate or prolonged embryonic duration) are temporary or permanent is particularly important in terms of overall impacts on fitness and survival of grasshoppers in the wild. Nymphal mortality in grasshoppers is normally very high, but further decrease in hatching and increase in embryonic duration can additionally result in major decreases in populations. In addition heavy metals could get magnified in these insects due to their phytophagous nature and the food (mainly the grass) growing on metal polluted soil which could join hand with the present findings and may cause mass mortality and may lead these species on the verge of extinction.

Acknowledgement

Author is thankful to the Head of the Department of Zoology, Suri Vidyasagar College for providing laboratory facilities. I am also thankful to the Director, Zoological Survey of India, Kolkata for his kind help regarding identification of the grasshopper species.

References

Andersen, A., Hovmand, M.F., Johnsen, I. 1978. Atmospheric heavy metal deposition in the Copenhagen area. *Environ. Pollut.*, 17 (2): 113-132.

Brumelis, G., Brown, D.H., Nikodemus, O., Tjarve, D. 1999. The monitoring and risk assessment of Zn deposition around a metal smelter in Latvia. *Environmental Monitoring and Assessment.*, 58 (2): 201-212.

Cervera, A., Maymo, A.C., Sendra, M., Martinez-Pardo, R., Garcera, M.D. 2004. Cadmium effects on development and reproduction of *Oncopeltus fasciatus* (Heteroptera:Lygaeidae). *Journal of insect physiology.*, 50:737-749.

Cortes, O.E.J., Barbosa, L.A.D., Kiperstok, A. 2003. Biological treatment of industrial liquid effluent in copper production industry. *Tecbahia Revista Baiana de Tecnologia.*, 18(1): 89-99.

Devkota, B. and Schmidt, G.H. 1999. Effects of heavy metals (Hg2+, Cd2+, Pb2+) during the embryonic development of acridid grasshoppers (Insecta,Caelifera) *Arch. Environ. Contam. and Toxicol.*, 36: 405-414.

Haldar, P., Das, A., Gupta, R,Kr. 1998. A laboratory base study on farming of an Indian grasshopper *Oxya fuscovittata* (Marschall) (Orthoptera: Acrididae). *Journal of Orthoptera Research.*, 8:93-97.

Hinks, C.F., Erlandson, M.A.1994. Rearing grasshoppers and locust: Review, Rationale and update. *Journal of Orthoptera Research.*, 3: 1-10.

Navarro, M.C., Pérez-Sirvent, C., Martínez-Sánchez, M.J., Vidal, J., Tovar, P.J., Bech, J. 2008. Abandoned mine sites as a source of contamination by heavy metals: A case study in a semi-arid zone. *Journal of Geochemical Exploration.*, 96 (2-3): 183-193.

Schmidt, G.H., Ibrahim, N.M.M., Abdallah, M.D. 1991. Toxicologycal studies on the long-term effects of heavy metals (Hg, Cd, Pb) in soil on the development of *Aiolopus thalassinus* (Fabr.) (Saltatoria, Acrididae). *Science of the Total Environment.*,107:109-133.

Vaalgamaa, S., Conley, D.J.2008. Detecting environmental change in estuaries: Nutrient and heavy metal distributions in sediment cores in estuaries from the Gulf of Finland, Baltic Sea. *Estuarine, Coastal and Shelf Science., 76*(1): 45-56.

Xu, J., Wang, Y., Luo, Y-M., Song, J and Ke, X. 2009. Effects of copper, lead and zinc in soil on egg development and hatching of *Folsomia candida. Insect Science.,*16: 51-55.

– Segment 3 –

Insect Taxonomy

Chapter 6

An Introduction to True Fly: Classification and Diversity of Diptera

☆ *Nilotpol Paul and Abhijit Mazumdar*

ABSTRACT

The Diptera or true flies (bots, flies, gnats, maggots, midges, mosquitoes, etc.) comprise 10-15 per cent (1,50,000) of existing animal species (1.5 × 10^6). The order is one of the four largest groups of living organisms. The living dipteran species have been classified into about 10,000 genera, 150 families, 22-32 superfamilies, 8-11 infraorders and 2 suborders. Of these, some 3,100 species are known only from fossils. There are more known flies than vertebrates. These insects are a major component of virtually all non-marine ecosystems. The major lineages within the order are now well established and summarized in phylogenetic relation for the Diptera. Studies concur that the traditional subordinal group Nematocera is paraphyletic, but relationships between the major lineages of these flies are not recovered consistently. The other major suborder, Brachycera, is clearly monophyletic, and the relationships between major brachyceran lineages have become clearer in recent decades. Phylogenetic tree will span the diversity of the order with increased quantitive rigor, which will enable a more clear understanding of the classification and phylogeny. The economic importance of the group is immense and are the most ecologically diverse order of insects, spanning ecological roles (ecological management) from detritivory to leaf mining, from vertebrate blood feeding to transmit pathogen. On the other hand, as food, biological pest control, pollination, forensic uses are other way by which human are benefited. Here in this review, a comprehensive idea of most important dipteran family with its characteristic features are summarized along with their diversity and biology. Many of them are directly influence in our daily life, and make it more comfortable and hazardous.

Introduction

The order Diptera (the true flies) is one of the most specious, morphologically varied and ecologically innovative groups of organisms, making up 10–15 per cent of known animal species. An estimated 150,000 species of Diptera have been described

(Thompson, 2005), however, the actual number of extant fly species is many times than that number. The living dipteran species have been classified into about 10,000 genera, 150 families, 22–32 superfamilies, 8–11 infraorders and 2 suborders (McAlpine and Wood, 1989; Yeates and Wiegmann, 1999; Thompson, 2005) (Figure 6.1), and around 3100 fossil species have been described (Evenhuis, 1994), although substantive information is available only for few families. Because their small size and delicate nature of many families, they are uncatchable. A phylogenetic tree gives the modern scheme of classification among infraorder and superfamily. With two significant morphological modifications (considered as synapomorphies), transformation of the hind wings into halters and sponging-sucking mouthparts, monophyly of Dipteran is well established in phylogeny (Hennig, 1973; Wood and Borkent, 1989; Griffiths, 1996). Such specious order has produced variation in biology, ecology, phenology and life history trait. These differences indeed came in this order because this order faced a great climatological changes or rather a shift during all these 250 Million years of their evolution. Biology of this order is an enigma with the most sophisticated and specialized life cycle which has enabled them to adapt in all possible weathered and clametic conditions; moreover it gives an advantage over pheno-ecological aspect that makes them probably the most

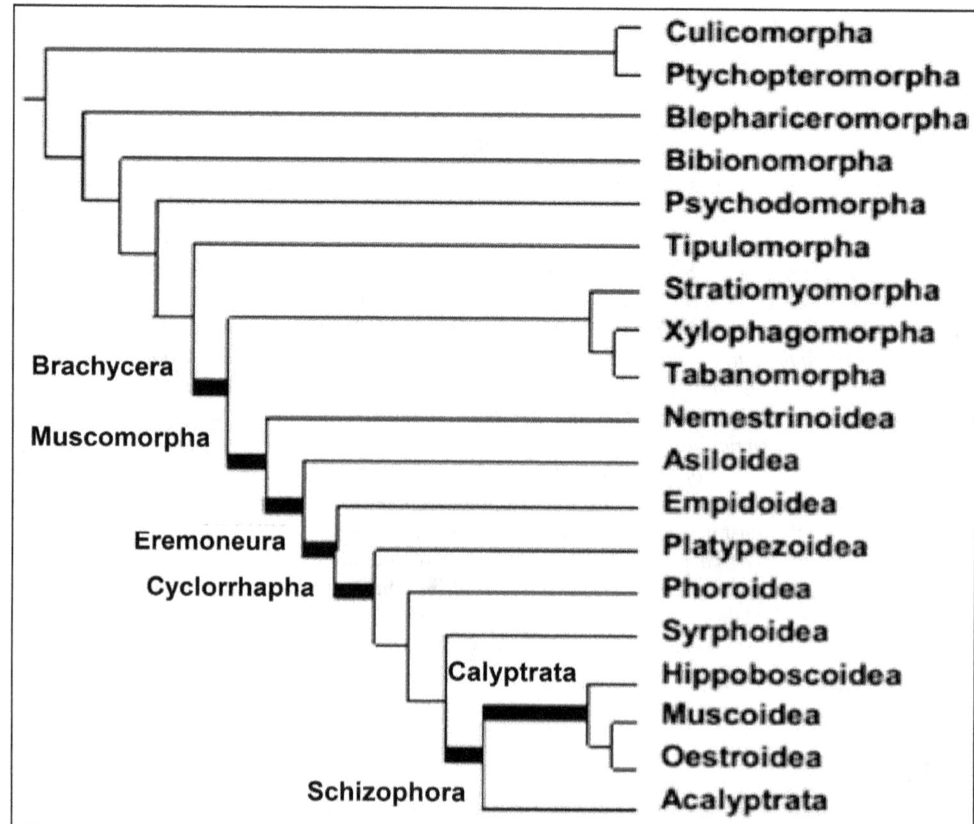

Figure 6.1: Relational Tree of Infraordinal and Superfamily Level of Order Diptera (Adopted from Yeates *et al.*, 2007).

ecologically diverse of the four megadiverse insect orders (Kitching *et al.*, 2005). Perhaps Diptera is the economically most important, in terms of both profits and loss. On one hand this insect order is the main causative agent of highest number of death per annum in human as well as other animals, while on other it blesses upon our modern society in terms of pest control, food security, interrogating criminal, cosmetics industry, chocolate industry, robotics and space sciences. In this review, almost all ecologically and economically important families (96) are briefly discussed with characteristic identification features and their eco-phenological aspect. In this review, we have highlighted the peculiarity of the life history trait of dipteran insects and how this features are related to our humanity. This review will also be useful as a manual for general Dipteran family as a whole with a revised classification scheme.

Description of Families

Perhaps classification scheme of Diptera is also controversial, according to Lindner (1949) Cyclorrhapha is a suborder but later many workers include it in the Brachycera. A peculiar family is Nymphomyiidae, placement of which has created much debate. According to Rohdendarf's hypothesis, it is a separate entity, which share primitive characters of dipteran (Hackman and Väisänen, 1982). Therefore, he created an infraorder nymphomyommorpha, and a suborder Archidiptera for this taxon, but later it has been placed under Nematocera. Thereafter the theme of classification and systematics changed during last few decades and here, general classification of Diptera is given following the sequence of the present phylogenetic tree, which is depicted in Linnaeus Tercentenary: Progress in Invertebrate Taxonomy (Griffiths, 1972; Hennig, 1973; Wood and Borkent, 1989; Yeates *et al.*, 2007 and McAlpine, 1989). Some of the families are still not included here because of scanty information, some others are also excluded, as they are not of human interest.

Order DIPTERA

Suborder NEMATOCERA

1. Infraorder CULICOMORPHA

Superfamily CULICOIDEA

Family DIXIDAE (Dixid Midges)

These midges are fragile, slender, yellow, brown or black, without scales on the wings or body. Eyes more widely separated above and without ocelli. Antennae long, 16-segmented, with 2nd segment large and globose; male antennae pilose. Wings large, with variable markings or hyaline; R_{2+3} strongly arched before it forks. Legs long and slender; hind tibia usually somewhat expanded apically. Adults do not bite. Short-lived midge and it is not certain whether they eat anything. During daytime they take rest, vertically and head up, on vegetation or on the ground. Males form swarms at dusk near vegetation along the edges of streams or ponds and females fly into the swarms to mate; mating can also occur in daytime without swarming. The slender larvae strain diatoms and other micro-organisms from the water with complex mouthparts. They typically rest in a U-shaped position at the surface film, usually against some substrate or plant near the water's edge; they swim by jerking the front half of the body from side to side. The family consists of approximately 200 described extant species worldwide.

Family CHAOBORIDAE (Phantom Midges)

Chaoboridae are morphologicaly closely related to Culicidae. Adults are delicate (1.5-10 mm), pale yellow to grey or brown in colour. Eyes separated in both male and female; ocelli absent. Male antennae plumose, 2nd segment large and globose. Proboscis short, extending only slightly past the clypeus. R 4 branched, 3 branches of vein Rs are almost straight and parallel; M and Cu are two-branched. Wing hind margin and most veins have scale-like setae; light and dark pigments in the membrane sometimes form wing pattern. Immatures truly aquatic. Adults are often found in large mating swarms. unlike mosquito, they do not bite. Most larvae are predators and show cannibalistic behavior, their antennae are prehensile and are used to grab prey. The larvae of *Chaoborus* are transparent (and thus called phantom midges); gas-filled air sacs at each end of the body help them maintain their position in the water of lakes and ponds.The Chaoboridae is a small cosmopolitan family, represented by 50 described extant species in six genera. The closely related family Corethrellidae is mainly tropical, but this midge is found in temperate zone also.

Family CORETHRELLIDAE

Corethrellidae is a family of tiny parasitic midges with a wing length of 0.6-2.5 mm. these midges are commonly known to parasitize frogs. Wing venation similar to Culicidae (R4, M2 and Cu2 branched) with branches of Rs and M nearly parallel. R1 closer to Sc or almost midway between Sc and R2.Adult female *Corethrella* is attracted to the mating calls of male frogs, their chosen host taxa. As obligate external parasites, the midges feed almost exclusively on the blood of these frogs. Because of this, *Corethrella* follow typical distribution patterns of external parasites and are restricted to only areas with abundant populations of their host frogs. Female midges most likely detect their hosts using a specialized organ called a Johnston's organ, a collection of sensory cells found on the second antenna segment. There is evidence of host specificity and selection of particular biting sites for some species.*Corethrella* species feed blood from individuals of the tree frog genus *Hyla* specifically; *H. avivoca*, *H. cinerea* and *H. gratiosa* were recorded as confirmed corethrellid hosts. They were until 1986 placed as a subfamily of Culicidae. A few, select species are known vectors of frog-specific species of the parasitic protozoan *Trypanosoma*. Corethrellid parasitism is thus a recorded cause of trypanosomiasis among host frog populations. The family contains members that date to the lower Cretaceous period some 110 million years ago. The family currently consists of just two genera, totalling around 97 species worldwide while few species are also known from fossile form. Most extant species are found in the lower latitudes, usually associated around the tropics.

Family CULICIDAE (Mosquitoes)

Mosquitoes are delicate (3-9 mm), long-legged, slender flies, and whole body clothed with scales. These scales vary in colour and often form patterns that are useful in species identification. Head globose and the compound eyes are concave medially where they meet at the bases of the antennae. Ocelli are absent. The 1st segment of the antenna small, 2nd large and spherical and rest 13 are slender and bear a whorl of setae. Mouthparts are elongate, stylet-like and enclosed in a sheath formed by the labium. In the male, the maxillary palp is about as long as the proboscis. Wings are narrow, long and lie flat above the abdomen when at rest; there is no discal cell and the single vein R_{4+5} lies between two branched veins – R_2 and R_3 in front and M_1 and M_2behind. Probably these are the most notorious and fascinating insects that is center of insect research due to their medical importance. Mosquitoes are common, widespread and well-known insects. Females of most species bite and feed on the blood of vertebrates, but not all biting species feed on human blood. Some mosquitoes transmit disease organisms to humans and other animals – malaria, filariasis, yellow fever, dengue, encephalitis and West Nile virus. Adult mosquitoes also pollinate many of the plants that they visit, especially various native orchids. Larvae and pupae are aquatic and live in marshes, ponds, pools, water-filled tree holes, man-made containers, and other places where water is get collected. They come to the water surface to breathe; most larvae eat organic debris and micro-organisms; a few are predators of invertebrates. At present, 3,540 recognized species, divided between 2 subfamilies and 112 genera are recorded in the world, while In India, 393 species in 49 genera and 41 subgenera are known. A total of 31 species are currently recognized in India for transmitting various mosquito borne agents of human diseases (Bhattacharyya *et al.*, 2014).

Superfamily CHIRONOMOIDEA

Family THAUMALEIDAE (Solitary Midges)

Thaumaleids are small (2-3 mm), stocky flies, and mostly shiny yellow to dark brown in colour. Head are small and globose; eyes meeting above in both sexes; ocelli absent. Antennae short, as long as head and project forward. 1st and 2nd segments large and spherical, while next 3 small and square; the last 7 are linear. Thorax robust; scutum strongly arched dorsally and without transverse suture, scutellum large and pointed. Wing broad, with the membrane covered with macrotrichia in *Trichothaumalea*, these setae only on the veins in *Thaumalea*. C reaches the wingtip; Rs usually forked to produce very short, nearly vertical R_{2+3} (meeting R_1) and a long, curving R_{4+5} reaching the margin. M_1, M_2, CuA_1 and CuA_2 normally strong and reaching margin. The legs are rather slender and lack tibial spurs and pulvilli and minute empodia. Adult flies are seldom seen; hense they are so named. They live around streamside vegetation. Larvae live on wet rocks (usually on vertical surfaces in the shade) in cold streams. They scrape diatoms off the rocks where the water flows as a film not thick enough to cover their bodies. Pupae lived in moss, wet leaves and muddy slime on the stream margins. Five genera containing about 120 described extant species are placed in the Thaumaleidae worldwide. Most of the species are Holarctic; two of the genera and 8 species live in the temperate parts of the southern hemisphere.

Family SIMULIIDAE (Black Flies)

Simuliidae are small (1.2-5.5 mm) stocky insects. They are completely black or dark brown, sometimes, grey, rust, orange or yellow. Head usually large and round, eyes meeting on top of the head in males, separated in females, and ocelli absent. Antennae consist of 9 to 11 bead-like segments, short and thick. Dorsally arched thorax found in males and is usually covered with short, dense, recumbent setae. Scutellum more or less triangular and densely clothed in long setae. Front tibia has an apical spur, the others have two; 1st segment of the hindleg tarsus elongate, which is often swollen in males. Anterior veins broad and weak posterior vein. Vein Rs simple or has a long fork; rarely the fork is short and obscure. A characteristic false vein (m-cu fold) is usually forked apically, but unbranched in *Parasimulium*. Female mouthparts are specialized for lecerate and sucking, while in some females and all males it is only for imbibing fluids such as water and nectar. However, not all species bite humans and some eat no blood at all. Through blood feeding, many species transmit parasitic disease organisms among birds and mammals; the worst is the filarial nematode that causes human onchocerciasis in the tropics of Africa and the New World. Larvae live in flowing water, attaching them by the tip of the abdomen to a pad of silk that they fix to submerged objects. Most feed by filtering food out of the water with fan-like mouthparts; some lack these labral fans (*e.g.*, *Gymnopais, Twinnia*) and graze off the substrate. The Simuliidae is homogeneous family of almost 1800 described living species.

Family CERATOPOGONIDAE (Biting Midges)

Biting midges are moderate (1-6 mm), and slender to rather stout. The compound eyes usually or mostly meet on top of the head (but well separated in Leptoconopinae); they are usually bare, but sometimes are finely setose. There are no ocelli. Antennae have 8 to 15 segments, although vertebrate feeders have 13 to 14 (*Leptoconops*); females have the last five segments elongated and most of males have plumose antennae. The proboscis is about as long as the head; most females have serrate mandibles. A pair of humeral pits often occur near the front edge of the mesonotum. The wing typically has one to three compact radial veins close behind the front edge of the wing and reaching the wing margin before the wing tip; two median vein branches reach the wing margin; crossvein r-m usually strong. The wings overlap over the abdomen when at rest, often patterned with dark or light spots or patches. The females of predatory Ceratopogoninae usually have at least one pair of raptorial legs with swollen, spiny femora or with enlarged tarsal claws. Larvae of biting midges live mainly in moist habitats around the aquatic or semi-aquatic environments. Most species of *Culicoides* fly at dusk, but those of some genera, such as *Leptoconops*, fly during the day. Females of many species suck blood to provide protein for egg maturation, and many are notorious for both their biting and for transmitting disease. Their small size enables most to crawl through the mesh of screens and, when large numbers of biting females are present, avoiding them outdoors is difficult. *Culicoides* species are vectors of filarial nematodes, blood protozoans and viruses such as bluetongue in livestock. *Culicoides, Leptoconops, Austroconops* and some *Forcipomyia* are the only genera that feed on vertebrates, usually mammals or birds, but also reptiles, amphibians and even the amphibious mud skipper fish of Southeast Asia. Many *Forcipomyia* drink the blood of large insects such as dragonflies, katydids and butterflies, usually feeding from wing veins. Other species kill small swarming flies and mayflies; some females eat males of their own species. *Dasyhelea* and some other genera feed only on nectar, and many others supplement their diet at flowers. Some are important pollinators of cocoa (from which chocolate is derived), rubber trees and other plants. The larvae of the most diverse subfamily, the Ceratopogoninae, run the gamut from burrowers in wet soil and manure to active swimmers in the waters of large lakes and rivers. Many are carnivorous. Those of the Leptoconopinae live in the soil and sand of arid habitats or the beaches of oceans and inland waters, feeding on microorganisms. Forcipomyiinae crawl in moist places such as moss mats or under bark, eating algae and fungi; Dasyheleinae wriggle in the fluids of sap flows, tree holes and other small, wet habitats. The fossil record of the Ceratopogonidae is rich, especially in amber, and reveals that the family was abundant and diverse at least 120 million years ago. Till date there are 6700 described extant species under 103 genera, but this is probably much less than half of the actual total (Borkent, 2016).

Family CHIRONOMIDAE (Chironomid Midges)

Chironomid midges are delicate, small to medium-sized flies (1-10 mm) with, slender legs and narrow wings. Various coloured, *viz.*, brown, black, green, reddish, yellow species occur; some have banded and patterned wings, abdomen and legs. Eyes usually separated, may be bare or setose. Ocelli are absent, although the frontal tubercles on few species may be modified ocelli. Antennae 3 to 17 segmented, usually with more segments in males than in females; plumose in males and pilose in famales. Mouthparts reduced, mandibles absent. Scutellum hemispherical; postnotum large, and usually bare and marked with a median longitudinal furrow. Legs often have the tarsus of the foreleg elongated and sometimes strongly setose. Wings lie flat when at rest. Costal vein usually is fused with R_{4+5} near the wingtip, but rarely reaches the wingtip. Subcosta usually ends before reaching the costa. Radius three-branched, normally more strongly sclerotized than the posterior veins; R_{2+3} often weak, sometimes absent, and branches in the Tanypodinae. The medial vein is straight and unbranched, meeting the wing margin near the wingtip. Vein CuA forked at, or past the r-m crossvein. These non biting midges are most ubiquitous and abundant among the freshwater insects. Adults are crepuscular, active swarms could be seen for mating. Although they do not bite, their enormous presence may create nuisance, pest in rice ago-ecosystem and also creat allergy. They thrive in almost all types of fresh water including brackish habitats or even in marine environment. They are found from the deepest lakes to 3 Km underground cave, tunnelling in rotting wood or mining the tissues of aquatic plants even as parasitically. Most chironomid larvae eat detritus and microscopic plants and animals; most live on or in the substrate, usually in tubes made from substrate particles bound with salivary secretions. Others, especially species in the subfamily Tanypodinae, prey on macroinvertebrates such as other chironomid larvae. The major groups of midges generally prefer certain types of environments: thus, the Podonominae and some other subfamilies are cold-adapted, living mainly in rapid streams; species of the Orthocladiinae, too, mostly prefer cool waters. The Chironominae are warm-adapted and most live in still waters, often where oxygen levels are depleted, most species of this subfamily live in or on the substrate and have oxygen-binding hemoglobin dissolved in their blood (the so-called "bloodworms"). Tanypodinae are free-living predators, and when oxygen levels drop, they can swim to surface waters where oxygen is more abundant. Through detailed analyses of these groups and their preferred environmental conditions, species composition of chironomid larvae has been used to classify lakes ecologically, to determine levels of pollution and to examine the historical and prehistorical changes in lake systems. Chironomidae is the largest and cosmopolitan family of over 7000 species worldwide including Antarctic and Arctic (Paul, 2016).

2. Infraorder PTYCHOPTEROMORPHA

Family PTYCHOPTERIDAE (Phantom Crane Flies)

Ptychopteridae is moderate sized, slender, long-legged flies. The head is transverse, tightly held against the thorax and the antennae are long, with 15 (*Ptychoptera*) to 20 or 21 segments (*Bittacomorpha, Bittacomorphella*). Mesonotum with transverse suture forming a strong loop rearward.Tibiae spurred. The wings with Rs short and R_{2+3}very close to R_1; R_4 and R_5 present. In Ptychopterinae, M_{1+2} forked but not in Bittacomorphinae; outer part of Cu A_2strongly sinuous; only one anal vein present. The halter has an unusual prehalter. The most striking members of the family are the two species of *Bittacomorpha*, the so-called 'phantom crane flies', which have the first segment of the tarsi inflated and filled with tracheae. This feature enables the flies to sail on the breeze, black and white banded legs outstretched. The striking habit gives them a ghost-like appearance, especially when seen in the dappled shade of the forest. The larvae live in wet mud and organic debris at the edges of forest streams and pools; they have long, posterior breathing siphon. The Ptychopteridae consists of over 60 described extant species in only three genera. There are two distinct subfamilies. The Ptychopterinae occurs in all regions except the Australian and Neotropical; its sole genus, *Ptychoptera*, has many species. The Bittacomorphinae contains *Bittacomorpha*,with two Nearctic species, and *Bittacomorphella* with four Nearctic and three eastern Asian species.

3. Infraorder BLEPHARICEROMORPHA

Family BLEPHARICERIDAE (Net-winged Midges)

These net-winged midges are delicate, slender (3-13 mm), long-legged flies and generally grey or dull yellow-brown coloured. Eyes large, finely pubescent and divided into two portions; the upper part normally reddish and is made up of larger facets, while the lower part is black. Eyes of male usually meet on top of the head along with three stalk ocelli. Antennae are usually short, with 11 to 13 barrel-shaped segments. Most females have slender, serrated rasping mandibles. Wings are broad with a prominent anal angle; the radius has only ≥ 3 branches reaching the wing margin in most species; M_2 is completely detached basally. A net-like pattern of fine folds present throughout the wing membrane, hence the family's common name. Net-winged midges are denizens of mountain streams, where the larvae cling to rocks in rapids and waterfalls, scraping water micro and mesofauna from the rock surfaces. The larvae are well-adapted to fast-flowing water, having a cephalothorax composed of the fused head, thorax and first abdominal segment; six ventral suctorial discs attach strongly to smooth surfaces. The pupae are also attached to rocks, often in the hundreds, all aligned in the same direction. Adults usually live near streamside rocks or in riparian vegetation and survive only a week or two. Female adults of many species are predators of soft-bodied insects such as mayflies, stoneflies and chironomid midges. Feeding habits of males unknown and females lacking mandibles but it consider that they might feed on nectar. Eggs are cemented to streamside rocks although some species oviposit while entirely submerged in the water. They are distributed worldwide; the 300 described species are placed in 27 genera. The family is usually divided into two subfamilies, the Edwardsininae in the southern hemisphere and the Blephariceriniae in both northern and southern hemispheres.

Family DEUTEROPHLEBIIDAE (Mountain Midges)

Mountain midges are delicate flies, approx 3 mm long, with silvery-blue wings and reduced mouthparts. Thorax dark brown or black, broad and arched, projecting over the head; the tapered abdomen is pale brown in male and dark green in female. Head small, transverse and rather flattened, with the small eyes below the antennae. Eyes lack pubescence and ocelli. Antennae with only six segments, but the terminal segment is greatly elongated in males, antenna = 4 × body length. Wing veins are much reduced however, a fan-like network of vein-like lines that help in wing folding. Legs are slender, the empodia of males are flattened or circular disc shaped with very long seta; one tarsal claw is slender, the other is reduced; the empodia of females are linear and shorter than the two stout claws. They live along rapidly running streams where the larvae cling (with seven pairs of finely-clawed abdominal prolegs) to smooth, light-coloured rocks near the water surface. Pupae prefer depressions and cracks in dark-coloured rocks. High-altitude species have a single generation a year whereas some species at lower elevations may produce several; species probably overwinter in the egg. Adults live only a few hours. The family Deuterophlebiidae is monogeneric, having only *Deuterophlebia*, found in widely separated regions of the temperate northern hemisphere. The most common species are *D. personata*, *D. inyoensis* and *D. coloradensis*.

Family NYMPHOMYIIDAE

Nymphomyiidae is a family of tiny (2 mm) slender, delicate flies. Under an alternative classification, they are considered the only living representatives of a separate, suborder called Archidiptera (or Archaediptera) which includes several Triassic fossil members. Modern classifications put all the species in a single genus *Nymphomyia*. Based on larval morphology, the family is suggested to be close to Deuterophlebiidae. The antennae are shortened as in the Brachycera and these flies are long, having a snout with vestigeal mouthparts, non-differentiated abdominal segments with large cerci. The wings are narrow and hair-fringed and have very weak venation. They are known to form cloud-like swarms in summer and the short-lived non-feeding adults have wings that fracture at the base shortly after mating. Nymphomyiidae are neotenic, retaining various larval features. Larvae are found among aquatic mosses in small, rapid streams in northern regions of the world, including northeastern North America, Japan, the Himalayas, and eastern Russia. They have strap-like wings with a highly reduced venation, and the wing margins have long fringes like those of the Thysanoptera. The wings break at the base after mating. The antennae are also very much reduced. Species in the genus *Nymphomyia* have atrophied mouthparts. Nymphomyiidae are unusual in that the adults are ventrally holoptic, meaning they possess two eyes that meet on the underside of the head. Adults form large swarms above water. One or two generations may breed in a single year depending on the region and climate.

4. Infraorder BIBIONOMORPHA

Family PACHYNEURIDAE (Pachyneurid Gnats)

Globose and small-headed Pachyneurid gnats are slender, long-winged, have more or less spherical eyes lothed in short setae.Three ocelli placed each on its own tubercle. Antennae have 15 to 17 cylindrical or bead-like flagellomeres, generally shorter than the thorax, except in the males of *Cramptonomyia,* where antennae longer than abdomen. Mesoscutum shallowly convex; transverse suture weak; dorsocentral setae pronounced. The wing has a distinct pterostigma. Vein Rs branches near the r-m crossvein; R_{2+3} is branched in *Pachyneura,* but is unbranched and connected to R_{4+5} by an extra radial crossvein in other genera; discal cell usually present. Abdomen is slender and longer than the wing. Known larvae of pachyneurid gnats bore in rotting wood. For example, larvae of the genus *Cramptonomyia* develop in logs of "red alder" along the Pacific coast; larvae tunnel under the bark or just below the surface of bare wood, taking at least a year to mature. Adults fly from February to April. The Pachyneuridae contains four described extant species in four genera. The Japanese genus, *Haruka,* the eastern Siberian one, *Pergratospes,* and the North American genus, *Cramptonomyia,* are more closely related to each other than they are to the fourth genus, *Pachyneura,* from Europe and Asia and are sometimes placed in a separate family, the Cramptonomyiidae.

Family BIBIONIDAE (March Flies)

March flies are dark, stocky, setose flies and moderate (5-12mm) long. Females' heads are rather flattened, more elongate than males; the eyes are small and widely separated. The compound eyes of males meet on top of the head, often covering much of it; the upper two-thirds have large facets, the lower third has smaller ones. Ocelli are present and prominent. The antennae are usually short with 9 to 12, normally rounded, compact segments; Only *Hesperinus* possess the long and elongated segment. The scutum is prominent, domed. Front femurs are swollen and front tibiae with apical spurs or rings of spines in *Bibio, Bibiodes* and *Dilophus.* In the wing, R_1 ends just beyond the end of Sc; Rs simple or forked; crossvein r-m at or beyond the half-wing. Medial vein forked, CuA_1 and CuA_2 reach the wing margin and A_1 is usually weak. There is frequently a pterostigma. These short-lived flies generaly emerge in late winter or spring (thus the name March flies) with huge mating swarms. Some genera visit flowers and may be important pollinators of fruit trees. Females, aided by spurs or spines on the fore tibiae, burrow into moist earth to lay eggs. Larvae scavenge in decaying organic material, rich soils and forest litter and eat the roots of cereal crops and vegetables, known as a serious pest. Almost 700 described extant species belonging six genera, among them *Bibio* is predominant. *Hesperinus brevifrons* is the sole member of its genus. *Bibiodes aestivus* Melander is a western species and *D. caurinus* is a widespread one. The family apparently reached its zenith in the Tertiary Period; *Plecia* is now predominantly tropical in distribution, but in the Eocene (50-60 Mya), it was abundant in North America (Rice, 1959).

Family MYCETOPHILIDAE (Fungus Gnats)

Fungus gnats are slender to moderate (2.2-13.3 mm) to robust flies. Body usually dull yellow, brown or black, but sometimes brightly coloured. Head usually flattened front to back, and inserted well below level of upper margin of strongly arched thorax. Eyes usually densely setose, situated on lower part of head, and never meeting above antennae; usually with three ocelli; frons between ocelli and antennal bases usually bare; antennae inserted at middle of head, and with length varying from slightly longer than head to several times length of body. Antennal segments usually cylindrical basally thickened and tapering to apex, and usually composed of 14 segments. Labella usually large and fleshy. Thorax varying from compressed and deep to depressed and low; thoracic vestiture variable, consisting of moderately strong bristles with apex bifed or otherwise modified, scale-like setae, or very fine appressed or erect setae; setae or bristles always present on pronotum, scutum and scutellum, but only occasionally present on other thoracic sclerites. Wing veins and cells often clothed with microtrichia and macrotrichia setae. R with three or fewer branches; usually with fork of M much longer than stem, and lanceolate rather than bell-shaped. Legs with coxae long and stout; femora and tibia usually slender, sometimes swollen, short setae arranged irregularly or in regular rows, and with bristles variable; tibiae with long, strong apical spurs and tarsi usually slender, sometimes with modified setae ventrally, or with some segments swollen below in female; tarsal claws rarely simple, usually with one or more teeth below. Abdomen usually broadest in middle; terga and sterna 1-6, 1-7 or 1-8 in male, and 1-7 in female well developed, but sternum 1 often reduced in size. Male with sclerites of segments 7 and 8 short and telescoped into segment 6; terminalia usually symmetrical. Mycetophilids are usually found to be most abundant in humid or moist habitats in wooded areas. Many larvae live in fleshly or woody fungi, on or in dead wood, under bark, or in nests of birds or squirrels. Most or all of these are probably myctophagous, hence the common name, but some members of the subfamily Keroplatinae seem to be predominantly predaceous. The family Mycetophilidae, is now generally thought to be paraphyletic with respect to the Sciaridae. There are approximately 4300 described extant species in over 225 genera worldwide.

Family SCIARIDAE (Dark-winged Fungus Gnats)

Dark-winged fungus gnats are small delicate (1-11 mm) flies, usually dull to yellowish in colour. Head usually ovoid, higher than long; compound eyes usually forming a dorsal bridge above antennal bases; frons with three ocelli. Pedicel and scape of antenna globular, rest 14 flagellomeres cylindrical, sessile or stalked; palps generally one to three segmented, basal one usually reduced; 2nd distinct segment often with a sensory pit.Thorax with anteropronotum setose; posterior pronotum sometimes with a few setae; scutum with bristles of variable length.Wings hyaline or smoky, costal vein ending between apices of R_5 and M_1. Tibiae with one or two apical spurs; tarsal claws simple or toothed. Abdomen cylindrical, in females, usually strongly tapered posteriorly. Male terminalia exposed, often broader than rest of abdomen, and not usually rotated, but sometimes rotated up to 180° during copulation. Adult dark-winged fungus gnats are usually found in moist places wherever fungus grows. Larvae generally feed on decaying plant material, animal excreta, or fungus. The family is worldwide in distribution with the exception of Antarctica. Some cosmopolitan species appear to have been spread synanthropically. There are over 1800 described extant species in about 90 genera worldwide with about 170 species in 18 genera known from the Nearctic.

Family CECIDOMYIIDAE (Gall Midges)

Gall midges are very small (1-5 mm) and delicate flies. Head with large eyes, holoptic or nearly so in both sexes; antennae usually long, with 12 or 14 segments (except scape and pedicel); mouth parts with generally fleshy labella, one to four segmented palps and labrum, the labrum and labella occasionally enlarged and styliform. Thorax squared; mesonotum convex, usually with two median and two lateral rows of setae. Wing covered with microtrichia, someties with scales; veins generally weak and reduced, costa continuous around wing, usually with a break just beyond insertion of unforked R_5. Legs usually long with coxae conspicuous; tibial spurs absent; claws toothed or untoothed. Abdomen elongate-cylindrical type in male, elongate-ovoid in female; posterior end of female abdomen often protrusible. Adults of the subfamily Lestremiinae often fly in cool weather, and can be found at lighted windows at night. The larvae are terrestrial and mycetophagous, usually found in decaying vegetation and wood, in plant wounds, and in mushrooms. The subfamily Cecidomyiinae contains numerous gall-makers, hence the family's common name. However, adult of other species are phytophagous or mycetophagous, while some larvae that are predaceous or parasitoids. As predators or parasitoid, species of this family feed or attack on mites, aphids, psyllids and coccoids. There are over 5000 described extant species of gall midges worldwide.

5. Infraorder PSYCHODOMORPHA

Family PSYCHODIDAE (Moth and Sand Flies)

Psychodidae possess moth-like appearance, and are small and setose with characteristic velvate wings. Head with 12 to 16 segmented antennae (sometimes longer than body); each segment usually with dense cupuliform whorls of setae, and with membranous thin-walled sensilla that may be broad or slender and branched or unbranched. Eye bridge is incomplete or absent; mouth part palps with 3 to 5 segments, the next to last segment with a sensory pit or a compact group of sensoria; proboscis usually very short, but long in blood-sucking species. Pronotum bare or setose, postnotum bare and metanotum large and projecting over abdomen; pleural sclerites variously setose or bare; transverse suture of scutum irregular shaped. Wings usually broad, longitudinal veins well developed, crossveins absent or restricted to basal half of wing and costal vein continuing around wing. Abdomen with sternum 1 sometimes unsclerotized; sternum 2 entire and divided into several sclerites, or unsclerotized. Males have inverted terminalia. Adult flies are nocturnal, during daytime adults usually rest in shaded moist habitats. Adult food habits are unknown, except for the blood-sucking habit of female *Phlebotomus*. In the tropics, species in this genus are the vectors of several diseases, such as leishmaniasis. Blood-sucking adults, commonly called sand-flies. Larvae live in moist or subaquatic habitats, with a few species often found in compost heaps and sewage disposal systems in soil, while other in semi-desert areas, streams and waterfalls. There are several thousand described extant species of moth flies worldwide.

Family TRICHOCERIDAE (Winter Crane Flies)

These flies are small to medium sized, fragile flies with long legs. Head with three ocelli, but with labrum reduced. Antenna 16 segmented (except scape and pedicel), elongate and setaceous, segmentation not clear in distal portion. Scutum flat with a V-shaped suture, mid part incompletely developed. Two strong anal veins reached the wing margin; A_2 half as long as A_1, and strongly curved at apex. Legs do not fall off readily, and females if they have an elongate cercus associated with the ovipositor, downwardly curved.Adult Trichoceridae are found in the colder months of spring (so the common name). Male fly swarms occur on sunny weather; otherwise, adults occur in dark places. Larvae are scavengers, and found in a wide variety of habitats, especially those with decaying leaves and vegetables. They also occur in manure, fungi, stored roots and tubers, as well as burrows of rodents. Approximately 120 described extant species belonging four genera found worldwide.

Family ANISOPODIDAE (Wood Gnats)

Wood gnats are small to medium sized (2-10 mm) flies. Body slender and elongate, with long and slender legs. Head small, rounded and usually somewhat flattened; eyes moderate, rounded to ovoid; males with eyes separate or meeting dorsally; frons with three ocelli, glabrous or with a few short, fine setae; antennae as long as length of head+thorax; scape and pedicel short, but other segment uniformly cylindrical; mouth parts with palps short and 3 or 4 segmented. Thorax convex, with reduced pronotum. Wings lying flat over abdomen when at rest; costal vein ending just beyond insertion of last branch of R, usually near wing tip; M with two or three branches; anal lobe well developed. Legs without strong spines; fore coxae long; tibia with apical spurs; claws simple. Abdomen elongate and cylindrical, slightly convex dorsally, but flattened ventrally. Male terminalia with rotation up to 180°. Adults can commonly be found on near larval habitats, often on tree trunk. They feed on nectar and other liquids. Males can form small to large mating swarms to attract females. Females oviposit on moist surfaces. Larvae occur in decaying organic matter, and are common in fermenting sap and mammal manure. There are about 100 described extant species in six genera worldwide.

Family SCATOPSIDAE (Minute Black Scavenger Flies)

Black Scavengers are moderate flies (0.6-4.1 mm). Usually black, dark gray or brown in colour, dull or shiny. Generally, head laterally compressed and setose; eyes occupying anterior half of head, usually holoptic, and with sparse or dense setae; frons with three ocelli. Scape and pedicel short, other segments (5-10 in number) short, often much wider, usually pedicellate type, covered with setae and microtrichia. Mouthparts reduced and palp single-segmented but labella larger. Thorax elongate and laterally compressed; scutum with sparse short setae. Wings with rather reduced venation; R_5 usually unbranched, and veins posterior to R faint; wing membranes covered with microtrichia, sometimes thickly; membrane and veins sometimes with obvious setae. Tibial spurs absent on legs. Abdomen with seven obvious pregenital segments; male terminalia sometimes rotated through 180°. Biology of Scatopsidae is sparse. The larvae of *Scatopse notate* (Linnaeus) have been found in decaying plant and animal matter. Adult biology unknown. There are about 250 described extant species of scatopsids worldwide.

Family CANTHYLOSCELIDAE (Canthyloscelid Flies)

The family Canthyloscelidae, includes the flies formerly referred to as the family Synneuridae. These are heavily sclerotized flies, shiny dark brown to black in colour and 2.0 to 3.5 mm long. Eyes meeting dorsally, ventrally narrowly separate to almost contiguous; frons with three ocelli. Antennae 10 to 14 segmented (except pedicel and scape), bead-like, distal segment largest; mouthparts reduced but palps 4 segmented. Thorax almost square; pronotum reduced medially, but with conspicuous postpronotal lobes; scutum narrow, strongly arched, slightly tapering and more narrow anteriorly, sparsely covered with short stiff setae; scutellum triangular and sparsely covered with short setae. Wings long and slender, with anal lobe at most scarcely developed; membrane and veins covered with microtrichia; costal vein strong, not reaching tip of wing, but extending beyond posterior branch of $R_{5.}$ Legs moderately long and stout; fore tibia with a single spur; middle and hind tibia with two spurs; claws small and simple, at most with a basal tooth. Abdomen long and slender, somewhat flattened dorsally, narrower anteriorly and widening posteriorly. Information is very little about the biology of these flies. Larvae usually found living in various kinds of decaying wood permeated by mycelia of various fungi.

6. Infraorder TIPULOMORPHA

Family TANYDERIDAE (Primitive Crane Flies)

Tanyderidae is a family of moderate to large-sized flies with long legs, no ocelli, and with a well-developed anal angle. The 5-branched radius is a hypothetically primitive condition rarely found in real flies. The antenna has 15 to 25 segments, usually the last segment being shorter than the second last segment. The compound eyes bear erect setae between the ommatidia. These dark-patterned wings are considered some of the most archaic among living Diptera. Adults are mostly lived in forests and the immatures are commonly aquatic, burrow in to the sandy sediments or rotten logs of streams. The primitive crane fly family is a relict group, scattered in the temperate areas of the world. The family contains 38 species in 10 genera. There are two genera and four species in North America; the sole eastern species is *Protoplasa fitchi*. The only western genus is *Protanyderus* with three species.

Family TIPULIDAE (Crane Flies)

Crane flies are normally slender, very long-legged flies with wingspans ranging from 5 to 85 mm. The head is variable in shape, often expanded forward into a snout-like rostrum. The antenna is usually short or moderately long, but sometimes reaching four times the body length and forming branches in males of some species. The number of antennal segments 5 in some *Chionea*, 13 in Tipulinae, 14 to 16 in the Limoniinae and 39 in some exotic species. Large, separate eyed insects (joined in *Limonia*) with absece of ocelli. Metanotal suture present in thorax. The long legs are unusually brittly, easily breaking between the trochanter and femur. Wings are elongate, rather narrow, reduced or lost in females of some groups, sometimes in both sexes (*Chionea*). The venation is normally characterized by two anal veins; 9 to 12 veins reaching the wing margin, the basal cells extending at least half the length of the wing, and a distinctive region near the outer third of the wing where the branching points of Rs, M and CuA often occur together in a transverse line. Crane flies occur almost everywhere insects live, from lowland deserts and tropical forests to the high Arctic islands and High Mountain. Most live in moist temperate forests, especially in cool, damp places near water. Larvae develop in fresh water, especially fast streams, in the intertidal zone, in mosses, decayed wood, wet leaf litter, organic soils and mud, dry soils, fungi, vertebrate nests and in the leaves of terrestrial plants. They are herbivorous, saprophagous or carnivorous. Crane flies comprise approximately 15,270 described species. *Tipula*, with well over 2000 named species and *Limonia*, with about 200, are among the most speciose genera of Tipulidae. The family includes largest flies, at least in wingspan *Holorusiahespera* Arnaud and Byers, measuring up to 85 mm. *Pedicia magnifica* Hine was first described from Vancouver Island, and have a striking brown triangular mark on the wings, which span at least 50 mm. Larvae develop in wet organic soil,while larvae of *P. vittata* burrow in decaying hardwood stumps and logs. *Tipula paludosa* Meigen (Marsh Crane Fly), a root-eating larvae, called leatherjackets, damage lawns, the adult of *Chionea* frequently are found striding purposefully over the snow, especially when the temperature hovers around freezing.

7. Infraorder AXYMIOMORPHA

Family AXYMYIIDAE (Axymyiid Flies)

These are medium sized (5-6 mm), stout flies often confused with *Bibio* (Bibionidae). Scutum with 2 shiny spots and radial vein 4 branched. Head large, bears eyes divided into upper and lower portion with a line or groove; the eyes in males meet dorsally, but are widely separate in females. There are three large ocelli on a strong convex tubercle. The mouthparts are vestigial and antennae short composed of 16 small flagellomeres. The thorax is strongly arched and bears a pair of shiny oval spots near the middle of the scutum. Thorax is covere only with short pale setae without any bristles or long stout setae. The legs are short and lack spines or long setae. The wing is longer than the body and, some species possess a faint pterostigma. Vein R_2 is short, nearly perpendicular to R_{2+3}, ending in C at or just past the end of R_1. The medial vein is two-branched. Haltere stalk is longer. The larvae live head-down in cavities that they excavate in waterlogged, bark-free logs that are continuously in contact with water or wet mud, apparently feeding on microorganisms living within the cavity. The Axymyiidae is restricted to the temperate northern hemisphere, where other lives in both tropical and temperate regions.

Suborder BRACHYCERA

8. Infraorder STRATIOMYOMORPHA

Family XYLOMYIDAE (Xylomyid Flies)

Xylomyids are slender, wasp-like or sawfly-like flies, 5 to 15 mm long, coloured in red, yellow and black or with pale markings on a dark background. The body has no bristles and is only inconspicuously setiferous. The head is hemispherical with the vertex and face more or less flush with the eyes. The eyes are bare of setae and are separate in both sexes. The antennae are 10-segmented, the eight segments of the flagellum tapering apically. The thorax is short, oval and rather flat. The legs are slender although the hind femur is sometimes thickened. The front tibia lacks apical spurs, the others have one or two; the empodia are pad-like. The wings are always clear and are folded flat over the abdomen when at rest. The costal vein does not extend past vein M_2; the medial vein is 3-branched and cell m_3 is closed. Adults are seldom seen, but fly in wooded areas where they appear around rotting logs and stumps. The larvae live under loose bark and in decaying wood where they are scavengers or predators of small invertebrates. As in the Stratiomyidae, pupation occurs within the last larval skin. The Xylomyidae is a small family closely related to the Stratiomyidae, with only about 130 described extant species in four genera worldwide. Two genera *Solva and Xylomya*, are predominant.

Family STRATIOMYIDAE (Soldier Flies)

The adults are slender to robust (2-18 mm); body ranges from rather bare to densely setose, without bristles. These are usually, green-blueish black. Head generally broad, thorax hemispherical and sometimes protruding. The head behind the compound eyes is often expanded, especially in females; ocelli present. Eyes are bare to densely setose, widely separated in females, joined or narrowly separated in males; the upper portion facets largest. The antennae are variable 2nd segment largest (often creating elbowed antennae) and the rest (5 to 8 segments) varies from simple and annulate to aristate. The proboscis is usually fleshy, either elongate or atrophied. The thorax is often characterized by pairs of spines on the scutellum; occasionally spines also occur on the notum. Legs are simple with padded empodia. The wing venation is distinctive, radial veins toward the costa so that R_5 reaches the wing margin well before the wing tip and costa does not reach past the wing tip. The veins arising from, and posterior to the discal cell are weak and tend to fade away. Viewed from above, the abdomen varies in shape from almost cylindrical or stalked to ovoid or almost round. Adult soldier flies rest and feed on flowers and are frequently found beside human habitat as well as wild. They are usually sluggish and slow but others often hover. Larvae of the subfamily Stratiomyinae are aquatic such as hot springs and saline pools. The Pachygastrinae live under tree bark; many are apparently predators, others scavengers. The other subfamilies develop in mostly terrestrial habitats and feed on rotting fruit and decaying vegetable matter, grass roots and vertebrate dung, the larvae are elongate and flattened; the cuticle contains small calcium carbonate plates, producing a shagreened surface. The pupa is enclosed within the final larval skin; this foreshadows the puparium of the

Muscomorpha, but is actually an example of convergent evolution. Stratiomyidae is widespread, with over 2650 species described worldwide, largest genera *Nemotelus* and *Stratiomys* are most specious, distributed through Holarctic and Neotropical.

9. Infraorder XYLOPHAGOMORPHA

Family XYLOPHAGIDAE (Xylophagid Flies)

Extremly varied flies, from stout and robust to slender and wasp-like, and range from 2 to 25 mm long. The body colour also varies from black, brown, reddish to yellow. Conspicuous setae are lacking, although the scutellum of *Coenomyia* bears two strong spines. Head looks spherical; eyes separated in females, but narrowly separated or meeting dorsally in males. The surfaces of the vertex and face are even with the compound eyes or only slightly depressed; in some species the latter is sunken deeply below the compound eyes. The antennal flagellum is 7 to 8 segmented in most genera, but has 20 to 36 saw-like or comb-like segments in *Rachicerus.* In *Dialysis* the flagellum, apical to the first segment, forms a thin arista. The legs are slender, each tibia has one or two spurs; the empodia are pad-like. The wings often have dark patches or spots. Except in *Rachicerus*the costal vein is continuous around the wing and cell m_3 is open. The medial vein is usually 3 branched and vein Rs arises before the base of the discal cell. The abdomen, with 7 to 9 visible segments, is tapered to the rear; the ovipositor is telescopic. Adult Xylophagids live in forest habitats where adults mainly feed on plant sap and nectar. The larvae are predators or scavengers in soil rich in organic matter (*Coenomyia*), under tree bark (*Xylophagus*) or in decaying logs (*Rachicerus*).The Xylophagidae contains over 120 described extant species world-wide.

10. Infraorder TABANOMORPHA

Family RHAGIONIDAE (Snipe Flies)

Medium size (4-15 mm) Snipe fly are mostly slender leged elongate fly with tapering abdomen; body covered with short setulae, sometimes with bristles. Colours range from grey to brown to black, and sometimes yellow or orange markings occur. Hemispherical head with vertex and compound eyes in same plane; convex clypeus with marginally deep grooves. Male's eye narrowly separated containing large facets, but female's eye widely separarted. Antennae vary, the primitively 8-segmented, often enlarged first segment bearing a slender, usually unsegmented stylus or seta-like arista. The proboscis is fleshy, sometimes developed for blood sucking. Legs have spurs on middle and hind legs and have peded empodium. The wings are broad and elongate and a small calypter; many have spotted one. Vein R_{2+3} is short, meeting the costa close to the tip of R_1. In both sexes, the abdomen has seven visible segments; they possess a telescopic ovipositor. Adult snipe flies are commonly perch in wet woods, wet areas, foliage and grasses, but few are predators of small insects. Genus *Symphoromyia* is known as biting pests of human, and showing ancestral dipteran habit of blood-feeding. Vertebrate-biting habits that arose later in the evolutionary history of flies rely on modified structures of the labium. The Rhagionidae is a cosmopolitan family still in need of considerable study with 22 extant genera and about 500 described species.

Family PELECORHYNCHIDAE (Pelecorhynchid Flies)

Moderate to large (4-18 mm) flies, mostly brown to black, and is well covered with setae, sometimes colourful (*e.g. Pelecorhynchus*). Head is large, with a strongly convex face; the compound eyes are large, unicolored, meeting above in males, separated in females. Three ocelli are present. First flagellomere is largest and elongated, while others (up to seven) reduced in size towards the apex. Scutum is broad, almost as wide as long; vein A_1 somewhat sinuate, calypter is large. The legs are strong; Leg with padded empodium. Biology of this insect is not known properly, adult found feeding at flowers while larvae live in wet soil in swampy and marshy habitats where they eat earthworms and other invertebrates. Pelecorhynchidae contains about 46 extant species in three genera. The largest genus is *Pelecorhynchus*, with 34 species confined to Australia and Chile. The most distinctive is *Pseudoerinna jonesi* (Cresson), a velvety black fly with orange antennae and palps and smoky wings. Genus *Glutops*, with seven Nearctic and four eastern Asian species.

Family OREOLEPTIDAE (Oreoleptid Flies)

Latest discovered family in Diptera, from the Canada in 2005. Oreoleptidae is a family closely related to the Athericidae and Tabanidae. These are dull grey pruinose flies, 5 to 7 mm long. Eyes meet on top of the head in males, but are widely separated in females; the upper two-thirds of the eyes have enlarged facets. Six segmented antenna having last three segments progressively narrower and comprising of an apical stylus. The top of the thorax is covered with long, semi-erect setae; a prescutellum is absent, but the subscutellum is present. The hind coxa bears a blunt peg on the front face; in the wing, cell r_1 is open. The first abdominal tergite is undivided on the mid-line. In the female, the abdomen is telescoped, with the genitalic segments not clearly differentiated from the others; the cercus is 2-segmented, the first one strongly lobed below. The few adult oreoleptids ever seen were reared from collected larvae and pupae, and nothing is known of the biology of the adult stage. The absence of mandibles in the female indicates that these flies are not blood suckers. The larvae, which bear two pairs of long, slender prolegs on abdominal segments 2 to 7, are predators of immature aquatic insects. Their flexible bodies allow them to crawl through the abrasive substrates of torrential streams; they pupate in sand and gravel at the high water line after spring run-off. The Oreoleptidae is known from a single genus, *Oreoleptis,* which lives in fast-flowing streams and groundwater wells in the mountains (Zloty, 2005).

Family ANTHERICIDAE (Athericid Flies)

Athericids are moderate (7-8 mm), heavely setulose, but without enlarged bristles.Generally yellow to black in colour and dark headed with spotted or banded abdomen and yellowish legs, faintly banded wings. Eyes widely separated in the female, almost meeting dorsally in the male; the three ocelli borne on a tubercle. First two antennal segments globular, but the third is kidney shaped and bears a thin, elongate arista. Vein R_{2+3} is short, ending together with R_1; the branches of R_{4+5} are splayed and enclose the wing tip; cell d is large, giving rise to three separate branches of M. Mid and hind tibiae bear two apical spurs, and bear paded empodia. Some female (*Suragina*) feed blood from humans and other mammals, other feed sap of vegetation while *Atherix* apparently feeds on honeydew. Female *Atherix* lays eggs on leaves and branches overhanging the water and the flies remain there and die; in some species, other females often lay eggs in the same mass, resulting in a bulky lump of eggs and dead flies. Hatching larvae fall into the water, becoming fierce predators of midge and mayfly larvae in stream riffles and underwater vegetation. The Athericidae is a small cosmopolitan family with 8 genera and about 70 species worldwide. Athericids, until recently, were considered snipe flies, but are probably more closely related to the Tabanidae.

Family TABANIDAE (Horse Flies and Deer Flies)

Tabanids are medium sized to large (5-30 mm), rather stout, large-headed flies, black, grey or brown, often orange or yellow coloured; bodies are moderately covered with sitae, but lack enlarged bristles. Eyes large, often brightly coloured, iridescent, striped or spotted, are separated in females, but meet dorsally in males; antennae with 5 to 11 flagellomeres, basal flagellum large and apically 2 to 8 annular. Proboscis is strong and rigid, with knife-like mandibles and maxillae in females of biting species. Large thorax bears stout legs; middle tibiae have spurs and are present or absent on the hind one. The empodium is pad-like. The wing bears large calypters. The venation is rather primitive and uniform; the costa extends around the wing margin and the radius has four branches. A distinctive feature is the splayed veins R_4 and R_5, which widely straddle the wingtip. The wings are often darkened and patterned distinctively. The abdomen is broad, often strikingly patterned; seven segments are visible. Horse flies and deer flies have fierce biting habit; females feed blood of humans, domestic and wild animals, and vectors of microorganisms that cause tularemia, anthrax, anaplasmosis and other diseases. Normally, adults are diurnal, active on warm, windless days; both females and males visit flowers to feed on nectar. Some species require no blood meal to mature the eggs, but most females suck blood from mammals, at least after the first batch of eggs is laid. The compact egg masses are laid on plant stems or leaves above the larval habitat. The larvae of most species live in the wet soil of marshes, fens, bogs, and along the margins of ponds and streams. A few live in the beds of fast flowing streams and some develop in dry soil. Apparently, most are predators of invertebrates, although *Chrysops* larvae may feed on plant matter in mud. Worldwide, the family consists of about 4,200 named species in 201 genera. There are three subfamilies: the Chrysopsinae, with hind tibial spurs and an antennal flagellum with five or fewer segments; the Pangoniinae, with hind tibial spurs and eight or nine apparent segments in the flagellum; and the Tabaninae, lacking apical spurs on the hind femur. *Chrysops* is a speciose and familiar genus. Deer flies are insects of open woodland, most diverse in north temperate regions, where members of the deer family are common. When biting people, they usually attack the head and neck. *Haematopota* is a speciose genus of over 300 species that has evolved with the mammal family Bovidae most occur in Africa and Asia, and North America has only five species (Teskey, 1990).

11. Infraorder MUSCOMORPHA

Superfamily NEMESTRINOIDEA

Family NEMESTRINIDAE (Tangle-veined Flies)

Tangle-veined flies are moderate and stocky flies look like bee, often strongly setulose, but without enlarged bristles. The colours are variable, often black, brown, yellow or white and frequently the thorax and abdomen are striped and banded, respectively. Head is large, almost as wide as the thorax; the eyes meet dorsally in most males, but are separated in females. In males, the upper compound eye facets are enlarged. Three ocelli are present. Antennae small with the first 3 flagellomeres about equal in size and a terminal stylus, usually of three segments. Generally proboscis is vestigial, or short and fleshy (*e.g. Trichopsidea*), or longer than the head, slender and sclerotized for flower feeding (*e.g. Neorhynchocephalus*). Legs are slender and lack tibial spurs; the empodium is pad-like. The wings are rather narrow, sometimes tinged with brown; the complex venation gives the family its common name. The veins in the apical half of the wing are parallel to the front and hind margins of the wing and usually end at the wing margin before the wing tip. The so-called diagonal vein is present, obliquely crossing the middle of the wing; it consists of parts of Rs, R_{4+5}, crossvein r-m, M_1-M_3, and CuA_1. The abdomen is conical or widest near the middle; the ovipositor is either a telescoping tube or a sabre-shaped structure formed from the elongated cerci. Adult tangle-veined flies are fast fliers and are often seen at flowers, hovering or motionless, with a high-pitched hum. Tropical species are known to be pollinators of orchids. The few species like *Neorhynchocephalus* and *Trichopsidea* adapted themselves as parasites of grasshoppers and beetles. *Hirmoneura* species of the US attack the larvae of scarab beetles; elsewhere it also apparently feeds on long-horned beetles. Thousands of eggs are laid in crevices in posts and trees; these hatch into tiny, active larvae, which search for hosts, and apparently, they are often dispersed by wind. Once the larva enters a host insect it develops into a maggot (hypermetamorphosis), producing a respiratory tube (lacking in *Hermoneura*) that links the spiracles at the larva's rear end to the outside air. The larvae parasitizing grasshoppers feed mainly on the fat body and ovaries of the host. Mature larvae overwinter in the soil. Approximately 300 named species of Nemestrinidae worldwide are arranged in 23 genera. The majority of genera and species live in the southern hemisphere.

Family ACROCERIDAE (Small-headed Flies)

Adult small-headed flies are bizarre, as the common name suggests, with tiny heads dwarfed by a grotesquely inflated thorax and abdomen. Variable in size (2.5-30 mm) and boast a range of colours from metallic blues, greens and purples to browns and blacks mixed with yellows and whites. Body almost bare, but usually is covered with fine setulae and without enlarged bristles. Globose and small head, fixed low on the thorax, consisting 3 ocelli, and head consisting mostly of the compound eyes, which meet dorsally in both sexes. The antennae are 3segmented; the first 2 are short and round, the 3rd (flagellum) varies in size, Mouthparts variable, vestigial, partly developed (short tube like), or well developed into a long sucking proboscis. Thorax is often humped, sometimes flat; the legs are normally simple, but some species have swollen femora. The empodium is pad-like, but is sometimes reduced or absent. The wing is variable in shape and venation; it is usually clear, but sometimes brown. The lower calypter is unusually well developed. Few males develop (especially *Pterodontia*) a striking costal thickening and projection occurs at the junction of Sc with $R_1$and R_{2+3}. The Sc reaches the middle of the wing or farther; R_{2+3}, when present, usually ending in the wing margin; an extra crossvein r-m sometimes present, forming cells $R_{4+5}$and R_5. The venation may be so reduced (as in *Ogcodes*) that there are no longitudinal veins behind R_1. Abdomen is usually globose, but some possess elongate (*e.g.Eulonchus*), tapered, wasp-waisted or laterally compressed. Acrocerid genera that have a well-developed proboscis (*e.g., Eulonchus, Lasia*) visit flowers for nectar, and species of *Eulonchus* are important pollinators of some wild flowers. Acrocerid larvae are internal parasitoids of spiders. An adult female lays several thousand eggs in the vicinity of a host population the oviposition site depends on the genus, for example, *Ogcodes* lays eggs on dead twigs, *Acrocera* on grass stems, *Eulonchus* on the ground. Like the first instar larvae of the Nemestrinidae, those of acrocerids crawl and jump around, searching for a host. Finding a spider, the larva burrows inside and takes up residence in the book lungs where it can breathe outside air. Usually, the larva rests there until they are almost mature, and then moults twice; the third and final larval stage is maggot-like and rapidly devours the liquid contents of the spider. Emerging, the larva pupates nearby, often in the host's web. The Acroceridae contains about 520 described extant species in 50 genera.

Superfamily ASILOIDEA

Family THEREVIDAE (Stiletto Flies)

These flies are slender to rather robust, small to medium (2.5-15 mm) sized flies. Silvery or gray white to black or yellow body often setulose, and with banded or spotted wings. The head is hemispherical, without a depression between the compound eyes; the compound eyes are separate in females, but meet dorsally in most males. There are three prominent ocelli. The antennae sometimes arise from a prominent protuberance and the first segment is often enlarged (*Tabuda*) and largest, while 2nd and 3rd form a small stylus bearing a bristle, often tiny. Prominent bristles are usually present on the thorax and legs, which are usually long and slender; the empodium is bristle-like or absent. Uniform wing venation present; vein $R_4$is long and usually sinuate; cell d is elongate with M_1, $M_2$and M_3arising from the apex; cell m-cu present. Pterostigma and calypter usually well developed. The 8 segmented abdomen tapers rearward and is convex to flattened above; acanthophorite spines are usually present in females. English name perhaps relates to the pointed abdomen of many species. Some therevids resemble small asilids in form and behaviour, although the structure of the head is completely different. The deep cleft between the eyes of robber flies is absent here and, because the proboscis is built for sponging rather than piercing, stiletto flies do not prey on other insects. Other species look like rhagionids and some mimic wasps. Adults frequent a wide variety of habitats, like forest openings, meadows, stream margins and dry areas. The elongate, snake-like larvae are predaceous in sand and loose soil, under dead tree bark and in decaying fruit or fungus where they attack invertebrates, like the beetles.The Therevidae is a cosmopolitan family containing over 1000 described species.

Family SCENOPINIDAE (Window Flies)

Scenopinidae are sturdy, small (5 mm) sized flies. The body is bare or setulose, but lacks enlarged bristles; it is usually dark in colour, but is sometimes marked with white or yellow. Most part of head consisted with compound eyes, which usually meet dorsally in males (Upper facets large), but are separated in females (Upper facets small); the vertex is in level with the compound eyes; a prominent flange is often present behind the eye, especially in some females.Three ocelli are present. Antennae 3 segmented, 3rd flagellum bears a peg-like stylus and varies from long and slender to short and oval, often with a slightly forked tip. Mouthparts portion of facial area is deeply recessed. Thorax with moderate convection and pruinose, and sparsely coverer by short to long or scle-like setulae; scutellum bare. Shortened legs dos not have empodia. The wings have prominent radial veins, R_{2+3} is usually short, ending close to R_1; R_{4+5} forks and R_4 ends well before the wing tip; cell r_5 is open to the wing tip and is narrowed, or is closed and stalked. The abdomen is broad and flattened or cylindrical, with seven easily seen segments in the male, eight in the female; most segments are subdivided by a transverse groove. Larvae are predators in nature. Few species of *Scenopinus*, prey on the larvae of household pests of cloth, wood, stored food and pets, including clothes moths, dermestid beetles, powder-post beetles and fleas. They prefer to live in household areas and probably in bird and mammal nests while others recorded in decaying wood, in fungi, in the galleries of wood-boring beetle larvae, in termite nests, and in the nests of woodrats and birds.The Family Scenopinidae is world wide in distribution, with about 414 species described. The most familiar are *Scenopinus fenestralis* (Linnaeus) and *S. glabrifrons* Meigen, originally from Europe, but now widespread over most of the globe, having been distributed everywhere by commerce because of their association with household goods.

Family MYDIDAE (Mydid Flies)

Probably the lagest flies (in wing span) of the world are Mydids (*Mydas heros*of Brazil, 60 mm length and 100 mm wingspan), sparsely covered in short setulae without large bristles except on the legs. Colour is variable, and usually is some combination of black, yellow, red or white. Wider head with usually swollen and setose face, only flatter face seen in the ancestral one. Most species have long, 4 segmented antennae with the two segments of the flagellum especially elongate; the second flagellar segment is normally clubbed and usually bears a tiny apical spine. The more ancestral species have a 1 segmented flagellum. Large compound eyes with always have uniformly small facets; there are three ocelli. The proboscis is well-developed, usually short, but can be up to five times longer than the head (for example, in *Rhaphiomidas*). Hind leg is frequently longer and stouter, with hind tibial spur and usually swollen and bearing spines femur. Unlike in the Asilidae, there is no empodium and the pulvilli have one rib instead of two. The wings can be clear or washed with yellow, orange or brown. The wing venation is distinctive, with most veins usually ending in the front margin of the wing. Vein Rs is very short and cell br is extremely long; all radial veins, M_1, and sometimes even M_2 end before the wing tip. Cells r_4, r_5 and m_1 (when present) therefore curve forward, parallel to the hind edge of the wing tip; cell m_3 lies parallel to the hind margin of the wing. The abdomen is cylindrical and taper rearward in males, but in most females is widest in the middle. Male tergite 8 almost always has a deeply concave hind margin. In most females (except *Mydas*), tergite 10 is developed into acanthophorites bearing a circlet of strong blunt spines. Mydids live mostly in warm habitats such as dry, open woodlands, grasslands and deserts. Adults, with their long antennae and colourful bodies, often resemble wasps and probably mimic them. These flies suspected as predators, although there is no evidence yet; probably most feed at flowers and those with atrophied mouthparts may not feed at all. Some species of larvae are known to prey on beetle larvae in rotting wood and sandy soil. The family Mydidae is widely distributed, especially in dry tropical, subtropical and Mediterranean climates. It is an old and scattered family of about 360 species and 54 genera that probably suffered a high rate of extinction. This fly family dates back at least to the Jurassic, and like the Apioceridae, the ancestral subfamilies of the Mydidae (Raphiomidinae and Megascelinae) show relationships and geographical distributions related to the break-up of Gondwanaland. These ancestral groups have only recently been transferred from the Apioceridae to the Mydidae.

Family APIOCERIDAE (Flower-loving Flies)

Flower-loving flies are robust (12-35 mm) and bare to moderately setulose with black or brown colour, often with extensive grey or tan tomentum and pruinosity creating patterns, especially on the abdomen. These short and wide headed flies have widely separated compound eyes (both sexes) along with three ocelli. Four segmented antennae, 1st flagellum pear-shaped and 2nd is short, cylindrical and acute. The mouthparts are sponging with large fleshy proboscis and 2-segmented palps. The stout mesothorax bears bristles on the sides and rear. The legs are unspecialized, but usually there are a few bristles on all segments; the empodium is bristle-like. Apiocerid wings are relatively short with venation resembling that of mydid flies. Veins R_{2+3} join R_1 before the wing margin; R_5 and M_1 curve forward to join the margin before the wing tip. Crossvein r-m lies at about the midway point of the discal cell. Elongate, tapering rearward abdomen has eight visible segments; male genitalia are enlarged and club-like while female with acanthophorite spines. They are inhabitants of dry, hot places, especially where there is a nearby source of water. They especially like the edges of sand dunes where there is a sparse cover of plants. Females lay eggs in sandy soil at the base of plants. Although vividly used in the books, the term 'flower-loving flies' is a misnomer. In fact most of these insects seldom visit flowers. They spend much of their time walking or running on open sand or soil; when the temperature is high they make fast runs and short flights between shaded spots under plants. Flight is often noisy. They are attracted to water in the soil and imbibe honeydew from beneath aphid-infested plants. The slender larvae apparently prey on invertebrates in the soil. The ancestral mydid subfamilies Raphiomidinae and Megascelinae have recently been transferred from the Apioceridae to the Mydidae, leaving *Apiocera*, with 137 described species, as the only apiocerid genus. This group is divided into four subgenera from four discrete geographical regions: western North America, southwestern South America, South Africa, and Australia. Evidence suggests that the family, like the Mydidae, arose in Pangaea before the middle of the Jurassic; its modern disjunct distribution resulted from the subsequent break-up and movement of the continental plates. The North American subgenus is the most primitive.

Family ASILIDAE (Robber Flies)

Robber flies are named after their predatory habits – they attack and devour other insects. The body form varies widely, from delicate and slender to heavy and stout, from almost bare to bristly or setulose. Some are tiny flies only 3 mm long, but others are gigantic, over 50 mm long. Colours range from browns, greys, silvers and blacks to colourful patterns of contrasting blacks, yellows and reds. Males of some species have additional ornamentation such as the expanded silver abdominal tip in *Nicocles*, the striking white abdomens of *Efferia* and the decoration on the tarsi of many *Cyrtopogon* species. The compound eyes are large usually rather flattened from front to back; in most species the broad forward-facing area has enlarged facets. The compound eyes are well separated dorsally in both sexes by a distinctive hollow. The ocelli are prominent, usually placed on a tubercle. Varying from flat to strongly protuberant, the face bears a characteristic tuft of setulae or bristles, the mystax. The antennae are erect, usually 4-segmented. The first segment of the flagellum is elongate or oval, the stylus is normally 2-segmented, but may be 1-segmented or apparently absent; it is prominent and seta-like in the Asilinae. Both sexes have stabbing and sucking mouthparts developed into a proboscis paralyzing saliva is injected through the needle-like hypopharynx (tongue), which is sheathed in the prominent labium. The thorax is prominent and powerful and usually bears distinctively arranged bristles. The legs are strong and raptorial, frequently with numerous bristles; the empodium is bristle-like, but is lost in some Leptogastrinae; the pulvilli are also lost in the latter subfamily and a few other genera. The wings are sometimes coloured or spotted. The venation is not much modified and vein R always has four branches, with R_{2+3} unbranched. However, R_{2+3} joins R_1 before the wing margin in some subfamilies (*e.g.* Laphriinae, Asilinae), closing cell r_1. Cells m_3 and cu *p* are often closed. The abdomen varies from cylindrical and tapering to short, broad and rather flattened; in the Leptogastrinae the abdomen is slender, elongate and club-shaped. Acanthophorite spines are present in females of several subfamilies and a prominent knife-like ovipositor is formed from the terminal abdominal segments in many genera of the Subfamily Asilinae. Robber flies are predators that as adults pursue other insects, seize them with powerful legs and kill them with a paralyzing stab of the hypopharynx. The liquefied contents of the prey are then sucked-up by the proboscis. They are mostly opportunistic predators, feeding upon any insect that they can subdue and kill. Some species, especially in the subfamily Laphriinae, are effective mimics of bees and wasps. Robber flies usually hunt in open areas where there is plenty of light and warmth; grasslands, scrub, deserts and open woodland are the best places to find them. Larvae are predators of the eggs, larvae and pupae of other insects in the soil (most groups) or in rotting wood (subfamily Laphriinae), although in a few species studied the immature larvae, especially, are ectoparasitic on their hosts.The Asilidae have 6700 described extant species worldwide. Genus *Cyrtopogon* is the most diverse among the Subfamily Stenopogoninae. *Efferia* is the largest robber fly group.

Family BOMBYLIIDAE (Bee Flies)

Bee flies are the main flower-visiting nector feeding flies, are small to large (1-25 mm), with the usually stout body normally clothed in delicate setulae or scales, or both, ranging in colour from black and brown to white, silver and gold. The wings are often patterned and colourful. Eyes globular or transverse, bare, and often meeting dorsally (subfamily Anthracinae, hind margin of eye is sharply indented, with a horizontal seam dividing the facets; flat, swollen or deeply concave black-headed) in males. There are three ocelli.The mouthparts are adapted for sucking from flowers; long and slender proboscis (Bombyliidae) or short and fleshy labella (Anthracinae) seen. Antennae 3 to 6 segments; 1st segment enlarged and others form a stylus. Thorax is flattened or humped. The legs are slender, with or without bristles, but normally with bristles at the apex of the tibiae. The front legs are often thinner, shorter and weaker than the other two pairs, especially in anthracine genera such as *Exoprosopa* and *Poecilanthrax*. The wing venation is variable. Vein Rs normally has three branches (R_{2+3}, R_4 and R_5), with R_{2+3} usually ending in the wing margin toward the wing tip (and along with R_4 often bent sharply forward), but sometimes short and joining R_1. Vein M usually with two branches reaching the wing margin; occasionally only one branch is complete because M_1 ends in R_5 or because M_2 is absent. Cell dm is usually present and cell cu *p* is open or closed. The abdomen is short and broad, elongate or cylindrical, consisting of six to eight visible segments and seldom bearing bristles. The females of many bombyliid groups have acanthophorites and spines on the 10th segment, but many have lost these egg-laying aids. Females in advanced groups, including the subfamilies Bombyliinae and Anthracinae, have a sand chamber, developed through modification of segment 8, in which eggs are coated with sand from the substrate before they are laid. Bee flies are diurnal; drive in hot, dry climates where sand and stony ground prevails. They have a strong, hovering flight. They feed on nectar and even some pollen. When perched, they hold their wings outstretched and swept back. Some of those that have been studied are parasitoids of the immature stages of bees, wasps, moths, beetles, flies and other insects or they prey on the egg pods of grasshoppers. Eggs are laid near the host insects and the larvae are similar to those of the related Nemestrinidae and Acroceridae– an active first-instar larva seeks out the food source and moults to a grub-like form in subsequent instars to feed voraciously on the host. The family consists of about 4700 species in 230 genera. Probably the most familiar species is *Bombylius major* Linnaeus, a fly that spans throughout the northern hemisphere. The largest *T. magnus* (Osten Sacken) is about 20 mm long and has the front of the wings dark shaded. Among the total describe bee flies, subfamily Anthracinae covers half the number.

Family HILARIMORPHIDAE (Hilarimorphid Flies)

Moderate (1.8-7.2 mm), dark flies having suboval head with a concave occiput and bare compound eyes which not emarginate laterally to the antennae. Males are holoptic, eyes contiguous from the vertex to the base of the antennae, and with the facets in the lower third of the compound eye smaller than those in the upper two-thirds. Females have small and dichoptic compound eyes with uniform facetssize.

Ocelli are present and appear prominent on a subtriangular vertex pad, usually with several short, fine, erect setae. Antennae have 2-segmented stylus and subconical scape. Moderately arched thorax with distinct vittae; scutellum short, relatively large, and subtriangular with short setae arranged in a row across the posterior margin. The wings are hyaline to pale brown, covered with microtrichia, and usually with a pterostigma. Veins are brown, with R_{4+5} and M_{1+2} similarly forked. Vein M_1 not curved forward, and vein CuA_2 reaches the margin of the wing near A_1. Legs are elongate, with tibiae lacking apical spurs. Tarsal claws are simple, the pulvilli are large and distinct, but empodia are absent. The male abdomen ends in swollen, globose terminalia, but the abdomen in the female is gradually tapering. Immature stages are unknown, but adults have been collected on species of *Salix* growing along the narrow gravel-bottomed streams. Generally, the adults are active from mid June to the end of July.The family contains a single, north temperate genus *Hilarimorpha*. Worldwide there are approximately 30 species.

Superfamily EMPIDOIDEA

Family EMPIDIDAE (Dance Flies)

Small to medium (1-12 mm), elongate, Dance flies have dark coloured body occatinally yellowish to pale brown. Head with large compound eyes, males often holoptic, females in the subfamily Hybotinae also holoptic; proboscis often elongate. The antennae have 3 or fewer segments, and usually have a stylus or arista. The thorax is usually somewhat rectangular in dorsal outline. Wings have highly variable venation, but the costa extends to at least the apex of the wing. Vein Sc usually joins C or ending freely, never abruptly joining R_1. Vein Rs originates well distal to the level of crossvein h, and with vein CuA_2 usually absent or vestigial. The posterior veins in the wing are not setose. Legs slender, long, and often show sexual dimorphism. If sexually dimorphic, males with basal one or two tarsal segments enlarged with silk glands. The femora may or may not be thickened, and the middle coxae lack a strongly developed prong. The hind tarsus also has the basal segments not expanded and flattened. The abdomen is usually more or less elongate and cylindrical, with the terminalia in the male symmetrical or asymmetrical, much larger than preceding segments or turned forward over these segments. Empidid larvae are either aquatic or terrestrial, living in soil, leaf-litter or rotten wood. Adults are often found on vegetation in moist habitats, on tree trunks, or on the surface of water. Both sexes of most species feed on protein sources and nectar. Empidids feed on live insects, especially swarming Diptera. Many species of empidids also swarm, in spring and early summer. Males in many species present nuptial gifts of food to females as part of the mating process. This food is either presented to the female as captured, or may be wrapped in silk or a frothy 'balloon' or ball of silk by the male prior to presentation. It is thought that such silk or silk-wrapped food, as well as insect food by itself serves to prolong the feeding response of the female, and so distract any predaceous attack by the female on the male. The higher classification of this group is in flux, but it is generally thought that the Empididae, as currently defined, is a paraphyletic group with respect to the Dolichopodidae. Worldwide, there are some 4,000 described species.

Family DOLICHOPODIDAE (Long-legged Flies)

Long-legged flies are moderate (1-9 mm), typically shining metallic green in colour, but also be brown or black, and somewhat pruinose. Head possess broad and anteriorly narrowed frons. Arista are2 segmented. Thorax with scutum usually strongly bristled. Wings oval in shape, with costa usually continuous to juncture with M. Vein R not strongly thickened, but Rs arising at or very near level of crossvein h, and R_{4+5} unbranched. Vein M often straight beyond crossvein dm-cu, the cells dm and bm united. Vein A_1 and cell cu *p* sometimes rudimentary or lacking. Legs usually with large bristles on tibiae. Male legs often highly modified, but never with one or more of basal tarsal segments of hind tarsus expanded. Middle coxae without a prong. Tarsi usually with bristle-like empodia and broad pulvilli. Male abdomen with tergum 5 sometimes modified. Female with usually posterior abdominal segments retracted into segment 5. Most larvae are predaceous and occur under the bark of trees, or in decaying vegetation. A few are aquatic. Larvae pupate in a cocoon made by the larva from pieces of wood, sand or soil. Adults are also predaceous, and are found on foliage, tree trunks, or damp earth, usually in swamps or along partially shaded streams, where they prefer small areas of sunlight. Males of species of *Dolichopus* in particular have complex mating dances. As noted in the discussion of Empididae, it is generally thought that the dolichopodids form a monophyletic group within the empidids in the broad sense. Worldwide more than 6,600 described species assigned in 200 genera.

Aschiza Section of Muscomorpha

Superfamily PLATYPEZOIDEA

Family PLATYPEZIDAE (Flat-footed Flies)

Flat-footed flies are slender to robust (2-10 mm) with black, gray, yellow, orange or brown, or a combination of these colours, although sometimes marked with shining blue or green. Head as broad as thorax, and rounded in front. 3-segmented bare arista situated terminally on flagellomere. Eyes of males (holoptic type) are upper facets large and lower facets small, and females (dichoptic type) with all small facets. Males have ocelli on prominent tubercle. Mouthparts are short and fleshy and thorax with transverse sutures usually visible only at the sides. Wings clear or brownish, the veins in the posterior half not setose. Vein C extending to at least apex of wing; subcosta reaching wing margin; radial veins not strongly thickened. Characteristically, the wings of platypezids have a relatively large cu*p* cell which always ends in an acute angle. Legs stout and with middle coxae lacking a prong. Often the hind tarsus of males, and in some genera also the females, are expanded or dilated, often appearing as flattened plates in dried specimens. The abdomen is elongate; and cylindrical or somewhat flattened. Larvae of all flat-footed flies appear to feed in damp woods (on fungi), where the adults are often encountered running erratically in a zigzag, stop-and-go, fashion on leaves of bushes in filtered sunlight, likely feeding on honeydew and other deposits on the leaves. Adults may also be seen hovering or running across the damp sand of stream beds. Males can form aerial swarms, into which females enter to select mates. Adults of *Microsania* are attracted to smoke; *Microsania occidentalis* is a widespread western species. There are about 20 described genera and 250 described species all over the world.

Superfamily PHOROIDEA

Family LONCHOPTERIDAE (Pointed-wing Flies or Spear-winged Flies)

Lonchopteridae are small and slender, yellowish to brown flies, 2 to 5 mm long. The head is almost wide or wider than the thorax; the frons is short, broad, and bare except for a pair of strong and divergent lower interfrontal bristles medially above the antennae. Antennae short, with bases widely separated, and with all segments small. Scape and pedicel each with a row of short bristles on the distal margin. The antennal arista is exposed, arises apically and has short pubescence. Eyes are moderately large, bare, prominent and broadly separated in both sexes; the ocelli are small and equidistant from each other. The proboscis is short, with palps short and clavate. The thorax is subrectangular and convex dorsally with the scutum rather strongly arched anteriorly and the scutellum moderately large, triangular, and rounded posteriorly. The legs are large and slender, with claws and pulvilli small, but the empodia are absent. The hind femora are slightly swollen, as are the fore tarsi of the male. The wings are elongate, slender, and somewhat pointed apically, with linear venation which shows sexual dimorphism. The main veins except Sc and R3 have black setulae on the dorsal surface, vein M is branched, and the crossvein r-m is situated near the base of the wing well before the middle of the wing. Vein CuA2 does not reach the wing margin as a free vein, but in the male terminates as A1+ CuA2. In the female it ends as A2+CuA2+CuA1. The abdomen is rather short, subcylindrical to oval.The flat, broad larvae live under leaves or in decaying vegetation, feeding on micro-organisms and fungal hyphae. A few species live in aquatic habitats such as seeps and beaches. Adults are found in moist, grassy, shady habitats. They have a characteristic jerky movement and feed on nectar, pollen, fungi and dead insects.The family is small, containing about 55 described species in a single genus, Lonchoptera. Many species remain to be named. The family is distributed worldwide, although *L. bifurcata* (Fallén) is the sole species in the Neotropical and Australasian regions, where it is almost certainly introduced by humans. However, it is probably native to at least the Holarctic region, although males are extremely rare there. Most populations are parthenogenetic. Seven Lonchoptera species live in the Nearctic (Klymko and Marshall, 2008).

Family PHORIDAE (Phorid Flies)

Minute to small (0.5-6.0 mm), inconspicuous, blackish, brownish or yellowish flies, with major bristles on head, legs and other parts of the body characteristically feathered. The head is small, sometimes rather short and flattened, with ocelli present except in apterous forms. The thorax has a characteristic hump-backed appearance. Hind femur often enlarged, and more or less laterally compressed. The tibiae have one or more apical bristles, with one or more seams often on the hind, and sometimes also on the middle tibia. The wings are usually large, usually hyaline to pale brown, and rarely with dark markings, but females are sometimes short-winged or apterous, branches of R strongly thickened and crowded into the antero-basal portion of wing, but with four other weak and peculiarly defined veins in the remainder of the wing blade. The costa ends near the middle of the anterior margin of the wing, and the veins in the posterior half of the wing are not setose. The abdomen is somewhat conical, and more or less tapering posteriorly, but often it is membranous (in case of wingless). Adults move about with a characteristic quick, jerky movement. They are common around decaying vegetation or animal matter, and sometimes in and around the nest of ants, termites and bees. Larval habits are varied, with many having been reared from fleshy or woody fungi. Some are scavengers, and some are parasites. The parasitic forms are often found in the nest of ants, termites, bees and wasps, and on beetles, caterpillars, millipedes and land molluscs. Worldwide there are about 245 described genera and over 3000 described species.

Superfamily SYRPHOIDEA

Family SYRPHIDAE (Flower or Hover Flies)

Slender to robust (4-25 mm) flies have usually black and strikingly marked with yellow or orange head, thorax and abdomen. Many of them mimic bees and wasps. Less commonly, species are brown, yellow or metallic green or blue, or with various combination of such colour. Integument usually smooth, but sometimes totally or partly punctate, sculptured or rugose, and may be somewhat pruinose. Body usually covered with dense, short setae, but rarely may be almost bare, or with long setae or stout bristles. Setae are sometimes flattened or scale-like, forming a dense tomentum. Eyes usually holoptic in males, but can be very narrowly or broadly dichoptic. Females moderately to broadly dichoptic.Upper part of facets in eye may be enlarged in male, and usually unicolourous, but rarely with dark spots or bands, or irregular markings, and with or without setose, if setose then short or long, sparse or dense. Three ocelli present, and the antennae with a short or long frontal prominence. Lower facial margin usually with a distinct median notch.Wings usually hyaline, but somewhat darkened or with distinct markings. Characteristically, theses flies with costa ending at the apex of R_{4+5}, and there is an unattached longitudinal vein, the spurious vein (*vena spurea*), running most of the length of cells br and r_{4+5}, posterior to Rs. The apex of vein M is bent strongly forward near the wing margin, and joins near the end of R_{4+5}, thus forming an apical crossvein. Cell c*up* in the wing is closed near the wing margin, and a pterostigma is usually present. The legs are usually slender, but can be somewhat modified, especially in males. The abdomen is usually suboval, but can be elongate or even petiolate.Subfamilies Syrphinae and Eristalinae are habitual visitors of flowers for obtaining pollen, nectar and honeydew, and are important pollinators: they are often considered second only in importance to some bee species in cross-pollinating many economic plants. Males are often seen hovering, almost motionless in the air, but dart swiftly aside when disturbed. Adults lay oval, chalky-white, sculptured eggs on or near the food of the larvae, those with aphidophagous larvae being laid singly, but others may lay eggs in masses of over 100. Larvae have a wide variety of habitats and food. Some are predaceous (Syrphinae, Pipizini), but others are phytophagous (*Cheilosia*, Merodontini), saprophagous (most Eristalinae), or scavengers (*Volucella*, Microdontinae). The mouth parts of such larvae are thus quite different. Predaceous types have four stylets for piercing and sucking, while phytophagous forms have strong mandibles. Saphrophagous species have a complex comb-like mandibular lobe, and like the scavengers, have a muscular and contractile cibarial chamber. The predaceous larvae of the Syrphinae feed primarily on aphids or on immature thrips, beetle, or lepidopterous larvae. Larvae of the Pipizini feed on aphids, preferring woolly or root aphids with a waxy integument, and attack these living both above and below ground. Larvae of the Syrphinae and Pipizini are thus important in natural biological control. The phytophagous larvae of *Cheilosia* feed on fungi or vascular plants, while the larvae of the Merodontini live in monocotyledonous bulbs, and sometimes in other plants, and may cause considerable economic damage. The saprophagous larvae of most species of Eristalinae live in tree holes, ulcerated tree wounds, or rotting wood, but larvae of species of the tribes Eristalini and Sericomyiini are aquatic, living in water with a high organic content. The larvae of the Eristalini are the rat-tailed maggots, so called because the end of the abdomen is drawn out into a protrusible rat-tail-like respiratory siphon. Among the scavengers, the larvae of the Microdontinae are known to live only in ant nests, and the larvae of *Volucella* likewise are scavengers in nests of colonial Hymenoptera. One of the most common syrphids in British Columbia is the cosmopolitan *Eristalis tenax* (Linnaeus). Although larvae of this species are aquatic, they have also been found living in the urogenital opening of cows. Three species of Syrphinae, namely *Eupeodes perplexus* (Osburn), *E. venablesi* (Curran) and *Syrphus opinator* Osten Sacken have been reported as important predators in the natural control of the woolly aphid *Eriosoma langerum* (Hausmann) in British Columbia. The Narcissus bulb fly *Merodon equestris* (Fabricius), is a pest of *Narcissus* and other bulbs, and most years causes considerable loss to some bulb growers. Larvae of *Heilosia* are called bark maggots, and cause a blemish in timber of western hemlock and other conifers, known in the lumber industry as 'black check'. Worldwide there are some 180 recognized genera, and about 6200 species with 4 subfamily and 15 tribe. Different types of ecosystem in India attracts 357 species of Syrphidae or hover flies or flower flies belonging to 14 tribes under three subfamilies of the family Syrphidae (Mitra *et al.*, 2015; Young *et al.*, 2016).

Family PIPUNCULIDAE (Big-headed or Big-eyed Flies)

Big-headed flies are small, dark bodied, 3 to 15 mm long, with characteristic globose or semiglobose eyes covering almost the entire head. Males have usually contiguous eye front, but dichoptic in the subfamily Chalarinae. The occiput is very narrow in the Chalarinae, and has ocellar bristles. In the Pipunculinae the occiput is generally swollen, clearly visible and lacks ocellar bristles. The thorax is shiny black, and pruinose to a varying degree, but usually somewhat setose. The femora of the legs usually have rows of small spines ventrally towards the apex. The hind tibiae often have strongly, erect anterior setae medially. The claws and pulvilli are moderately large. The wings are typically long and slender, hyaline to faintly infuscate, and iridescent in direct light. The costa ends at the wing apex, and R_{4+5} unbranched. The abdomen is typically subcylindrical. Adults are frequently found hovering in or over vegetation. Females search out immature leafhoppers and other Homoptera for ovipositon. Females snatch the prey with their long legs and hold them in their strong tarsal claws. While in flight, females then insert an egg into the host through weak areas of the abdomen. The larvae are endoparasitic, and mature in the host and then emerge, drop on the ground and pupate in the soil or ground litter. Worldwide there are about 1380 described species.

Schizophora Section of Muscomorpha

CALYPTRATAE (Schizophora, which possess calyptra)

Superfamily MUSCOIDEA

Family SCATHOPHAGIDAE (Dung Flies)

Scathophagids are slender, small to medium-sized flies (about 3 to 11 mm long); bristling ranges from weak to strong and some species are densely setulose. The ground colour is usually black, brown or yellow, but some are strikingly bicoloured; some are densely grey or yellow pruinose. The head is normally higher than long, with the frons of equal width in males and females. There are 1 to 3 orbital bristles and 1 to 6 frontal bristles; the latter are curved inward. Also present are ocellar, postocellar and inner and outer vertical bristles. Parafacial area bare; the vibrissa is weak or strong and there are few to many weaker bristles and setulae nearby. Eyes bare, 2nd antennal segment with a distinct seam above; the 3rd segment 2-5 times longer than wide, usually rounded at the tip and sometimes with a long bristle near base of the arista. Arista 3 segmented, straight or rarely elbowed, and bare or feathery. Bristles of the thorax are normally rather strong, but those on the disc of the scutum are sometimes short and setae-like. Postpronotum usually has one or two bristles; there are normally two presutural and three postsutural dorsocentral bristles. Scutellum rarely has more than two pairs of marginal bristles. The wing venation is rather constant; the costa has costagial, humeral and subcostal breaks, the subcosta is complete and vein R_1 joins the costa before the middle of the wing. Vein R_{4+5} meets the costa at about the wing tip and the costa ends at vein M_1. Vein A_1 usually complete, but sometimes ends before the wing margin. The wings are normally clear, but sometimes spots or bands occur. Legs slender; bristles vary from weak to strong. Front femur and tibia sometimes bear short black setae below and are sometimes highly modified in males. Abdomen slender, but in males it frequently enlarged at the tip; the 1st and 2nd tergites are fused. Only some species of dung flies develop in dung. Larvae of the large and common genus *Scathophaga* mostly do so, although some larvae feed in rotting seaweed on ocean beaches. Other Scathophaginae feed on a wide range of plants, especially monocots including *Scirpus* and *Juncus*. Still others are predators under water, wet soil or plant tissues. Larvae of the subfamily Delininae mine in the leaves of plants in the lily and orchid families. Adults prey on insects and other small invertebrates. The Family Scathophagidae has often been treated as part of the Anthomyiidae or Muscidae. Its approximately 260 named species are almost completely restricted to the Holarctic region.

Family ANTHOMYIIDAE (Root-Maggot Flies)

Anthomyiids are small to medium-sized flies (2 to 12 mm long), usually yellow, brown, grey or black and without any metallic sheen. Eyes usually well-separated in females, often meeting above in males; they are bare to densely pilose. Frons bears frontal, vertical and ocellar bristles; flies with separated compound eyes normally also have orbital bristles. Vibrissae present. The antenna with a quadrate or oblong 3rd segment and the arista is bare to plumose. The scutum normally has paired postsutural acrostichal bristles, 1-2 pairs of presutural dorsocentral bristles and 1-4 pairs of postsutural dorsocentrals; notopleural bristles 2. Scutellum with a pair of basal and a pair of apical bristles; there are normally setulae on the underside. Costa reaches the tip of M_1. Cell r_{4+5} broadly open at the wingtip and A_1 usually traceable to the wing margin. Legs are slender and bristly. First segment base of the hind tarsus bears a strong bristle underneath. Abdomen usually cylindrical and more or less conical. Sternite 5 in males is bilobed; female ovipositor usually tubular. Adult root-maggot flies can be common on flowers; the larvae mostly feed on the roots, stems, leaves and flowers of living or decaying plants. Subfamily Fucelliinae contains marine beach species whose larvae feed on seaweed washed ashore; adults swarm in large numbers around these piles of wrack. The rest of the family comprises the subfamily Anthomyiinae (cosmopolyton in nature), a diverse group living in many varied habitats. The most obviously economically important species are pests attacking crops like *Pegomya hyoscyami* (Spinach Leaf Miner), *Delia antiqua* (Onion Maggot), *D. platura* (Seed Corn Maggot) and *D. radicum* (Cabbage or Radish Maggot =*Hylemya brassicae*). Other species attack many cultivated plants. Many more species, however, are certainly important in pollination and in the recycling of organic matter. Some are scavengers living on the dung of mammals and birds, others are parasites or inquilines in the burrows of tortoises and rodents, and species of *Leucophora* live in the nests of solitary bees and wasps, especially in dry, sandy habitats. Many species, such as some *Hydrophoria* are common in and around wetlands of all sorts and some larvae are aquatic. Still others breed in the fungi of forests and meadows. The family often has been treated as a subfamily of the Muscidae and many authors include the Scathophagidae as a subfamily within the Anthomyiidae. There are 41 genera and over 700 described species in the Nearctic.

Family MUSCIDAE (Muscid Flies)

Muscid flies are slender to stocky (2-14 mm), and with a colour that ranges from yellow to grey or black, but some are metallic blue or green. Sometimes flies brightly setulose. Wings usually unmarked, but some with clouded crossveins. Head usually higher than long with the frons in males narrow to broad and its central plate sometimes strongly reduced; frons in females is at least 25 per cent as wide as the head with the central plate always distinct and normally wider than the fronto-orbital area. Frontal bristles 1 to many and curved inwardly. Parafacial area usually bare, but the vibrissa normally strong with bristles or setulae. Usually flat or concave face, rarely with a medial ridge or tubercle on the upper part. 3rdantennal segment at least twice as long as broad and usually rounded at the tip; the arista 3segmented, bare to plumose or rarely comb-shaped. Thoracic bristles usually prominent; normally presutural 1-2 and postsutural dorsocentral bristles 3-4. Scutellum usually bears 2 pairs of marginal bristles; Underside of the apex, an isolated group of setulae rearly present. C has costagial, humeral and subcostal breaks; C usually ends where M_1 meets the margin. M_1 more or less parallel to

R_{4+5} or bent forward; A_1 never reaches the wing margin. Slender legs with varied bristling; base of 1st tarsal segment of the hindleg lacks the distinctive ventral bristle characteristic of most anthomyiids. Both sexes have 5 exposed abdominal tergites usually bearing strong marginal bristles. Larvae develop in many habitats, from dung and decaying plant matter to carrion and fungi, and also found in the nests of bees, wasps, birds, mammals and other animals. Others live in fresh water or soil; a very few develop in living plant tissue. Most evidently feed on excrement, decaying organic matter and the micro-organisms that inhabit this material; some are known as predators of insect larvae or other invertebrates. A few species such as *Fannia canicularis* (Lesser House Fly) and *F.scalaris* (Latrine Fly) can invade the human body and cause intestinal and other myiasis. The food of adults is also varied. Adult feed on dung or decomposed matter, plant sap, honeydew and pollen while some others are predators of insects; the remainings suck vertebrate blood or feed on the exudates of mammals and other animals. Those species that feed on human feces and food, such as *Musca domestica* Linnaeus (House Fly) can spread human diseases such as typhoid fever, cholera and dysentery. Muscid flies are widespread around the world, occurring on all continents and most oceanic islands. Often now the subfamily Fanniinae is considered to be a separate family (Fanniidae). About 4300 species in about 180 genera are described with 4 genera and about 270 described species of those in the Fanniinae and the rest in the Muscidae in the strict sense. The male species of *Fannia* hover in small swarms near the lower branches of trees. Although it is most commonly found breeding in excrement and decaying garbage around human dwellings, these species also breed in bird and rodent nests and wasp and bumblebee colonies. Other *Fannia* species develop to breed in mushrooms, leaf litter, decaying vegetation and animal carcasses. Holarctic common *Neomyia cornicina* is a striking, bright metallic green muscid can be mistaken as calliphorid. Its larvae exhibits an unusual blue colour, live mostly in cow dung. *Musca domestica* is the notorious house fly, common now over much of the earth and certainly widespread. *Musca autumnalis* commonly feeds on fluids from the eyes and nostrils of cattle and called the face fly. *Muscina stabulans* can be a parasite of insect pupae and of nestling birds and is known to cause intestinal myiasis in humans. Several muscids suck the blood of vertebrates, like *Stomoxys calcitrans*, the stable fly, a cosmopolitan pest that lays its eggs in mouldy hay bales and rotting vegetation of various sorts.

Superfamily OESTROIDEA

Family CALLIPHORIDAE (Blow Flies)

Generally blow flies are stocky, medium to large (4-16 mm) with partly or completely metallic blue, green, black or brassy body, sometimes small, slender and without metallic coloration. Sexual colour- dimorphism present. Head higher than long and eyes in males rarely meet above, male frons narrower. Lunule bare and shining; frontal setae reach forward to 2nd antennal segment; frons usually with finely setulose. Vibrissa strong and thickly setulose gena present. Females with 1 backwardly bent orbital bristle above and 2 forwardly bent ones below. Ales lack frontal and outer vertical setae. Long arista, basal ¾ plumose. Thorax with 2 notopleural setae; usually 2-3 anterior and posterior dorsocentral bristles present.

Scutellum with 1-3 pairs of lateral bristles, apical one usually strong. Subscutellum weakly developed or absent. Wing vein M with acutly or right angular bend; cell r_{4+5} almost always open at the wing margin. Blow flies are predominantly flesh-eaters. The most common species called bluebottles and greenbottles, lay eggs on carcasses of all kinds and also are attracted to fresh and cooked meats and dairy products indoors. The restless flying around in rooms, so commonly seen, is mainly a search for places to lay eggs. The term 'blown' refers to meat that has had blow fly eggs laid on it. The development of some species, such as *Phormia regina* and various species of *Calliphora*, on human corpses is monitored or investigate in forensic sciences. Some species are also attracted to excrement and can transmit digestive system pathogens. Screw worms lay eggs in body orifices, on wounds or soiled hair and wool, especially on domestic animals and create severe economic damage on livestock. Some species of *Lucilia* attack amphibians. Larvae of *Protocalliphora* and *Trypocalliphora* suck the blood of various nestling birds and can be troublesome in nest boxes. Species of *Pollenia* (cluster flies) often overwinter in the attics and they lay eggs in the soil and the larvae attack earthworms. Some genera, such as *Angioneura*, are internal parasites of land snails. Many calliphorids are easy to rear and are regularly used in laboratories studying insect physiology and biology. This family has wide distribution, with about 1000 described species under 150 genera. About 80 per cent of the species are from the Old World. This family contain 9 subfamily, 30 genera and 119 species from India (Bharti, 2011).

Family OESTRIDAE (Bot and Warble Flies)

The Family Oestridae contains medium to large (9-25 mm) flies with stout, bristleless, but often thickly setulose bodies and they often resemble bumble or carpenter bees. Broad head, higher than wide, and flat face shield-shaped broad or strongly narrowed below antennae. Antennae small and sunk deeply pitted; 3rd segment globular bearing an arista. Arista generally slender and bare with a thickened base or, in *Cuterebra* Clark, feather-like and lacking the broad base. Mouthparts small and compact, or often atrophied.Thorax covered with short setulae or densely long-setulose; weak bristles rarely present. Short legs, stocky and setulose; femora and tibiae often bear some weak bristles and tibial tip and tarsal segments usually with short, spiny bristles. Wings with vein M highly variable, straight or weakly curved, ending at the wing margin well behind wing tip, or curving forward to join vein R_{4+5} just before margin, or meeting margin immediately behind tip of R_{4+5}, or strongly angled with prominent stump vein beyond crossvein dm-cu. Abdomen globose to conical with shiny cuticle and long, coloured setulae or partly or completely pruinose. Species of Oestridae are obligate parasites of mammals and most are strongly host-specific. Adults have tiny or atrophied mouthparts and apparently do not feed on sugars or protein as do other calyptrate flies. Some species of *Cephenemyia* and *Cuterebra* evidently drink fluids by pressing the underside of the head against a damp substrate. In subfamily Cuterebrinae laied eggs on the host mammals and the larva enters the host through any orifice, eventually migrating to subcutaneous tissues where it feeds in the warble, forming a pouch in the connective tissue of skin. The larva breathes through the posterior spiracles. Females of *Dermatobia hominis*, which attacks a wide range of hosts (including humans) in the Neotropics, first catches a blood-sucking fly (such as a mosquito), on which she lays her eggs and then releases. When the fly lands on a host to bite, the eggs hatch and the larvae burrow into the host's body. Most species of the Subfamily Gasterophilinae lay eggs on hairs around the mouth or on the legs of the host like horses, zebras, elephants and rhinoceroses. In the Hypodermatinae, which attack rodents, lagomorphs and ungulates, the larvae of *Hypoderma* Latreille burrow into the skin, migrate through the body by varoius routes to the skin of the back where warbles are formed. Larvae of the Oestrinae live in the respiratory passages of a wide range of hosts. The eggs are retained in the female until they hatch; the young larvae are expelled within droplets of fluid onto the face of the host. In some *Cephenemyia*, at least, the larvae crawl into the mouth and enter the sinuses and throat via the palate. Bot fly larvae feed on blood and mucus and can cause severe damage to the tissues; mature larvae are coughed out of the mouth or sneezed out of the nostrils. *Oestris ovis* infests the noses, sinuses and throats of domestic sheep, but evidently has not transferred to native species.The Family Oestridae, containing over 150 species in 28 genera, is widespread on all the major continents, but is most diverse in Africa and central Asia. The African Elephant alone supports five species.

Family SARCOPHAGIDAE (Flesh Flies)

Species of the family Sarcophagidae are robust, mostly grey flies with variable range (2.5-23 mm). Thorax with usually 3 dark stripes on top and abdomen striped, banded or spotted (sometimes red) with markings that shift tones depending on the angle of the light. Head wider than high, usually little higher than long. Eyes bare; fronto-orbital area usually with sparse setulae but central part normally bare. Frontal setae 4-10 and orbital setae 1-3 curving backward, while 2-4 curving forward (except males Sarcophaginae). Concave face without a central ridge. Vibrissae present. Arista bare or finely setulose (most Miltogramminae) to plumose, especially on basal half or two-thirds (most Sarcophaginae).Thorax: 3-4 anterior and 2-4 posterior dorsocentral bristles; scutellum with 2-3 pairs lateral setae and 1 pair of discal setae; sometimes a pair of apical small setae occurs. Origin of R_{4+5} setulose above and below, with setulae continuing toward crossvein r-m above. Any extension at the bend in vein M is usually not developed, but merely present as a short, darkened fold. Flesh flies lay larvae or eggs that are ready to hatch. Larvae of the tribe Miltogramini are deposited by adult females at the entrance to the nests of burrowing solitary bees and wasps; the larvae feed on the provisions in the cell and sometimes kill the host's egg or larva. *Eumacronychia* attack lizard and turtle eggs and *Macronychia* parasitize adult tabanid flies. Species of the tribe Paramacronychiini (subfamily Miltograminae) and the subfamily Sarcophaginae are obligate parasites or scavengers that become facultative parasites in a wide variety of animals mostly arthropods and some myriapods and snails, while others can develop in excrement. A few species are implicated in myiasis of the human flesh (as secondary invaders) exception is species of the genus *Wohlfahrtia* which lay larvae on the skin of the young mammals, including human infants, although has not gained the notoriety of the blow flies or house fly. A few species are economically beneficial. *Sarcophaga aldrichi* attacks the pupa of *Malacosoma disstria* (Forest Tent Caterpillar) and is a major control agent of this defoliating pest. Adults feed on plant sap. There are over 2500 named species of sarcophagids worldwide, placed in 108 genera. Most speciose genus is *Sarcophaga* Meigen, and it contains almost 800 known species.

Family TACHINIDAE (Tachinid Flies)

They are tiny to large flies, often strongly and densely bristled. Head usually higher than long with sloping frons and small antennae (in holoptic males), but ranges to box-like with a horizontal frons, face and antennae long. Holoptic males' forns narrow and usually lacks lower orbital bristles curving forward; but 2 occur in females and dichoptic males. Frontal bristles usually bend inward. Ocelli rarely absent and ocellar setae usually curve forward. Antennae variable in size; 2nd segment bears setae on tip; arista normally bare, but sometimes feather-like. Face usually concave or flat, but sometimes convex; normally with strong vibrissa. Labium sometimes extremely long and slender. Postpronotum usually with 4 strong bristles, but vary 2-5. Scutum with highly bristled, especially in subfamily Goniinae which is almost devoid of them (many species in the subfamily Phasiinae); presutural dorsocentral bristles 3 and postsutural 3-4; scutellum usually with 3/4 pairs of marginal bristles and 1pair on disc; subscutellum well-developed, looking evenly convex in side view. Meron with a vertical row of bristles.Wings normally transparent, but some marked or spotted and all other dark. Vein M bent forward, ending in wing margin just behind R_{4+5} or in R_{4+5} itself, thus closing cell r_{4+5}. Abdomen variable from petiolate to broad, from convex above, flattened to globose. Usually more or less covered with strong, erect bristles, but these are lacking in some species. All larvae of Tachinidae parasites of arthropods. They have controlled populations of other insects many species now being used in the biological control of insect pests. In the subfamilies Phasiinae and Exoristinae stick a conspicuous, undeveloped egg directly on the cuticle of the host, considered the most primitive system of attack on an exposed host and modifications to speed up the penetration of the larvae into the host have developed. The females of most Phasiinae have abdominal structures for inserting eggs into the host's body. Generally Tachinidae, store the eggs in an ovisac until embryonic development is complete. Unlike sarcophagids, which deposit active larvae that have hatched within the female, all tachinids lay eggs that hatch (but sometimes within seconds) after deposition. A female in the tribe Goniini lays thousands of minute eggs where the host will eat them along with its food; the larvae burrow into the host through the gut wall. However, most tachinids broadcast their eggs on damaged foodplants or other places likely to support their host; the larvae lie in wait for a suitable host to pass by, board it and burrow inside. Still others (tribe Dexiini) have larvae that burrow into the substrate where they hatch and actively search for the host in soil, rotten wood and other places. Most tachinids, including most species in the subfamilies Tachininae and Exoristinae, attack the larvae of Lepidoptera. The Phasiinae parasitize true bugs, especially the Pentatomidae, Coreidae, Nabidae and Lygaidae. Beetle larvae of the families Scarabaeidae, Cerambycidae and Elateridae are hosts for the Dexiini in the subfamily Dexiinae; adult scarabs are attacked by species of *Cryptomeigenia* and adult carabids by those of *Zaira*. Certain members of the tribe Blondeliini parasitze larval and adult chrysomelid beetles. The Orthoptera are hosts for several genera. There are a very few records of tachinids parasitizing centipedes, spiders and scorpions. Adult tachinids are active and eagerly search for sources of sugar; they are particularly attracted to honeydew. Some, in search of nectar, often visit flowers, especially those of the Asteraceae, and others feed on tree sap. Males of many species congregate on sunny hilltops and ridges to wait for females. Species aggregate at specific sites in these places. In the Diptera, the family Tachinidae is second only to the Tipulidae in the number of described species, at least 8000 species are known worldwide. The related family Rhinophoridae contains tachinid-like flies that share characters of primitive tachinids, calliphorids and sarcophagids; species parasitize woodlice (terrestrial isopods). Perhaps the most known subfamily is Phasiinae, rotund and almost bristleless flies that attack big, green *Chlorochroa* Stål stink bugs. Genus *Cylindromyia* Meigen are also distinctive parasites of pentatomid bugs. *Compsilura concinnata* is considered the most polyphagous of all tachinids, it has over 200 different recorded hosts. It was introduced from Europe, mainly to control Gypsy Moth (*Lymantria dispar*) in eastern North America, but has spread west to the Pacific coast.).

Superfamily HIPPOBOSCOIDEA

Family HIPPOBOSCIDAE (Louse Flies and Keds)

Flies of the family Hippoboscidae are flattened, rather tough and leathery looking flies, ranging from 1.5 to 12.0 mm long. They are usually setulose. The head is broad and somewhat flattened; the mouthparts thrust forward. The compound eyes are normally well developed and horizontally elongate; ocelli are either present, vestigial or absent. The inner vertical bristles are long, the outer ones are absent and the orbital bristles are few to many. The lunule is usually bare, shiny and conspicuous. The antennae are strongly modified and are more or less immovable; they lie in deep pits. The first segment is usually present, but sometimes is fused to the lunule and invisible; the rounded second segment is the largest and sometimes bears a third, flattened, segment or a spatulate or branching arista. The one-segmented palps form a sheath for the blood-sucking, retractible labium. The thorax is flattened; the scutellum is usually large, often broadly rectangular. The wings are usually fully developed, but in some genera that parasitize mammals, such as *Lipoptena* and *Neolipoptena*, the wings break off after the fly settles on its host. In others, the wing is reduced to a small flap or, as in *Melophagus ovinus*, which lives in the fleece of sheep, is a tiny knob. This species has lost its halteres, but they are normally present in other species. The wing veins are usually crowded forward with the posterior veins often fading towards the wing margin. Some species have greatly reduced venation. The legs are strong, rather short and often well-bristled. The coxae and femora are usually swollen; the apical tarsal segment is the largest. The tarsal claws are large and strong, simple or forked; the empodium is setulose or feathery and the pad-like pulvilli are often long and soft. The abdomen is largely membranous with tergites and sterna mostly reduced. Hippoboscids are ectoparasites and feed on the blood of birds and mammals. The majority attack birds and are often called louse flies; those on mammals (mostly ungulates) are frequently called keds. The Sheep Ked (often erroneously called the sheep tick) is probably the only species in North America that causes any economic loss; heavy infestations of sheep sometimes result in anaemia and staining of the wool. This species was introduced from Eurasia to sheep-producing temperate environments worldwide and, apparently, no truly wild populations exist today. Those populations found on bighorn and thinhorn sheep are thought to have transferred from domestic sheep. Several species have been recorded biting humans, but this is a rare and accidental phenomenon. Most species are not host specific, but rather seem to inhabit a variety of hosts from a particular type of habitat. Together, the Hippoboscidae, Nycteribiidae and Streblidae are sometimes called the Pupipara, because they appear to give birth to pupae. This is inaccurate, in that the fully grown larva is extruded by the female and immediately pupates afterwards. One at a time, an egg and then a larva develop in the female uterus, nourished by secretions from the so-called milk glands. A female can produce 7-8 or more mature larvae during her life. The Family Hippoboscidae is cosmopolitan, but is most diverse in the tropics and subtropics; about 200 species are named in 21 genera. Most species but not all are in the subfamily Ornithomyiinae and parasitize birds. *Ornithoctona erythrocephala* ranges across Canada and into South America; it especially favours hawks and falcons. *Olfersia fumipennis* is a Holarctic species widespread over much of North America; it is collected regularly on Ospreys and Bald Eagles. Species of the Subfamily Lipopteninae normally parasitize even-ungulates, such as deer and sheep. *Neolipotena ferrisi* lives on deer. *Lipoptena* has three Nearctic species, but only the western *L. depressa* occurs in the province; it is especially common on coastal deer. The introduced *Melophagus ovinus* does not fly at all. It can be a problem for sheep flocks and is also known from native wild sheep. *Hippobosca longipennis* (Subfamily Hippoboscinae), an Old World species which attacks mammals of various types, including domestic dogs.

Family NYCTERIBIIDAE (Wing Less Bat Flies)

Wing less bat flies are most fascinating one among all Diptera, look like spider, about 1.5 to 5 mm long. Head small, sclerotized, which is retractile and rotatory in nature. Compound eye mostly absent, or is usually reduced to two ocellus-like facets. Antenna have 3 segment, possessing a branched arista. Proboscis is piercing-sucking and palps usually large. Scutellum is absent, postnotum rounded and raised. Unique to the Nycteribiidae, a movable crescentic row of spines (thoracic ctenidium), is present in front of coxa middle. The wings are absent, but halteres are present, lying in a groove, which may be covered by a flap arising from the inner edge of the groove. The legs are rather long and stout, variously armed with setulae and bristles. The first segment of the tarsus is usually longer than the others combined; the last segment is usually the broadest. The claws are simple, but large and stout; the empodia are absent. Normally, the abdomen has both the tergites and sterna bearing prominent rows of bristles and setulae on the hind margins. Nycteribiids are blood-feeding ectoparasites of bats. Like species in the related Hippoboscidae and Streblidae, those of the Nycteribiidae are viviparous, giving birth to fully developed larvae. One mating produces multiple larvae; the sperm is stored in the two spermathecae - the ovaries ovulate alternately, and one egg at a time enters the uterus after fertilization. The larva is nourished from special glands and development takes about nine days. The female normally deposits the mature larva on the vertical surface of the bat roost, pressing it into place; the larva pupates immediately. Most species of nycterbiids are apparently bat species or genus specific although there may be considerable variation in host-parasite relationships over the range of a species of fly. The family Nycteribiidae ranges worldwide except for the polar regions and contains about 260 described species in 13 genera. *Basilia* Ribeiro, the best known genus, contains over 40 species.

Family STREBLIDAE (Streblid Bat Flies)

These bat flies are small (0.75-5.00 mm), rather setose and bristly flies. They look like flea, dorsoventrally flattened, or have a convex body type. Head size usual, often compressed laterally or flattened, and bearing conspicuous bristles and setulae. Eyes are reduced, facets ranging from 0 to 36, but usually 7 to11; ocelli are absent. First segment of antenna fused to head and 2nd with a furrow above and mostly concealing the oval third segement bears a comb-like arista. Piercing and sucking type mouthparts. The thorax varies from convex and globose to slightly convex, laterally compressed or dorsally flattened; the pronotum indistinct. The scutellum is normally obvious, its hind edge angulate in New World genera and bearing one to ten (usually four) long bristles. Hindleg is often much longer. The tibiae lack apical spurs; the tarsal claws are large and adapted for clinging. Most species possess wings, but some have poorly developed wings while other lacks them. Wing veins run longitudinally; the costa ends at the apex of R_{4+5}; the subcosta is faint, fused with R or absent. Vein R_s forks at about one-third the wing length, vein R_{2+3} ends just before the wing tip and vein R_{4+5} ends at the tip. Crossvein r-m is usually at about the midlength of the wing and dm-cu, which joins the unbranched vein M to vein CuA_1, is near the wing tip.Veins CuA_1 and CuA_2 are fused for much of their lengths, with vein CuA_2 branching off at about two-thirds the wing length and joining A_1 near the wing margin. The abdomen is mostly membranous and sac-like, especially in egg-bearing females. Tergites 1 and 2 fused, with strong lateral lobes bearing bristles and setulae. Both sexes Streblids are obligate, blood-feeding ectoparasites of bats; *Trichobius* Gervais sometimes bite humans. Single larva develop in the female abdomen showing adenotropic viviparity, where they feed on special glandular secretions. When fully grown, the larva is extruded and immediately pupates. Adults must feed before mating and, apparently, mating is required each time a new egg is produced. Streblids frequently parasitize colonial, cave-dwelling bats, but most on forest-roosting bats. Solitary bats are seldom attacked. Flightless streblids mostly inhabit bats that eat fish or fruit. The adults of most fully winged NewWorld species, although highly mobile, stay close to their hosts; most are host specific or live on a few species of closely related bats. The Family Streblidae ranges subtropically and tropically on all the main continents and oceanic islands, mostly between 40 degrees north and south. It contains about 220 species in 32 genera.

ACALYPTRATA (Schizophora, without calyptra)

Superfamily CONOPOIDEA

Family CONOPIDAE (Thick-headed Flies)

Thick-headed flies are medium to large (3-20 mm), elongate and wasp-like who have lost prominent setae or bristles. They are coloured black and yellow, black, or red-brown, frequently with pruinescence; wings are often spotted or darkened along the front margin. Head large, broader than thorax; bare compound eyes are widely separated in both sexes. Ocelli are generally present except Subfamily Conopinae. The proboscis is long, elbowed at the base (Conopinae and the genus *Zodion*) or elbowed at both the base and in the middle of the proboscis. Antennae variable: the first segment of the flagellum is elongate with a short 3segmented stylus (Conopinae) or is short, bearing a 2/3segmented arista on the upper surface. Wing with R_{4+5} and M_{1+2} strongly convergent and sometimes fused near the wing tip, closing cell R_{4+5}. A short crossvein joins Sc and R near the end of Sc in some genera. Sometimes a spurious vein occurs, running outwards from crossvein r-m. Cell *cup* is always closed and petiolate. The legs are almost bare and lack distinct bristles; they are rather uniform, although the femora are thickened in some genera. Abdomen cylindrical and more or less club-shaped, broad or constricted at the base and often curving downward at the tip. In the female, the sterna of segment 5 and/or 6 are often enlarged and plate-like. Adult conopids are frequently wasp-like in form and colour and apparently mimic various Hymenoptera. They are notorious for their variability, especially in coloration. They feed at flowers, especially of the aster, mint and carrot families. The known larvae of most genera are internal parasitoids of aculeate Hymenoptera. Adult females deposit eggs on the hosts in flight. The Subfamily Stylogasterinae, parasitize cockroaches and calyptrate Diptera; females are known to hover over the front lines of army ant columns laying eggs as their hosts are flushed by the ants. About 800 described species assigned in 45 generaspread worldwide. In North America there are nine genera and 67 species described; many species range across the whole continent (Smith, 1959).

Superfamily TEPHRITOIDEA

Family LONCHAEIDAE (Lance Flies)

Lance flies are stout, setose, 3 to 6 mm long, with a large, wide head and a broad, flattened abdomen with shining blue-black or dull brown in colour. Wings clear, but often clouded with yellow or brown; haltere black. Eyes are large and round to oval. Frons setulose, narrower in male than in female; lunule large and exposed. One orbital bristle, one inner and one outer vertical bristle are present, and ocellar bristles are strong. No vibrissae are present, but some subvibrissal setae may be enlarged. 3rd antennal segment varied, short or long and narrow, black to orange, often hanging vertically; arista bare, pubescent or plumose. Thorax with scutum arched, black or brown and pruinose to shining; the setulae and bristles are usually dense and strong. The scutellum has two pairs of bristles, one apical and one basal. The wing is gradually tapered from base to tip; C extends to M and is constricted or broken at the end of Sc; Sc is complete. Vein A_1 short or continues toward wing margin as a sinuate fold. Upper calypter prominent, with white to brown margins. Legs stout and black; the tarsi often yellow. Femora often strongly setose and bristled; mid tibiae bear a bristle underneath at the tip. Lance flies are commonly secondary invaders of injured or decaying vegetation. Larvae of *Lonchaea* and *Dasiops* often live under the bark of dead or dying trees (mainly conifers) or in damaged or rotting fruits and vegetables. Some are associated with plant-attacking insects such as bark beetles, weevils and fruit flies, where they scavange or prey on larvae. Others may be primary invaders of plants; for example, most species of *Earomyia* infest the cones of conifer trees. Male adults of some species swarm in patches of light in woodland.The Family Lonchaeidae contains ten genera and about 700 described species.

Family PALLOPTERIDAE (Flutter Flies)

Flutter flies are small to medium-sized (3 to 5 mm long), with grey or yellow bodies and brown-patterned wings. Head higher than long with the frons broader than high and always yellow, at least on the front half. One pair of orbital bristles and strong ocellar and inner and outer vertical bristles present. Ocellar bristles pair, weaker and somewhat divergent. Yellow face slightly convex, usually with a median ridge and lacks setae and without oral vibrissae. Small antenna with 2nd segment notched at the tip and bears a single bristle; the third segment is oval with a bare or short-setulose arista. The thorax is yellow to black and normally pruinose, but sometimes shining or patterned in yellow and black; thoracic bristles are prominent. Wing is rather long and narrow, usually with brown markings. Subcosta is complete; cell sc is always dark. The costa has three weakenings or breaks and ends just beyond its junction with R_{4+5}. Cell cu*p* convex at the tip; vein A_1reaches or almost reaches the wing margin, at least as a fold. Tibiae and tarsi are usually yellow; the tibiae lack preapical dorsal bristles, but the middle tibia has an apicoventral bristle. The abdomen is elongate oval, yellow to dark brown and normally unpatterned. Flutter fly larvae are apparently phytophagous or carnivorous and some have been found in the flower buds and stems of plants in the aster and carrot families. Others live under the bark of dead trees and prey on the larvae of long-horned and bark beetles. *Palloptera claripennis* has been reared from the cones of Douglas-fir, where the larvae fed on the larvae of *Contarinia* midges (Cecidomyiidae) in Vancouver, Canada. Adults are usually seen on flowers or on the lower branches of trees and shrubs. Males vibrate their wings, giving the family its English name-flutter flies. This small family is a small family, closely related to the Piophilidae and distributed in the north temperate region, south temperate America as well as New Zealand, about 54 described extant species in 12 genera.

Family PIOPHILIDAE (Skipper Flies)

Generaly small (3-6 mm) and vary in colour from metallic black or blue-black to pale brown or yellow. The frontals often yellow. Bristles are usually strong and black and the body is sometimes densely setulose. The wings are usually clear and iridescent, but some genera (*e.g. Mycetaulus*) have brown-marked wings. Viewed from the side, the head is normally higher than long; the compound eyes are nearly round and lack setulae. The frons is parallel-sided, as broad as long in both sexes. There are two or three orbital bristles, the inner and outer vertical bristles are about the same size, the two ocellar bristles point forward and the postocellar bristles diverge. The oral vibrissae are usually prominent, sometimes there are several and in *Amphipogon* they form a thick beard. The face is convex, with the antennae seated in grooves. The second antennal segment lacks a dorsal seam, but normally bears a dorsal bristle; the segment is about as long as wide, but is strikingly elongate in males of *Prochyliza xanthostoma*Walker. The third segment is oval with a bare or short-setulose arista arising near the base. The scutum normally is shiny, but is pruinose in some *Mycetaulus*; it is densely setulose to almost bare. The scutellum is sometimes flattened and occasionally bears a pair of tubercles; there are four bristles. In the wing, the costa has a subcostal break. Subcosta is complete and not fused with R; vein A_1 faintly reaches the wing margin or ends abruptly just before the edge. The tibiae lack preapical dorsal bristles, but often have other strong bristles, especially ventrally and laterally near the tip. The abdomen is shiny black to yellow and lacks a colour pattern. Skipper flies usually develop in dead animal matter (especially dried carcasses advanced in decay) and rotting fungi. The larvae of one European species are ectoparasitic blood feeders on passerine bird nestlings. Perhaps the best-known species, *Piophila casei* (Cheese Skipper) infests cheeses, meats and hides and can be a serious pest in the food industry. This fly has also caused mild nasal and enteric myiasis when people have eaten maggot-infested cheese. The larvae are called skippers because they can flip into the air like little springs. The Family Piophilidae contains 69 described species, 11 of which contain Holarctic species. 23 genera world wide, but is most diverse in the northern temperate regions.

Family OTITIDAE (Picture-winged Flies)

The family Otitidae is small to medium-sized flies (3-12 mm) with often brightly coloured and frequently metallic body. The head in profile is normally higher than long; inner and outer vertical bristles, ocellar, postocellar and one or two orbital bristles usually present, but vibrissae absent. The face is normally fully sclerotized, broad and convex. The size and shape of the antenna is variable; the third segment often has a sharp tip and the arista is long and bare to plumose. The wing has Sc complete, its tip gently curved; R_1 is bare or setulose. Cell *cup* usually has an acute projection at the postero-distal corner. Females have abdominal segment 7 flattened and more or less triangular; the ovipositor is sword-like. The larvae of picture-winged flies live in decaying vegetable, dung even some feed under the bark of dead trees and others are phytophagous; a few, such as those attacking sugar beets and onions, are economically important. Adults can be common in moist places and meadows; they rest on low vegetation and visit flowers, fungi and tree wounds. Many vibrate their patterned wings during mating displays. The family Otitidae consists of two subfamilies, the Otitinae and Ulidiinae, which are sometimes considered separate families. The family consists of about 800 described species in 50 genera.

Family PLATYSTOMATIDAE (Platystomatid Flies)

Platystomatids are small to medium sized flies (2.5-12 mm), with brightly metallic body and banded wing. Dome shaped head with reduced bristles; there are 1/2 orbital bristles, an inner and outer vertical bristle and 1genal bristle. 3rd antennal segment is usually elongate, sometimes with a sharp point tip and slender arista, either bare or setose. The proboscis and palps are well developed. Thorax with maximum one pair of dorsocentral bristles; two or three pairs of scutellar bristles present. The wing is normally long and slender; C with break near base, but no subcostal break near end of Sc. Vein Sc is complete; vein R1 is setose above. Cell *cup* is always rounded at the tip, never with a pointed extension at the lower end. Little is known of the life histories of platystomatid flies, but the lives of adults and larvae are probably much like those of the Otitidae. In other parts of the world, adults frequently visit mammal dung; larvae live in logs and vegetation damaged by other insects or attacked by fungi. Closely related to the Otitidae (considered as a subfamily), most of the 119 genera and 1200 described species of Platystomatidae are found in tropical and subtropical Africa, Asia and Australia.

Family TEPHRITIDAE (Fruit Flies)

Brightly coloured,small to medium-sized tephritid flies usually have banded or spotted wings. The head is variable; in some exotic species the compound eyes are stalked. Usually one pair of inner and outer vertical bristles occur, along with one pair of postocellars and ocellars and one to several pairs of orbital and frontal bristles which are thickened or flattened. Vibrissae are absent. The second antennal segment sometimes bears a seam on top and the third segment is often pointed on the upper end; the arista is usually bare or finely setulose. The proboscis is sometimes long and elbowed. The scutellum is swollen and shining in some genera, with 1-4 pairs of bristles, normally on the margin. The bristles on the scutum are variable, but there is always at least one pair each of dorsocentral and acrostichal bristles. The wing has a distinctive Sc bent sharply forward toward the costa and weakened after the bend, often not reaching the costa. Vein R_1always bears short setae above. The cell *cup* usually has an acute projection on the hind margin. Colour patterns usually present, ranging from almost entirely dark brown or black to combinations of bands, stripes, spots or reticulations in black, browns and yellows. Adult female fruit flies lay eggs in living, healthy plant tissue and the developing maggots feed in a wide variety of plant parts, depending on the species. Some form galls on stems and roots; a few tunnel in leaves; others develop in the fruits, seeds and ovaries, especially in plants of the huge aster family. Those that attack fruits and vegetables can be severe agricultural pests, one of the worst is *Ceratitis capitata* (Mediterranean Fruit Fly). These are among the most economically important fly families, not only because of its destruction of useful plants, but also because many species are extensively used in the biological control of weeds. Adults are frequently seen on their host plants flower. The Family Tephritidae is a speciose cosmopolitan family of about 4350 known speciesin 481 genera.

Family PYRGOTIDAE (Pyrgotid Flies)

Pyrgotids are medium to large flies (wings 6-18 mm long), relatively slender, usually with strongly patterned wings. Head is usually prominent and rounded although sometimes tapering ventrally; face often broad with a medial pronounced ridge. Mouthparts, palpi, and compound eyes normal; vibrissae and ocelli absent. Head bristling often greatly reduced with the exception of the inner vertical bristle which is always distinct, ocellar and postocellar setae sometimes well-developed. Antenna generally large, with an elongate pedicel; flagellomere somewhat oval in shape, with a medial to near basal, bare arista. Dorsal thoracic bristles generally reduced giving thorax a setulose appearance. Wing long, with a humeral break and sometimes a subcostal break in the costa; subcosta usually reaching costa. Wing with either a distinct banding pattern, like some Tephritidae, or with a more diffuse mottled colouring. Cell dm long, cell cup with lower apical corner straight (1 North American species) to having a long pointed corner. Alula and calypteres well developed to virtually absent. Legs long, robust; strong bristles lacking; hind tibia of few species with basal third much smaller in diameter than apical two-thirds. The abdomen of the female is greatly elongate and highly modified to lay eggs onto adult scarab beetles while in flight. Larval pyrgotids are internal parasites of adult scarabs and some pyrgotid species are known to help control population levels of some species of pest scarabs. As the scarabs fly at night, generally it is most effective to collect pyrgotids at night using light traps when the flies are actively pursuing hosts.There are about 330 described species in 50 genera worldwide (Steyskal, 1978).

Superfamily NERIOIDEA

Family MICROPEZIDAE (Stilt-legged Flies)

Stilt-legged flies are small to medium (3.5-20 mm), slender, and almost without bristles or setulae. The common name, stilt-legged flies, is a reference to the strikingly long, slender legs. Colours vary from yellow through red to black and the body often banded or has pale pruinescence. The wings are clear or coloured, often with spots or brown bands. The head is usually globular or sometimes conical and pointed in front (*e.g. Micropeza*) with large eyes.Third segment of the antenna is oval and, on top near its base it bears with or without setose arista. Vibrissae or ocellar bristles absent. The thorax is elongate, with the front legs placed well forward of the middle pair; small scutellum with 1 pair bristles. Long and slender wing; vein C reaches the wing tip at vein M_{1+2} and has no breaks; Sc is complete. All longitudinal veins are rather straight and extend to the wing margin; R_{4+5} and M_{1+2} usually converge or fuse near wing tip. Cell dm is long and narrow. The abdomen is elongate, slender and sometimes constricted at the base; the genitalia of both sexes are flexed down and forward. Males often have processes under segments 5 and 6 and laterally on segment 7 (especially on left side). The larvae develop in decaying wood, fruit and other vegetable matter even in dung; one oriental species attacks growing ginger roots. Adults are usually found in marshes or wet woods, perched on leaves and tree trunks, often at wounds in the bark. Some are evidently predators of aphids and other small insects that they stalk in vegetation. Generally they are wingless and mimic ants.The Micropezidae is cosmopolitan, tropical family containing about 520 described species (Merrittand Peterson, 1976).

Superfamily DIOPSOIDEA

Family TANYPEZIDAE (Tanypezid Flies)

These flies are generally 4.5 to 7 mm long, slender, long-legged and with patches of silvery tomentum on the body. The legs are yellow, brown head with very large compound eyes. The frons is narrower than the eye and is much narrower in males than in females. The third antennal segment is oval, rather large and somewhat elongate, with a basal, setulose arista. Vibrissae are absent; the upper orbital bristle arises on the vertex. The scutellum has two pairs of bristles, but is otherwise bare. The wing is clear; the costa has a weak subcostal break and Sc is complete. Veins R_{4+5} and M almost meet at the wing tip; cell cup is closed by the L-shaped CuA_2.

Vein A_1 almost reaches the wing margin. The adults live in damp forests. This is primarily a New World family containing 3 genera and about 22 species. Only *Tanypeza* is the dominant one.

Family STRONGYLOPHTHALMYIIDAE (Strongylopthalmyiid Flies)

These slender and long-legged flies are about 3.5 to 4.5 mm long and with few bristles. They generally black with coarse yellow setulae while antennae, front of head and legs are yellow. They have faintly patterned wings with pale brown patches. The head is almost globose, slightly longer than high, with prominent compound eyes and antenna covered with fine setae; arista arises basally on the 3rd segment. They possess a wider (male and female equal sized) frons besides the anterior ocellus. There are three orbital bristles followed by a row of fine short setae reaching the level of the antenna. No vibrissae occur; inner and outer vertical bristles are present, both have one. Thorax elongate, almost twice as long as wide. A strong subcostal break near the end of vein R1 present; Sc fades just before this break; Rs fork strongly divergent while A_1 does not reach the wing margin. The legs are slender with fore coxa distant from middle coxa; bristles lacking except for a short apical one on the middle tibia. Biology unknown. The larvae have been collected from under the bark of trees.This family contains 33 known species in two genera: *Nartshukia* with one species in Vietnam and *Strongylophthalmyia* with 23 species in the Oriental and Australasian regions, eight in the Palaearctic region and one in the Nearctic. These flies have long been placed in the Psilidae, but the family is probably most closely related to the Tanypezidae.

Family PSILIDAE (Rust Flies)

Members of the Psilidae are small to medium in size (3-8 mm), rather slender and with only sparse bristles. They are brightly coloured, ranges from yellow and red to brown or black with usually clear wings often yellowish or smoky and with darkened crossveins on wingtip. The head is circular or triangular in profile with the frons projecting anteriorly and the face below the antennae strongly sloped backwards. Variable size eyed insect and, the back of the head below them is somewhat swollen. The antenna ranges varies from short to long; 3rd segment (and often 1st and 2nd) often elongate (*Loxocera*) and bearing a setulose arista on the basal half. Bristles of head are few and variable, but there is always at least an inner and outer vertical bristle on each side; vibrissae are absent. Wing has a strong subcostal break well before the end of R_1; a clear strip in the wing membrane obliterates the end of Sc and reaches or crosses R_1. The outer end of cell c*up* is squarely truncate and vein A_1fails to reach the wing margin. The legs lack bristles, but there are longer setae on the tips of the tibia. Underside femur of *Loxocera* species bears a pad of short dense setulae near the tip. Larvae feed in the roots and stems of many kinds of plants (recorded hosts include sedges, rushes and lupines); a few species are pests of crops. The most injurious is *Psila rosae* (Fabricius), the Carrot Rust Fly, which tunnels in the roots of carrots, parsnips and celery. Larvae also live under the damaged bark of trees. Adults rest on foliage, especially in the shade; some species are attracted to tree sap. The Family Psilidae is mainly Holarctic with a few species scattered in the southern hemisphere. It contains seven genera and about 200 species.

Superfamily LAUXANIODEA

Family LAUXANIIDAE (Beach Flies or Lauxaniid Flies)

Small delicate (2.5-5.5 mm) flies, having yellow, brown, black or a combination of these colours even dull or shiny body, sometimes with dark spots; wings are often tinged with yellow, and sometimes spotted or clouded. Head shape varies with bare or sparsely micropubescent eyes, and bare to long plumose arista. The vertex is not strongly excavate, but rounded or carinate; vibrissa are absent with distinctly convergent postocellar bristles. Four pairs of strong backward leaning bristles present between the eyes. The thorax has the metepisternum bare, and posterior thoracic spiracle lacks both bristles and outstanding setae on the lower margin. The legs usually have a preapical dorsal bristle on all tibiae, but sometimes absent on hind tibia while 1 to 3 apical spurs on mid tibia, the hind tarsus is sometimes ornamented, and often a ctenidium present on fore femur. Vein C without a subcostal break and ends at M_1, while Sc is complete and separate from R_1. Adults are normally found in woodlands and in dense vegetation by water in the shade. Larvae are saprophagous, and typically live in leaf litter, vegetable trash, in rotting tree stumps, dung and bird's nests. The family, often previously called the Sapromyzidae, occurs on all continents, except Antarctica. More or less1600 speciesfound worldwide, under 128 genera. Most predominant *Lauxania* Latreille with the arista thickly white pubescent is represented by the Holarctic *L. cylindricornis*(L.) and the Narctic *L. nigrimana* Coquillett.

Family CHAMAEMYIIDAE (Aphid Flies)

Small to medium (1-4.5 mm) flies with silvery gray-brown and densely pruinose predator of aphids. They can also be shiny black, often with brown stripes on the thorax, and black spots or bands on the abdomen; wings often uniformly milky hyaline, occasionally brown markings. Head generally as broad or broader than the thorax with bare eyes and with short pubescence. The ocellar triangle is occasionally enlarged and prominent. Vibrissae are absent; 0 to 3 pairs of fronto-orbital bristles present. The antennae are short and porrect. The thoracic scutum bears neatly arranged setulae and bristles, and the scutellum has just four long setae. Legs weakly bristled (except fore femur) and all tibiae without preapical dorsal bristle. The costal vein C is unbroken, and Sc is complete and free. The radial vein R is bent forward close to Sc, and ends close to the apex of Sc. The A_1vein is abruptly terminated in the basal third to half of the wing. Very little information is available on biology of most adults. The larvae of most species are free-living predators of adelgids, aphids, coccids and scale insects. About 20 genera are known worldwide.

Superfamily SCIOMYZOIDEA

Family COELOPIDAE (Seaweed Flies)

Strongly bristled, robust (3.7-8.9 mm), flattened, flies generally black or grayish-black in colour or shiny or subshiny. Head turns strongly concave, and with oblique compound eyes. Antennae are short, and appressed to the face. Thorax generally distinctly flattened, with metepisternum setose below the posterior thoracic spiracle and lost bristles on there lower margin. Strong and swollen fore femora, and all tibiae have a long dorsal preapical bristle. At rest, wings lie flat on abdomen, with the wing blade posterior to cell dm frequently folded under and closely appressed to the ventral surface, of the anterior portion of the wing. Veins generally bare, except upper and lower surfaces of the stem of R, and upper surface of apical third of R_1 are sometimes setose. C is without a subcostal break and without obvious spines; subcosta complete, and A_1+ CuA_2 extends to the wing margin. They often are found on sea coasts, associated with seaweeds. The larvae live in rotting seaweed, and breed in vast numbers, the adults often occurring in dense swarms around kelp. Worldwide there are about 30 described species in nine genera. *Coelopa frigida* (Fabricius), this Holarctic species evidently is confined to the Atlantic and Arctic coasts.

Family SEPSIDAE (Black Scavenger Flies)

Slender, moderate (2-6 mm) flies, mostly shiny black, but also can be dull black, brownish or yellowish. They possess a globular head, with large eyes and bare antennal arista. The thorax has a silvery pruinescence on at least part of the pleuron, and the posterior thoracic spiracle with 1or more fine bristles on lower margin. Legs slender, and males possess characteristic bristles, tubercles or emarginations on fore femora, and often also in fore tibiae for grasping the base of the wing of the female during copulation. Mid tarsus of male sometimes enlarged or varicolored, and hind tibia often has a slit or elongate area (osmeterium). Narrow and hyaline wings usually with a dark spot near the tip of vein R_{2+3}, and are also sometimes blackish base; C without a subcostal break. Abdomen is usually elongate, and often constricted near the base, giving these flies an ant-like appearance. Adults are scavengers, and most often can be caught by sweeping grass in meadows, woods, or around dung. When walking about these flies repeatedly and characteristically flip their wings outward. Larvae are often found in carrion or excrement. About 250 described species in 21 genera occur worldwide with numerous species often present on more than one continent like *Decachaetohora aeneipes, Meroplius sterocoraria, Sepsis biflexuosa, Sepsis puncta* and *Themira putris*. The latter four species are Holarctic in their distributions.

Family DRYOMYZIDAE (Dryomyzid Flies)

Moderately bristly to quite setose flies, 4 to 12 mm long, yellowish, brown or dark grey in colour. Head with face convex in middle, and with clypeus large and prominent in profile, bulging below lower margin of face.Vertex not strongly excavated, and the postocellar bristles vary from slight convergent to greatly divergent. Vibrissae are absent, and with short pedicel. Thorax longer, and with bare metepisternum. Scutellum has 2 or 3 pairs of bristles, and the posterior thoracic spiracle is lacking both bristles and outstanding setae. The tibiae with dorsal preapical bristles, and males may have an apicoventral projection on the first segment of the fore and hind tarsi. Wings hyaline or smoky or tawny, and sometimes dark brown spoted crossvein r-m and posterior crossvein seen. C without subcostal break, and subcosta is complete to the costa at some distance proximal to the tip of vein R_1. The costa may have costal spines, and crossvein bm-cu is always present. Species of *Dryomyza* have been reared from decaying organic matter, including decaying fungi, carrion, and dung. Members of the Helcomyzinae (often now considered a separate family) have been reared from seaweed. Dryomyzinae is restricted to the Holarctic region, with currently 20 described species in the two genera. Helcomyzinae are restricted to marine coasts and occur worldwide, with 13 species in 6 genera described till date.

Family SCIOMYZIDAE (Marsh Flies)

Slender to robust flies (1.8-11.5 mm). Colour varying from shiny black to dull gray, brown, reddish or yellowish. Head usually concave with wide frons (in both sexes), and the vertex not strongly excavate. The postocellar bristles are strong and parallel or slightly divergent. The compound eyes are prominent and bare, and vibrissae are absent. Characteristically porrect antennae with usually elongate pedicel found, even short or long antennae seen, but arista is always short, pubescent to plumose. The thorax lacks a precoxal bridge and the metepisternum is bare. Posterior thoracic spiracle also is without either bristles or setae. The legs typically have the femora well developed and usually strongly bristled. Preapical dorsal bristle present on one or more of the tibiae. Wings usually longer and are immaculate or heavily patterned or spotted. C lacks spines and is without a subcostal break. The subcosta is complete, being free from vein R_1distally, and ending in the costa. Adults can often be swept from vegetation along streams or ponds. Larvae are associated with freshwater or terrestrial molluscs, and are parasitic or predaceous even be saprophagous. Worldwide there are some 600 described species under 60 genera.

Superfamily SPHAEROCEROIDEA

Family HELEOMYZIDAE (Heleomyzid Flies)

Robust and bristly flies, 3.0 to 7.0 mm long, reddish yellow or reddish brown to black in colour, and often distinctly pruinose.Head with 1-3 pairs of orbital bristles, and ocellar bristles arising on ocellar triangle above anterior ocellus. Oral vibrissae ½ pairs, presence of strong convergent postocellar bristles. Short antennae, arista minutely pubescent to plumose. Thoracic scutellum with 2-3 pairs of strong bristles. Moderately bristled legs; usually a preapical dorsal bristle on mid and hind tibiae, although sometimes minute such as in *Borboropsis* Czerny. The wings are usually hyaline, but sometimes faintly yellowish or brownish. Wing surface may be mottled with contrasting whitish and dark gray areas, the crossveins are often clouded and the longitudinal veins fuscous. C extends to the end of vein M_{1+2} and has a distinct subcostal break, but no humeral break; C also usually strong and conspicuous spines. Subcosta is usually completely separate from vein R_1, ending in the costa sometimes close to R_1, and a pterostigma is often present. Vein CuA_2 short, and fuses with A_1, the combined vein often nearly reaching the wing margin. Adults are commonly collected in moist, shaded area. Many larvae breed in fungi, Heliomyzinae breed in decaying plant and animal matter, and are known to occur in bird's nests, mammal burrows, bat caves, carcasses of large mammals, and excrement. Often workers now consider the Heleomyzidae to include the family Trixoscelididae as a subfamily. Here, It maintained as separate families to allow for easy generic identifications using the Manual of Nearctic Diptera. Worldwide there are about 65 described genera and over 500 species in the Heleomyzidae, in the broad sense.

Family TRIXOSCELIDIDAE (Trixoscelidid Flies)

Very small (1.3-3 mm) flies with yellow legs, face, antennae and a portion of the frons, but in species of *Zagonia* Coquillett, the whole body is yellow. The body is mostly pruinose, but the abdomen is sometimes glabrous. Wings hyaline, sometimes fuscous along the veins and crossveins, conspicuous hyaline spots in back end. Trixoscelididae, ocellar bristles lie just outside the ocellar triangle. Face slightly concave, vibrissae present, and postocellar bristles strong and not divergent. Vertex not strongly excavated, and antenna may partly hidden in antennal socket. Thoracic scutellum with 4 bristles. Fore femora slightly swollen, a preapical dorsal bristle present in all tibia. Mid tibiae also have one or more apicoventral bristles. C with a subcostal break but lacks a humeral break, and extends to the end of M_{1+2}. The subcosta is complete and free from vein R_1, ending in the costa near R_1, near the basal third of the wing. Adults are frequently collected on flowers or vegetation. However, there is no information available on larval biology and habitats. This family is often now included within the Heleomyzidae, although in the Heleomyzidae are recognized by the ocellar bristles arising on the ocellar triangle, above the anterior ocellus.

Family SPHAEROCERIDAE (Lesser or Small Dung Flies)

Lesser dung flies are minute to small (0.9-5 mm), easily recognized by the short, thick basal segment to the hind tarsus. These are generally dull coloured flies, often black or dark brown, but sometimes brown or with head or legs yellowish often are highly coloured. Broad frons slightly narrowed anteriorly, and typically with two proclinate or lateroclinate bristles. Vibrissae present, and short antennae with preapical or subapical arista. Scutum usually setose, but is warty in some species of the subfamily Sphaerocerinae; bare anepimeron. Legs with the femora, especially the hind femora somewhat swollen, and with the hind tarsus distinctive with its swollen basal segment. Wings generally fully developed (macropterous), or reduced (brachypterous), or may be totally absent (apterous). If macropterous, the wings are rarely spotted, with costa ending at R_{1+2}, or at R_{4+5}, or between them, and with costagial, humeral, and subcostal breaks. Subcosta is incomplete, and A_1 never reaches the wing margin. Some lesser dung flies have larvae that are scavengers, commonly associated with decaying organic matter, including animal dung, carrion, cave debris, compost, conifer litter, dead vegetation, fungi, leaf litter, mammal nests and supralittoral seaweed debris. There are over 1340 described species of lesser dung flies in 111genera worldwide.

Family CHYROMYIDAE (Chromyid Flies)

Very small (1-4.5 mm), *Drosophila* like flies are usually yellow in colour. Head with anteriorly narrowed frons, and with ocellar bristles often relatively strong; postocellar bristles are usually present, and convergent. Vibrissae weak and short, antennae with microscopically pubescent arista. The thorax has yellow bristles and spines, and the proepimeral bristle is absent. Also, the proepisternal bristle is usually absent, while the anepimeron is bare. The legs are weakly bristled, with the fore and hind femora in the male frequently enlarged. All tibiae lack a dorsal preapical bristle. C with subcostal break only, and extends to M_1. Subcosta complete but weak on the apical 1/5, and joining the costa very close to the insertion of vein R_1. Cell cup present at posterior base. Adults have been swept from foliage of shrubs and herbs, especially at margins of creeks and ponds. *Chyromya* and *Gymnochiromyia* have been reared from bird's nests, mammal burrows, and wood debris of hollow trees. However, adults of *Aphaniosoma* appear to frequent grasses and sedges on seashores, and around alkaline or saline ponds or lakes. Only 3 genera and about 40 species are reported so far.

Superfamily EPHYDROIDEA

Family DROSOPHILIDAE (Pomace Flies/Vinegar Flies/Lesser Fruit Flies)

Small to medium (1-6 mm) flies with a wide range of body colouration from yellow to brown or black, and can be shiny or grey pruinose and frequently having stripes or spots on the thorax and abdomen. Eyes usually covered in distinct micro-pubescence and are often bright red. Head with 3orbital bristles; usually the back two curved rearward, the front one forward. Ocellar bristles large to small, postorbitals moderate and convergent; one to several strong vibrissae present. Arista plumose, sometimes bare or reduced in branching. Scutum with usually 2 postsutural dorsocentral bristles. Tibiae have apical and preapical dorsal bristles. The wing has both humeral and subcostal breaks of the costa; end of the subcosta usually vestigial; r-m and dm-cu are always present; cells bm and dm either separated or joined. The larvae of pomace flies mostly eat yeasts and other microorganisms in fermenting organic matter. Adults live around garbage, compost, rotting fruits and vegetables, decomposing cacti, sap from tree wounds, fungi and dung. A few species can be annoying pests in markets, breweries, bakeries and canneries. Some are used as laboratory animals for genetic and physiological research because they are small, fecund and so easy to rear and maintain. Many species of *Scaptomyza* are leaf miners; larvae of *Cladochaeta* are ectoparasites of cercopid nymphs; species of *Pseudiastata* prey on mealybugs. Currently this family has about 60 genera and 3000 species worldwide, many of them in the tropics. The family is dominated by the genus *Drosophila* Fallén, which has about 1600 described species worldwide (Grimaldi, 1990). *Drosophila melanogaster* Meigen is probably the most familiar species in the genus; it is certainly the best known scientifically because of its wide use in research laboratories. The European species *D. buskii* and *D. funebris* are known for being attracted to sour milk and rotting potatoes.

Family DIASTATIDAE (Diastatid Flies)

Diastatids are small (2.5-4.0 mm) flies, grey-brown, with patterned wings. Head higher than long and the frons narrows from the vertex to the antennae. The vertical bristles are strong; the inner pair is often longer than the outer pair. Postocellar bristles converge and the ocellar bristles arise prominently behind the front ocellus. Two orbital bristles of unequal size curve backward and a third one points forward. Face flat; vibrissa strong, lying in front of a row of 5-7 subvibrissal setae. Eye bare, 2nd antennal segment swollen above; the 3rd segment points downward and bears an arista that ranges from almost bare to feathery. Scutum with short and depressed setae; 2postsutural dorsocentral bristles present. Scutellum flat, lacks setulae, but bears two pairs of bristles. Wing usually twice as long as wide; the anal lobe and alula weak or absent. C weakly, but distinctly spiny and has both humeral and subcostal breaks. It extends to the end of vein M_{1+2} at the wing tip, but is weak past the end of R_{4+5}. Subcosta is incomplete. Crossvein r-m basal to the middle of cell dm and the cossvein dm-cu about as long as the part of vein CuA_1 apical to dm-cu. Vein A_1 weak and short. Wing usually spotted with brown or has white spots on a brown background. Legs generally slender; front femur always with a comb of short stout setae on the lower front margin of the apical half. All tibiae have a dorsal bristle just before the tip. Adults are encountered around rich herbaceous vegetation in moist woodland and on the edges of peatlands and other wetlands. The males of some species wave their spotted and reflective wings during courtship displays. The biology of the immature stages is unknown. The Diastatidae is a nonspeciose family, mostly Holarctic in distribution, containing about 40 described species in 4 genera worldwide; one genus is extinct and known only from Baltic amber.

Family EPHYDRIDAE (Shore Flies)

These are small to medium-sized flies (1-11 mm), usually dark and dull in colour, but highly variable in structure and difficult to characterize. Head has the frons usually wider than long; the face is variable in form, setulose and arched, often bulging, tubercles and ridges are often present. The subcranial cavity is often large and gaping and the mouthparts, if large, can be pendulous. Fronto-orbital bristles often prominent, curving forward or backward, often in many species, strongly projecting laterally. The antennae are short, the second segment often has a single bristle above; the third segment bears a dorsal arista, which is bare, finely setulose or comb-like, with any aristal rays almost always on the upper surface only. Thorax extremely variable in surface features shiny, dull, or densely pruinose, sculptured or smooth. Wing clear or spotted and with both humeral and subcostal breaks of the costa; the subcosta is incomplete and R_1 joins the costa before the middle of the wing. Cells bm and dm are not separated by a crossvein; cell cup absent. Fore and mid femora often with spines or bristles below; forelegs of *Octhera* are raptorial. A preapical dorsal bristle occurs only on the middle tibia. The empodia and pulvilli frequently reduced or absent in subfamily Ephydrinae. These flies typically live in aquatic and semiaquatic habitats along the muddy shores of marshes, ponds and streams, but few can live in marine marshes, tidal pools, and the alkaline lakes of arid and semiarid environments. It is in these harsh saline habitats that the most distinctive forms have evolved and where the family has reached its greatest biological importance, both immature stages and adults can be so abundant that they are significant food for wildlife, especially waterfowl and shorebirds. The family is pre-eminent in its ability to withstand the osmotic pressure of salt water. Most larvae feed by filtering microorganisms such as bacteria, algae and yeasts from the water and mud, but others feed on excrement in places such as cess pits and in the sludge found in sewage filters; from sewage it is a short step to feeding on decomposing carrion and carcasses. In salty aquatic habitats, especially, there is a strong trend to carnivory; *Octhera* larvae eat chironomid midge larvae. Others exploit the many invertebrates that fall into the water and cannot escape. The carnivorous larvae of *Helaeomyia petrolei* live in pools of crude petroleum that flow from the ground in California; they feed on organisms trapped in the oil. Some larvae of genera such as *Hydrellia* and *Psilopa* are leaf or stem miners and some develop in plants far from water. Other species of *Hydrellia*, when abundant, can damage crops of watercress, rice and other irrigated cereals. Species of *Hydrellia* mining the leaves of the aquatic *Potamogeton* can access oxygen by inserting their posterior spiracles (which open near the ends of sharp, hollow spines) into the leaf tissue. Most adults feed on unicellular algae and other microroganisms. Both adults and larvae of several shore fly species feed on the bacteria and cyanobacteria growing in algal mats in hot springs. Adults of *Ochthera* feed on small insects and still others eat nectar and leaf tissue. Family Ephydridae is speciose, with about 1800 described species in 114 genera worldwide. *Hydrellia* (70 species) and *Notiphila* (55 species) are found to be predominant. *Atissa pygmaea* is tiny species, only 1 mm long, and almost completely white pruinose. Few inhabit saline environments like *Hydropyrus hians.* In the Cariboo the larvae and pupae are found massing in huge numbers in almost completely saturated sodium carbonate solutions, the adults swarming on the salt crusts and on the water surface itself. *Clanoneurum americanum* is the commonest in salty habitats in Atlantic coast and *Coenia curvicauda* is frequent in coastal and inland salt marshes transcontinentally.

Superfamily OPOMYZOIDEA

Family ODINIIDAE (Odiniid Flies)

These flies are small (3-4 mm) stout insects. They are gray, marked with brown; the wings are spotted or mottled with brown. Head higher than long; the frons as broad as long, equally wide in both sexes. Postocellar bristles divergent; the inner vertical bristle usually stronger than the outer one; the lower of the three orbital bristles points inwards. Oral vibrissa strong, with adjacent bristles decreasing in size rearwards. Antenna mostly yellow; the 2nd segment has a dorsal bristle and the globular third segment bears a shortly setulose arista. Scutellum large and convex with two pairs of bristles; scutum usually strongly bristled, including four or five dorsocentral bristles. Wing short and broad, always darkened around the subcostal break in North American species. Costa broken at the subcosta only, and extends around the wing to R_{4+5} or M_1. Subcosta incomplete; R_1 lacks setae and joins wing margin near the subcostal break. Cells bm and dm are complete; the apex of cell *cup* convex. Vein A_1 does not reach the wing margin. Legs stout, often yellow and usually with brownly banded tibiae; hind femur enlarged in males. Legs moderately bristled; preapical dorsal bristles are present and are strongest on middle tibia. Abdomen short and broad, normally brown grey spotted. Adult odiniids gather at wounds in trees, on polypore fungi and rotting tree trunks and stumps. The larvae are apparently frequently associated with wood boring beetles and moths. They may feed on fluids of decay or on fungi in insect galleries or decaying wood; some may parasitize other insect larvae. Closely related to the Agromyzidae, the family Odiniidae occurs worldwide. It is divided into 10 living genera consisting about 60 described species.

Family AGROMYZIDAE (Leafminer Flies)

Agromyzidae is a large family of minute to small (1-6 mm) stocky flies ranging from yellow to brown, grey and black in colour; often they are black with yellow markings. Wings normally clear, but are patterned in some tropical forms. Eye vertical or slanting, bare or sometimes setulose.Orbital bristles 1-3 and frontal bristles 1-5, lower ones usually angled inwards; postocellar bristles diverge. A well-developed vibrissa present; sometimes several vibrissae fused (*e.g.* in males of some *Ophiomyza*). 3rd antennal segment varies from small to large and globular to elongate; arista bare or short-setose. Thorax with dorsocentral bristles 2-5 and scutellar bristles1-2 pairs. Costa (C) ending near R_{4+5} or M_1 and broken only at the end of the subcosta. Subcosta either distinct and joins R_1 or reduced to a fold that may or may not end in the costa. Cell *cup* present; A_1 fails to reach up to wing margin. All tibiae lack preapical dorsal bristles. Abdomen usually more or less depressed and tapering; six segments are visible in front of the genitalia. Agromyzid larvae eat living plant tissue. Most species feed between the upper and lower surfaces of leaves, making conspicuous mines, but others attack stems, roots and seeds. Most, being small and inconspicuous, are more easily recognized by the mines that they make than they are by the adult insects themselves. Mines can be blotch-like, linear or serpentine and their location in the leaf can be diagnostic. Even the distribution pattern of frass can be characteristic. Most agromyzids limit their development to particular plant species or groups of related species. Some are more polyphagous, including species such as *Liriomyza trifolii*, which attacks many legumes and other plants such as tomatoes, cucumbers, asters and chrysanthemums. Many species are pests of agricultural crops and ornamental plants. Family Agromyzidae ranges throughout the world, inhabiting all environments from arctic tundra to tropical forests. There are over 2700 species named in 29 genera, but certainly many hundreds of species remain undescribed (Spencer, 1969).

Family CLUSIIDAE (Clusiid Flies)

Flies of the family Clusiidae are slender, small to medium-sized (1.8-7.5 mm), yellow to black species. Bodies often marked, but seldom pruinose. Frequently smoky or brown marked wings, especially at the tips.Head normally higher than long and wider than high, flat or concave at the back. Antennae short: 1st segment tiny, 2nd with a dorsal bristle and often with a characteristic triangular projection on the outer margin. 3rd segment more or less globular, consisting a setulose arista near the tip. Ocellar bristles short; postocellars when present are divergent. Prominent vertical bristles, the inner ones are longer than the outer ones. Fronto-orbital bristles 2-5. Oral vibrissae strong and pointed forward and upward. Wing usually twice as long as wide with a broadly rounded tip. Costa normally with a subcostal break and usually rounds the apex and reaches M_1. Subcosta complete, running parallel to R_1. Cell bm complete, but A_1 does not reach the wing margin. On the legs, most species have well-developed preapical dorsal bristles on at least the mid tibia. Adomen slender and tapered. Clusiid larvae develop in rotten wood, mostly under the bark of dead and dying deciduous trees. They can jump like the larvae of Piophilidae. Adults feed on nectar, sap and decayed vegetable matter. In some species, lekking occurs, where males aggregate to interact and attract females. Males of many species use their patterned wings in mating displays and head-pushing in male-male interactions sometimes occurs. This cosmopolitan family Clusiidae consist on about 220 species under 25 genera. However, recent work suggests there are over 600 species worldwide and almost 400 from the New World.

Family ACARTOPHTHALMIDAE (Acartophthalmidae Flies)

Acartophthalmids are dull grey or black flies only 2.5 to 3.0 mm long; the wings are somewhat infuscate. Head higher than long and bears prominent bristles; eyes are round and finely setulose. Frons strongly narrowed from the top of the head to the antennae. Three fronto-orbital bristles; upper one longest and lower one shortest. The ocellar, inner and outer verticals and postocellar bristles are strong and all about the same length; the postocellars are widely separated and diverging. Vibrissae weak and accompanied by 4-5 additional bristles about the same size. Antennae short, 3rd segment globular with a shortly setulose arista near the base. Scutum with 3-4 dorsocentral bristles; 2 pairs of bristles on the scutellum. The subcosta is complete, meeting the costa well before the end of vein R_1; costa broken near the humeral crossvein. Cell bm complete; A_1 does not reach the edge of the wing. The legs lack strong bristles; preapical dorsal tibial bristles are absent. Habits are poorly known in this family, and the immature stages are undescribed. Adults sometimes are seen on logs, stumps and rotting fungi in wet forests. The males often flick their wings in the manner of otitid flies. Formerly treated as a subfamily of the Clusiidae, the Acartophthalmidae consists of only three known species, two Holarctic and one Palearctic in distribution.

Family OPOMYZIDAE (Opomyzid Flies)

Opomyzids are small, slender flies, 2.0 to 4.5 mm long; the body is yellow, red, brown or black and can be shiny or pruinose. At least 1 dark spot at the tip of the wing is present; usually other additional markings, especially on the crossveins present. Few species of *Geomyza* have reduced wings and nearly flightless. Head with frons somewhat narrowed towards antennae. One strong orbital bristle present, the inner and outer vertical bristles are strong and the postocellar bristles are usually lacking; if present, they are diverging. No vibrissae occur on the weak vibrissal angle, but a significant row of setae may be present under the main eye. Eye sparsely clothed with short setulae. 3rd antennal segment is oval, produced downward; the arista often has the upper setulae longer than the lower ones. Scutum strongly bristled; dorsocentral bristles 3-4. Scutellum with 4 bristles, the basal pair are often weak; without dorsal preapical tibial bristles. Wing moderately broad to unusually narrow; the alula and anal lobe are frequently absent. Costa extends to M_1 and broken only at the end of subcosta. Subcosta incomplete; cells dm and bm usually incompletely separated; cell cu*p* present; A_1 incomplete or absent. Adult opomyzids usually frequent in moist grassy habitats while larvae feed within the stems of grasses. Family Opomyzidae is small, with about 40 species in four genera; most live in the north temperate regions, but a few occur in eastern and southern Africa (Vockeroth, 1961).

Family ANTHOMYZIDAE (Anthomyzid Flies)

Small (2.0-3.4 mm) and slender flies with shining black to moderately pruinose, yellow to black in colour. Frons slightly narrowed anteriorly, and 1-3 pairs of orbital bristles, with anterior pair small or even absent. Ocelli present alonhwith strong ocellar bristles. Post-ocellar bristles short, weak and convergent, while a row of very weak genal setae present below the eyes, ending anteriorly in 1 or 2 distinct vibrissa-like bristles. Eyes bare; antennae short, turned downward, and the arista with short to long setae.Thoracic postnotum with one bristle, and scutellum bare with two short subbasal and two long, apical bristles. Prosternum and anepisternum bare. All tibiae lack a dorsal preapical bristle, and the fore femora usually have a strongly developed ctenidial spine. Wings usually slender and unmarked, but rarely can be reduced or even absent. Wings when fully developed have the costa extending to the apex of vein M, with a subcostal break, while the subcostal vein is incomplete, and not reaching the costa. Cell *cup* is present at the base of the wing, while the halteres may be well developed or rudimentary. The abdomen is more or less depressed. Adults are commonly swept from grass or low vegetation, especially in marshy areas. Larvae are associated with *Juncus, Typha, Elymus, etc.,* living between the closely fitting leaf blades of terminal shoots. Worldwide there are about 53 species in 14 genera.

Family PERISCELIDIDAE (Periscelidid Flies)

Periscelidids are small (3-4mm) *Drosophila*-like flies. Colour dull grayish-black or brownish-black. Head broader than high, narrowed below the antennae. The frons has one pair of orbital bristles, and the postocellar bristles are strong and divergent. Ocelli distinct and single pair of ocellar bristles distinct. Arista plumose.Compound eyes with short sparse pubescence.Thorax with propleura and scutellum bare, the latter with four bristles. Legs short and stout, with tibiae frequently banded. Fore and middle femora each with row of fairly strong posteroventral setae. Preapical dorsal bristles on tibiae undeveloped.Wings short and broad with milky, sometimes brownish markings.Costal vein without subcostal or humeral breaks. Subcosta is incomplete and not reaching costa, but R_1 joining costa near middle of wings, and veins R_{4+5} and M not convergent. Veins CuA_2 and A_1 atrophied. Abdomen rather broad and dorsoventrally flattened. Adults frequently wound and flux on trunks of deciduous trees. Adults of *Periscelis wheeleri* have been reared from larvae found in galleries of the cerambycid *Sternochetus lapathi* on willow (*Salix* sp.) in British Columbia. The Periscelididae has primarily a Neotropical distribution, with about 9 genera and fewer than 60 species described worldwide. *Periscelis* genus is the predominant one.F*amily ASTEIIDAE* (*Asteiid Flies*)

Asteiids are very small, delicate and rarely collected flies, 1.0 to 2.5 mm long. Head normally higher than long, with concave face. Eyes large and bare; Frons broad with one or two pairs of bristles. Ocelli present, and ocellar bristles short or seta-like, divergent or proclinate. Post ocellar bristles weak or absent. Oval vibrissae are well developed, but pale and inconspicuous. Antennae short and decumbent, arista with alternating short and long setae, giving the arista a characteristic zigzag appearance. Thoracic scutellum with two pairs of marginal bristles, the basal pair of bristles often seta-like, but posterior bristles well developed and obvious. Legs short and slender, with tibiae lacking preapical dorsal bristles.Wings long, hyaline and unspotted. Costa without a subcostal and humeral break, and with incomplete subcostal vein, not reaching the costa. R_1 joins costa in the basal third of the wing,

R_{4+5} and M distinctly convergent distally, and CuA_2 and A_1 atrophied. The abdomen is narrow, rather membranous and in most weakly sclerotized. Little is known about the biology and life cycle of these flies. Adults have been collected on windows and at bleeding wounds on trees or fungi, otherwise adult have been reared from dead plant remain or plant part. Worldwide there are about 100 species under 11 genera. Five genera with 18 species are known from the Nearctic.

Superfamily CARNOIDEA

Family CARNIDAE (Carnid Flies)

Small flies, 1.0 to 3.0 mm long, and more or less shining black. Face of head rather concave, and frons with at least two orbital bristles. Ocellar bristles strong, and postocellar bristles subparallel. The gena has a row of strong bristles in the middle. Vibrissae strong and a row of strong subvibrissal bristles present. Antennae often with a deep antennal groove. Proboscis with a bulbous base and short inconspicuous labella.Thoracic scutellum armed with 4 short bristles, and propleura bare. Legs slender, and tibia lack a preapical dorsal bristle. Wings when fully developed hyaline, and the costa with both humeral and subcostal break, while the subcosta is complete although weak and faint distally. R_{2+3} typically bisinuate. In *Carnus* Nitzcsch the wings usually broken off, leaving merely a short stub. Abdomen normally well developed, but in *Carnus* is rather inflated, lacks sterna, and the membranes with numerous setiferous sclerotized spots in females. Most species are saprophagous and associated with carrion or excrement, with many having been naturally reared in bird nest. The Holarctic *Carnus hemipterus* occurs in bird's nests and adults are often found on nestlings and may feed on blood. The genus *Carnus* was reviewed by Grimaldi in the year 1997, is mostly a Holarctic family, with over 40 described species (Grimaldi, 1997).

Family TETHINIDAE (Tethinid Flies)

Tethinids are small (1.5-3 mm), stocky flies, coloured yellow, grey or black and usually strongly pruinose. Head usually higher than long; frons and face strongly to weakly narrowed in front; face with a weak depression below each antenna with a slight median ridge. Ocellar bristles, inner and outer vertical bristles and 1 to 5 orbital bristles are strong (the latter curve backward or laterally); postocellar bristles absent; inner occipital bristles variable in size and convergent. Lower edge of the head below the eye bears a row of weak to strong setae; the foremost vibrissa-like. Antennae short, usually droop; the 3rd segment almost circular; arista bears extremely short setulae. Scutum with 1 presutural and 3 postsutural dorsocentral bristles; scutellum rounded, about twice as broad as long, without setae, but with 2 pairs of marginal bristles. Wings clear or sometimes slightly white to brown; rarely with brown clouds on the crossveins; anal and alula well-developed. C spines and broken only at the end of the subcosta. The subcosta complete or sometimes fused with R_1 near its tip. Veins R_{2+3}, R_{4+5} and M_1 typically rather straight and more or less parallel to the long axis of the wing. Cell cu*p* present, but small. Legs slender, femora sometimes slightly swollen. Coxae and femora bear, or with a few weak bristles; the mid and hind tibiae usually have an apical bristle below. Most tethinid flies live along ocean beaches or near alkaline ponds and lakes in more interior regions.

The larval biology is poorly known, although the larvae probably live in the soil or in masses of algae or seaweed. Some species in the southern hemisphere may be associated with seabird colonies. Tethinidae is a cosmopolitan family occurring on all the main continents and many oceanic islands. It is a nonspeciose group, only 120 described species in 14 genera, although many other undescribed species likely exist.

Family MILICHIIDAE (Milichiid Flies)

Milichiids are physically small, 1-7 mm long, acalyptrate flies, often largely brown or black, occasionally orange or yellow. Males of some genera with silvery pollinosity on abdomen. Milichiids usually have two or three pairs of medioclinate frontal bristles, and two or three pairs of lateroclinate to proclinate orbital bristles; rarely with many reclinate frontal and orbital setae in an indistinguishable series. Frons usually with two rows of interfrontal setae, sometimes on distinctly shining stripes; lunule often with one or two pairs of setulae; proboscis often elongate, geniculate. Wing with humeral and subcostal break; the region of latter is often modified into a costal lappet; cell cu*p* closed, small. Milichiids are generally recognizable based on the head bristling, humeral and subcostal breaks and the closure of cell cu*p*. Although the biology of the majority of species is unknown, many milichiids have general saprophagous or coprophagous larvae, developing in decaying matter ranging from rotting fish and animal dung to rotting plant material. In the Nearctic, larvae of *Eusiphona* have been found feeding on the pollen in megachilid bee nests. Some *Phyllomyza* species have been found in association with formicine and ponerine ants in the Palearctic and Oriental Regions; these groups of ants may also be hosts of *Phyllomyza*. Two Nearctic species of *Pholeomyia* have been recorded from the refuse heaps in leaf-cutting ant nests. Species in the genera *Pholeomyia, Milichiella, Leptometopa,* and *Phyllomyza* are known from caves and in association with bat guano, and may occur in similar habits. Many adult milichiid flies including the genera: *Phyllomyza, Desmometopa, Neophyllomyza, Paramyia, Milichiella,* and *Leptometopa,* all of which occur in British Columbia, have species with adults that are kleptoparasitic on a wide range of predaceous arthropods, such as spiders, assassin bugs and robber flies. The vast majority of milichiid specimens collected as kleptoparasites of spiders are female and it is suspected that the extra protein from kleptoparasitic meals may be necessary for egg maturation. Although *Leptometopalatipes* is documented as a spider kleptoparasite, the larvae are also commonly associated with bird nests, including records from bird nests. Adult male milichiids have equally well documented unique behaviors. Males of some species of *Milichiella,* and *Pholeomyia* have silvery abdomens and form mating swarms that can be seen from long distances. Worldwide there are about 20 described genera and 275 described species.

Family CHLOROPIDAE (Frit Flies, Grass Flies)

Chloropidae are minute to small, 1.5 to 5.0 mm long; with the number and size of the body bristles clearly reduced. Body black, grey, black and yellow, or black and red. Frons broad, usually with the ocellar triangle well-developed, plate-like, clearly demarcated, shining to pruinose and normally with a single row of setae-bearing punctures along the lateral margins. Frons usually projects only slightly and the face is normally somewhat concave with the vibrissal angle rounded. The third antennal segment is usually round, sometimes kidney-shaped or elongate and bearing an arista that is normally pubescent or sometimes setulose, but rarely is bare. Head bristles are usually short and weak; the inner and outer verticals, ocellars and postocellars are commonly present. The latter bristles are parallel, convergent or crossed. Fronto-orbitals are usually represented by short setae curving rearward. Vibrissae, when present, are usually fine and setulose. Scutum normally longer than broad, with fine setulae set in distinct rows and frequently with coarse, setulae-bearing punctures; one posterior dorsocentral bristle almost always occurs. Scutellum sometimes has marginal tubercles bearing bristles. Proleuron sharply ridged in front. Wings rarely absent or reduced and normally lack any colour pattern. The costa has a subcostal break; the subcosta is incomplete, usually faint. Veins R_{4+5} and M_{1+2} are long, the former ending before the wing tip, the latter ending behind the tip. Cells b-m and d-m are completely joined, forming a single long cell; CuA_1 often has a characteristic jog near the middle of cell bm+dm. Vein A_1 and cell cu*p* are always missing; the anal area of the wing is usually broadly rounded. Normally legs short, slender, and without bristles, except an apical or subapical spur sometimes occurs on middle or hind tibia. Many species have an elongate oval tibial organ on the upper surface of the hind tibia. Abdomen is broad, tapers to the tip, and each segment in front of the terminalia is about equal in length. The larvae of many species of frit flies develop in grass stems, shoots, and decaying plant matter. A few species damage crops; one of the best known is *Oscinella frit*; It also attacks rye, barley, lawn grasses and corn. The wheat stem maggot, *Meromyza americana* is also destructive. Some species produce galls; the European *Lipara lucens* forms galls on the giant marsh grass *Phragmites*. Predaceous chloropids includes: *Thaumatomyia glabra* which preys on root aphids, and the larvae of *Pseudogaurax* which feeds on egg masses of spiders, tussock moths and mantids. Adult chloropids are most common in grass and sedge habitats. Some are found on flowers and a few, especially in the *Hippelates*, are bothersome and even transmit disease by hovering around sweating faces and sipping liquid secretions from eyes, sores and wounds. The family Chloropidae ranges around the world, with over 2100 described species in more than 160 genera among them 155 species and 55 genera known from India (Cherian, 2002).

Conclusion

Diptera, the highest developed insects have made a long 250 Mya journey, which seems very eventful, as these age old (evolved in Upper Permian) fly have told many stories of convergence and divergence. Classification and phylogeny (based on morpho-taxonomy + molecular taxonomy) of Diptera are rightly demonstrated in Manuals of Nearctic Diptera and Manual of Palaearctic Diptera. Large in number as well as greatest variation (morphologically and genetically) make them the widely adapted in all probable habitat especially non-marine fresh water ecosystems. This diversity documents the importance of the group to man and reflects the range of organisms in the order and is considered as second to third largest groups of living organisms. The economic importance of the group is immense. One needs only to consider the ability of flies to transmit diseases. Mosquitoes and black flies are responsible for more human sufferings and death cases (by transmit pathogens) than any other group of organisms, at the same time these flies are pest of many grains and fruits. On the positive side of the ledger, outside their obviously essential roles in maintaining our ecosystem, flies are of little direct benefit to man. Some are important as experimental animals (*Drosophila*) and biological control agents of weeds and other insects. Others are crucial in helping to solve crimes or in pollinating plants. Without Diptera there would be, for example, no chocolate!

References

Bharti, M. (2011) An updated checklist of blow flies (Diptera: Calliphoridae) from India. *Halters*, 34-37.

Bhattacharyya, D. R., Rajavel, A. R., Natarajan, R., *et al.* (2014) Faunal richness and the checklist of Indian mosquitoes (Diptera: Culicidae). *Check List*, 10(6): 1342–1358

Borkent, A. (2016) World species of biting midges (Diptera: Ceratopogonidae). Ceratopogonidae Catalog. Available from:http://www.inhs.uiuc.edu/research/FLYTREE/(accessed 30 June 2014)

Charian, P. T. (2002) The fauna of India and the adjacent countries Diptera volume ix Chloropidae (part 1) Siphonellopsinae and Rhodesieijlinae, ZSI publication, 391 p.

Evenhuis, N.L. (1994) *Catalogue Of The Fossil Flies Of The World (Insecta: Diptera)*. Leiden, Backhuys, and updates at hbs.bishopmuseum.org/fossilcat/.

Griffiths, G.C.D. (1972) *The phylogenetic classification of Diptera Cyclorrhapha, with special reference to the structure of the male postabdomen*. The Hague: Junk.

Griffiths, G.C.D. (1996) Review of papers on the male genitalia of Diptera by D.M. Wood and associates.*Studia Dipterologica*, 3, 107–123.

Grimaldi, D. 1997. The bird flies, genus *Carnus*: species revision, generic relationships, and a fossil *Meoneura* in Amber (Diptera: Carnidae). *American Museum Novitates*,3190: 30pp.

Grimaldi, D.A. (1990) A phylogenetic, revised classification of genera in the Drosophilidae (Diptera).*Bulletin of the American Museum of Natural History*, 197, 1–139.

Hackman., W and Väisänen, R. (1982) Different classification system in Diptera. *Ann. Zoo. Fenniei.*, 19, 209-219.

Hennig, W. (1973) Diptera (Zweiflügler).*Handbuch der Zoologie (Berlin)*, 4, 1–200.

Kitching, R.L., Bickel, D.J. and Boulter, S. (2005) Guild analyses of dipteran assemblages, a rationale and investigation of seasonality and stratification in selected rainforest faunas. In D.K. Yeates and B.M. Wiegmann, eds, *The Evolutionary Biology of Flies*. New York: Columbia University Press, 388–415pp.

Klymko, J. and S.A. Marshall. 2008. Review of the nearctic Lonchopteridae (Diptera), including descriptions of three new species. *The Canadian Entomologist*, 140, 649-673.

Lindner, E. (1949) Handbouch. In: Lindner, E. (ed): Die Fliegen der Palaearktischen region. Stuttgart, Band 1: 1-422.

McAlpine, J.F. and Wood, D.M. eds. (1989) *Manual of Nearctic Diptera Volume 3*. Ottawa, Research Branch Agriculture Canada.

McAlpine, J.F. (1989) Phylogeny and classification of the Muscomorpha.In J.F. McAlpine and D.M. Wood, eds., *Manual of Nearctic Diptera Volume 3*.Ottawa: Research Branch Agriculture Canada, 1397–1518 pp.

Merritt, R.W. and B.V. Peterson.1976. A synopsis of the Micropezidae (Diptera) of Canada and Alaska, with descriptions of four new species.*Canadian Journal of Zoology*, 54, 1488-1506.

Mitra, B., Roy, S., Imam, I and Ghosh, M. (2015) A review of the hover flies (Syrphidae: Diptera) from India, *International Journal of fauna and biological studies*. 2 (3), 61-73.

Paul, N. (2016) Biosystematic studies of the subfamily Tanypodinae (Diptera: Chironomidae) from the eastern Himalaya, India.PhD thesis, University of Burdwan, 221 p.

Rice, H.M.A. 1959. Fossil Bibionidae (Diptera) from British Columbia.*Bulletin of the Geological Survey of Canada*, 55: 1-24.

Smith, K.G.V. 1959. The Conopidae (Diptera) of British Columbia.*Proceedings of the Entomological Society of British Columbia*, 56: 54-56.

Spencer, K.A. 1969. The Agromyzidae of Canada and Alaska.*Memoirs of the Entomological Society of Canada*, 64: 1-311.

Steyskal, G.C. 1978. Synopsis of the North American Pyrgotidae (Diptera).*Proceedings of the Entomological Society of Washington*, 80: 149-155.

Teskey, H.J. 1990. The horse flies and deer flies of Canada and Alaska (Diptera: Tabanidae). Part 16.The Insects and Arachnids of Canada.Publication 1838. Research Branch, Agriculture Canada, Ottawa, ON.

Thompson, F.C., ed. (2005) *Biosystematic Database of World Diptera*. Version 7.5, http://www.diptera.org/biosys.htm.

Vockeroth, J.R. 1961. The North American species of the family Opomyzidae (Diptera: Acalypterae). *Canadian Entomologist*, 93, 503-522.

Wood, D.M. and Borkent, A. (1989).Phylogeny and classification of the Nematocera. In J.F. McAlpine and D.M. Wood, eds., *Manual of Nearctic Diptera Volume 3*. Ottawa: Research Branch Agriculture Canada, 1333–1370pp.

Yeates, D.K. and Wiegmann, B.M. (1999) Congruence and Controversy: Toward a Higher-Level Classification of Diptera.*Annual Review of Entomology*, 44, 397–428.

Yeates, D.K., Wiegmann, B.M., Courtney, G.W., Meier, R., Lambkin, C. and Pape, T. (2007) Phylogeny and systematics of Diptera: Two decades of progress and prospects.*In*: Zhang, Z.-Q. and Shear, W.A. (Eds) (2007) Linnaeus Tercentenary: Progress in Invertebrate Taxonomy.*Zootaxa*, 1668, 1–766.

Young, A.D., Lemmon, A.R. and Skevington, J.H., *et al.* (2016) Anchored enrichment dataset for true flies (order Diptera) reveals insights into the phylogeny of flower flies (family Syrphidae). *BMC Evolutionary Biology*, 16, 143.

Zolty, J., Sinclair, B.J. and Pritchard, G. (2005) Discovered in our backyard: a new genus and species of a new family from the Rocky Mountains of North America (Diptera, Tabanomorpha). *Systematic Entomology*, 30, 248-266.

– Segment 4 –

Insect Behaviour

Chapter 7

Food-Induced Food-Transporting Strategies of the Ants *Pheidole roberti* and *Paratrechina longicornis*

☆ *K. Naskar and S.K. Raut*

ABSTRACT

Six different varieties of food viz., sugar cube, biscuit fragment, papad fragment, dry fish fragment, nut particle and mosquito were offered to the ants Pheidole roberti and Paratrechina longicornis in their natural foraging grounds at Garia, Kolkata, West Bengal, India to note the food transporting strategies in these ants. It is revealed that depending upon the weight, size and shape of the food materials offered 38.72 per cent, 45.33 per cent and 44.61 per cent, 39.22 per cent food particles were transported by pushing and pulling act by the ants individually, belonging to P. roberti and P. longicornis respectively. In contrast transporting of 15.95 per cent and 16.17 per cent food particles required cooperative action of 2 and/or more ant individuals depending upon the weight, shape and size of the food matters of respective species. Statistically there exists no significant difference in the food transporting behaviours exhibited by the two ant species. It is evident that cooperative transportation in ants is an induced impact of the characteristic features of the food to be transported by the ants.

Introduction

Ants forage almost in equal tempo during day and night time (Naskar and Raut 2014a,b,c, 2015a,b,c,d,e,f,g, 2016a, b). Depending on the size of the food matter they decide whether they would continue feeding on the said food at the site or would try to carry the same to the nest. If it is decided to carry the food matter to the nest then what kind of transporting strategy is to be applied must be decided by them. In fact, various authors (Hölldobler *et al.*, 1978; Traniello,1983; Traniello and Beshers 1991; Robson and Traniello,1998; Cerada *et al.*, 2009, Czaczkes *et al.*, 2010; Czaczkes and Ratnieks 2013; Naskar and Raut 2015) have reported about food-transporting behaviour in different ant species. The essence of such studies is that, the ants carry the food particle either individually (individual transport mechanism) or collectively (cooperative transport mechanism). Though lifting or pushing, pulling or dragging are associated with the food transporting mechanism it is still unknown which characteristic of the food matter stimulates the ant in expressing such behavioural manifestation. However, Traniello (1983), Traniello and Beshers (1991) and Robson and Traniello (1998) opined that the number of ants to be required to carry a food matter is decided with respect to the size of a certain food matter. But in fact architecture, weight and shape or size of a food particle may differ with the variation and type of the food in such cases the ants have no alternative but to apply effective transporting device to carry the foods to the nest. To verify the same, we offered six types of food to the experimental ant populations *i.e. Pheidole roberti* and *Paratrechina longicornis,* and the findings are described.

Materials and Methods

We selected six types of food *viz.*, sugar cubes, biscuit fragments, papad fragments, dry fish fragments, nut particles and mosquitoes (entire body of the freshly killed *Anopheles sp* and *Culex sp*) for experimental studies. The sugar cubes, biscuit, papad, dry fish and nut were purchased from the local market. The sugar cubes were of different size and weight, and were considered as such to use in the experiment while biscuit, papad, dry fish and nut were crushed and fragments/ particles of different sizes/shapes and weight were used in the experimental studies. The mosquitoes were collected by a net and were killed before considering themc for the present study.

The number of food particles/matters considered for each experimental trial varied from 15 to 60. Among the mosquitoes offered as food to the ants only 6 belonged to the genus *Anopheles* while the remaining specimens were of *Culex* variety. The food particles were offered to the ants on a floor of a room located at the ground floor of a house which is surrounded by flower and vegetable gardens at Garia (Latitude 22°46′ North and Longitude 88°4′ East), Kolkata, West Bengal, India. A total of 29 trials were performed in different seasons of the study year. Each trial was performed by supplying only one type of food. In each case observations were continued for a period of 3 hours following the time of supply of the food materials at the experimentation site. Except mosquitoes the other food materials offered were of 15 mg and more in weight while the shape and size varied to a great extent with respect to the type of the foods offered.

In the course of observations, attention was paid to note the number of food particles transported individually or collectively (*i.e.* by 2, 3, 4 and more ants cooperatively). Also, in case of individual food-transport mechanism, data on 1) pushing (lifting) and 2) pulling (dragging) were noted regularly with respect to the food materials offered and the ant species involved in food transportation. Likewise, in case of cooperative transporting device data on the number of ant individuals acting as pusher or puller (cooperation of at least 5 individuals in carrying a food material) in effecting a cooperative transport act, have also been noted regularly. To avoid any kind of confusion, as well as considering regular occurrence of the two almost similar sized (workers) ant species *Pheidole roberti* and *Paratrechina longicornis* were taken into account in course of data collection on food-transporting behaviour. The trials which were interfered by the ants other than these two species were discarded. Also the number of food particles which were not carried by the ants within a period of 3 hours from the offered sites were recorded.

One-way analysis of variance (ANOVA) was employed (Campbell 1989) to ascertain if any differences do exist between the food-transporting behaviours of the two selected ant species.

Results

In all the 29 trials, 1000 food particles belonging to the six above mentioned types were offered to the ants. Five trials were discarded because of interference by ant species other than the ones under present investigation. The ants *Pheidole roberti* and *Paratrechina longicornis* transported a total of 848 food particles/elements from the supplied spots to the nest (Table 7.1). A comparative account of the food-transportation device applied to transport each of the six types of food materials offered to the ants have been shown in terms of percentages in Figures 7.1–7.3.

Table 7.1. Number of Food Particles Transported by the Ants *Pheidole roberti* and *Paratrechina longicornis* by Applying different Food-transporting Strategies

Food	*Food-Transporting Strategy*					
	Pulling/Lifting		*Pushing/Dragging*		*Cooperative*	
	Pheidole roberti	*Paratrechina longicornis*	*Pheidole roberti*	*Paratrechina longicornis*	*Pheidole roberti*	*Paratrechina longicornis*
Sugar cube	40	19	43	36	6	18
Biscuit fragment	6	2	118	10	5	8
Papad fragment	13	16	58	41	12	9
Dryfish fragment	86	19	6	26	33	10
Nut particle	40	21	4	16	22	7
Mosquito	14	72	4	2	4	2

Irrespective of food-types both the ant species were seen to carry a food material cooperatively by the joint efforts of 2, 3, 4 or more individuals. In case of transportation of a food material by two ant individuals, one was found to be in the pushing act, while the other was involved in the pulling act. Likewise, the cooperative act exhibited by the three individuals was in most cases pulling of the food matter was effected by one individual while two were involved in pushing. In case of transportation of a food matter by four individuals, two were seen in pushing act and the remaining two were in pulling act. However, transportation of a food matter when effected by five ant individuals, two or three were seen either in pulling or pushing of the food matter. The role of individuals when more than

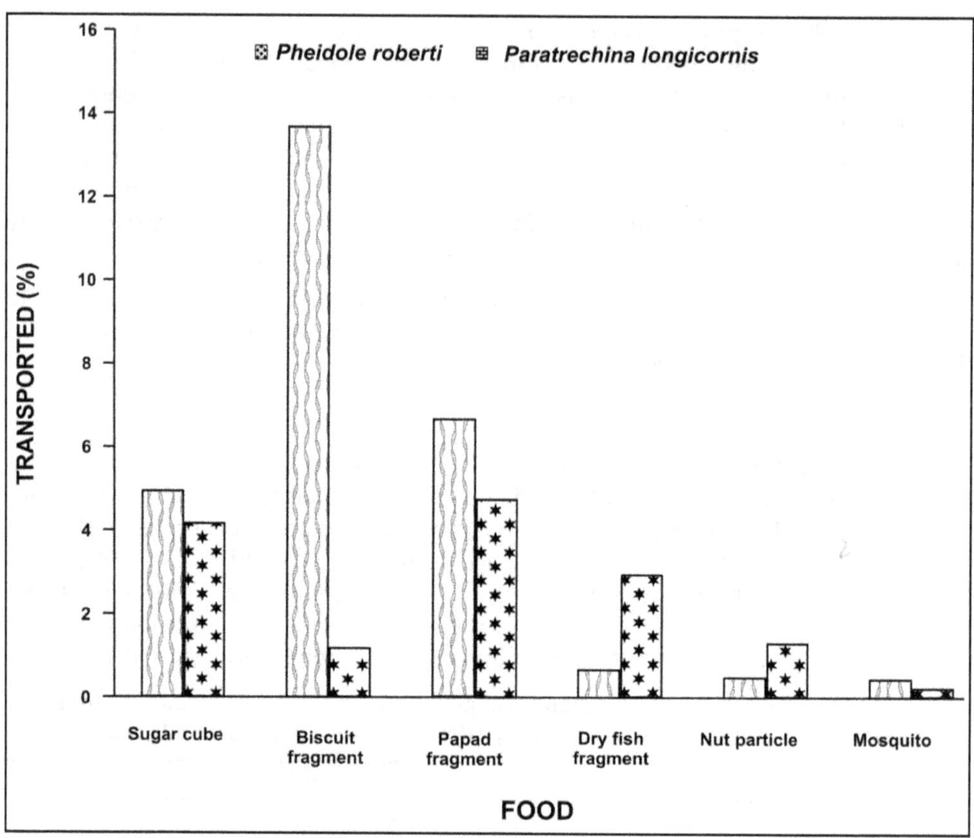

Figure 7.1: A Comparative Acocunt of the Food Transported (Per cent) by the Ants *Pheidole roberti* and *Paratrechina longicornis* through the Application of Pushing/Lifting Strategy.

five were involved in transportation varied to a great extent as the same individual was seen to act as puller or pusher as per need.

Though there exist variations in the percentages of food-materials transported by *Pheidole roberti* and *Paratrechina longicornis* such variations are not statistically significant (ANOVA results: for pulling/lifting transportation F= 0.28, df=1, Fcrit = 4.96 ; for pushing/dragging transportation F= 0.76, df=1, Fcrit = 4.96 ; for cooperative transportation F= 0.8, df=1, Fcrit = 4.96).

Cooperative transport exhibited by the ants varied to a great extent with respect to the number of food materials carried to the nest. Of the six sugar cubes (Table 7.1) each one of the 5 was carried by two ants of *Pheidole roberti* jointly, one acted as the pusher and the other, as the puller. The remaining one was carried by 3 workers where one acted as the puller and the two performed as pushers. *Paratrechina longicornis* transported 18 sugar cubes with the help of two or more individuals. Of these, each one of the 17 sugar cubes was transported by 2 ant individuals where one

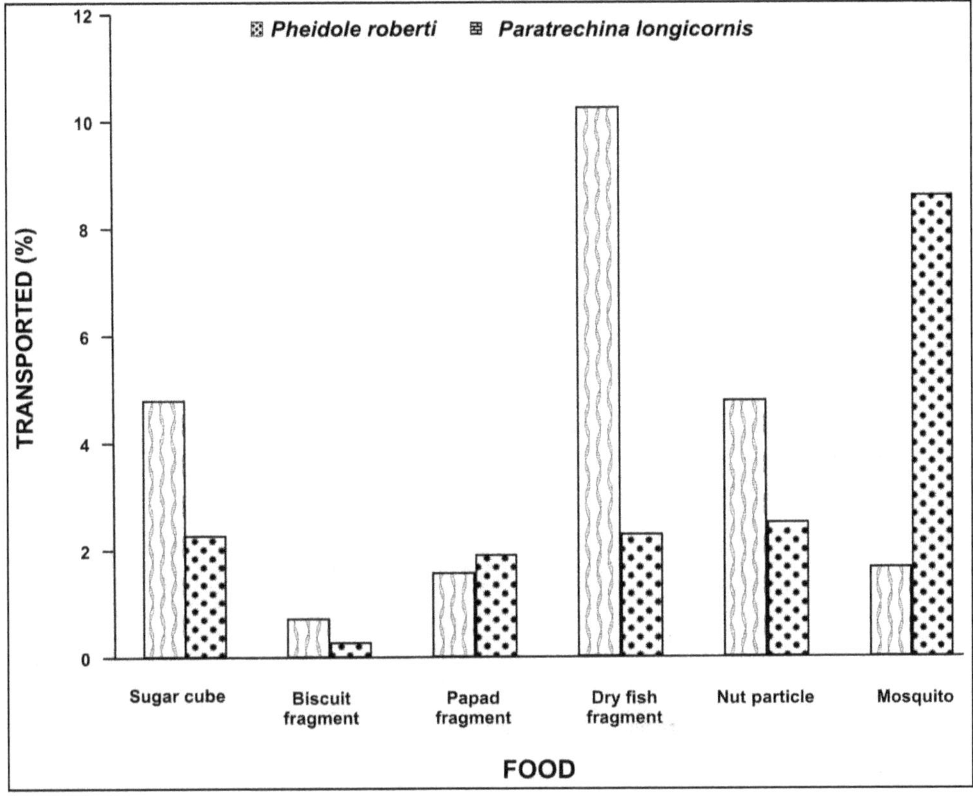

Figure 7.2: A Comparative Acocunt of the Food Transported (Per cent) by the Ants *Pheidole roberti* and *Paratrechina longicornis* through the Application of Pulling/Dragging Strategy.

was puller and another one was pusher. The remaining food matter was transported by the active participation of 5 ants where 3 were engaged in pushing and 2 were involved in pulling. Each of the 5 and 8 biscuit fragments was transported by 2 individuals of *Pheidole roberti* and *Paratrechina longicornis* respectively, where one acted as the puller and the other one as the pusher. *Pheidole roberti* transported 12 papad fragments cooperatively. Of these, each one of the 9 fragments was carried by 2 workers, one served as puller and the other one served as pusher; each one of the 2 fragments was carried by 3 workers where one acted as puller and the two served as pushers, the remaining one was transported by 4 workers of which 2 were pullers and 2 were pushers. *Paratrechina longicornis* transported each one of 8 papad fragments with the help of 2 workers, one pusher and one puller. In one case, of papad fragment transportation 2 individuals served as pullers and 2 individuals served as pushers. Cooperation of 2 *P. roberti* workers, one as puller and the other as pusher was effective in transporting 32 dry fish fragments though transportation of one dry fish fragment needed 2 pullers and 1 pusher. Ten dry fish fragments were transported by *P. longicornis* cooperatively. Each one of 9 fragments was carried by

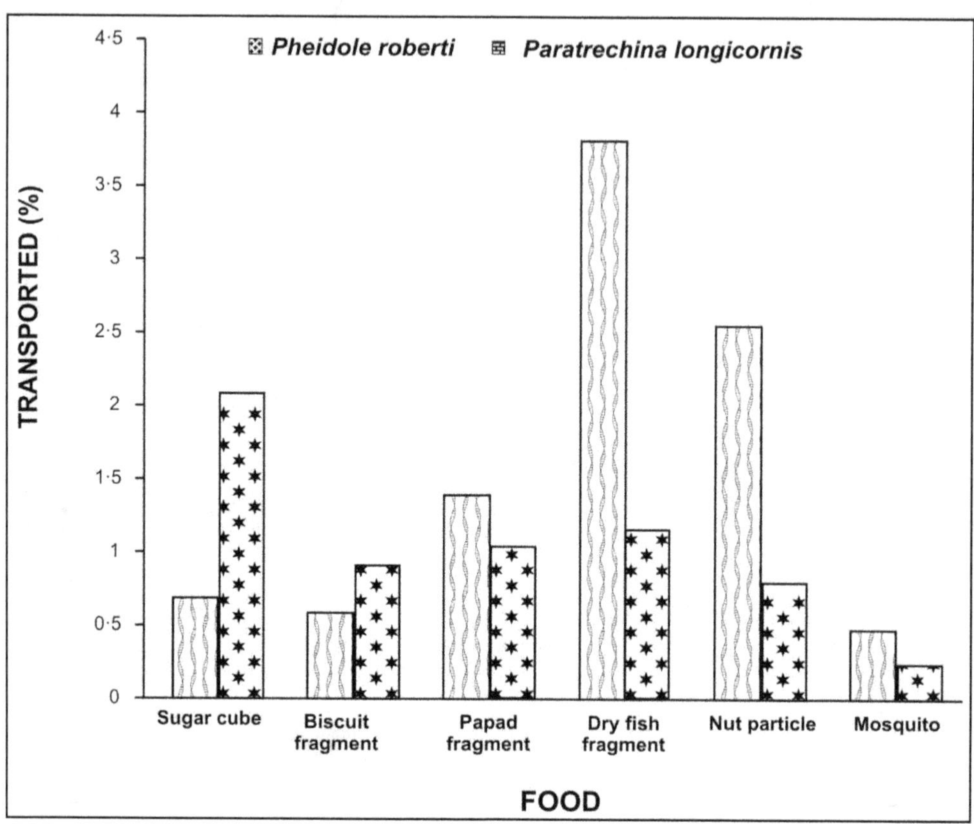

Figure 7.3: A Comparative Acocunt of the Food Transported (Per cent) by the Ants *Pheidole roberti* and *Paratrechina longicornis* through the Application of Cooperative Transporting Strategy.

1 puller and 1 pusher. The remaining one was carried by 3 pullers and 2 pushers. *P. roberti* procured 22 nut particles cooperatively and each particle was transported by 1 puller and 1 pusher. Of the 7 nut particles each one of the 5 particles required 2 *P. longicornis* individuals, 1 as puller and 1 as pusher. One particle needed 1 puller and 2 pushers while another particle was transported by 2 pullers and 2 pushers. *P. roberti* transported 4 mosquitoes (*Anopheles sp*) cooperatively, each one of the 2 mosquitoes was carried by 1 pusher and 1 puller while each one of the remaining 2 mosquitoes was carried by 2 pushers and 1 puller. On the other hand, transportation of 2 mosquitoes (*Anopheles sp.*) required 2 workers of *P. longicornis* where 1 acted as puller and the other one acted as pusher.

Discussion

From the results it appears that the food transporting strategy of ants is influenced by the weight, size and shape of the food materials to be transported to the nest. For the food particles manageable by an individual ant, the mode of transportation is either lifting (pushing) or dragging (pulling) ; however, it is most likely that the pulling act is applied when the food particle is little bit heavier than the one usually selected for transportation by the pushing act. This may be because in course of pulling, the ant individual has to overcome the stress not only because of the higher weight of the food particle, but also the strain to further carry the same. Perhaps, for the same reason involvement of two ants to carry a heavier food matter became inevitable. Thus, cooperative food transportation device by means of active transportation of two ant individuals (one as puller and another one as pusher) ultimately became a strategy for food transportation in the ants. This sort of decision leads the ants to develop cooperative transportation of heavier food materials by active participation of required number of worker individuals depending on the strength to be needed to transport the food matter through the application of lifting-dragging device. This is well judged from the fact of cooperation of 3, 4, 5 or more ant individuals, some as pullers and some as pushers, with respect to demand of the situation in transporting a food matter.

McCreery and Breed (2014) in their review clearly demonstrated the phases of transport in generalized process of cooperative transport of ants. Czaczkes and Ratnieks (2013) have discussed at length regarding adaptations for cooperative transport in ants. Indeed, cooperative transport is a behavioural adaptation in ants *P. roberti* and *P. longicornis* at least to avail the opportunity of utilisation of foods available in their foraging ground. Because, animals depend on the food sources occurring in their foraging area, and also there exists no guarantee of availability of sufficient amount of food matters to be collected by the worker ants individually to satisfy the need of the colony members. Therefore, these two ant species being foragers of the same foraging area have adapted themselves for all sorts of cooperative food transportations to ensure their survival.

In the present study the offered foods were of different nature with regard to weight, shape and size. It appears that the food materials which are manageable for individual transportation were transported only by one worker individual either as pusher or as puller, irrespective of species. But, in cases of heavier food particles cooperation of two individuals, one as pusher and the other as puller became inevitable for transportation of the said food matter on the way of lifting and/or dragging the same. The said transportation behaviour was intensified further by joining more worker ant individuals in the transportation act with respect to increasing weight and/or changes in shape of the particles involved. As the fragments or particles of sugar cubes, biscuit, papad, nut and fish were of different nature regarding weight and size/shape the ants had no alternative but to adopt cooperative transport in cases of transportation of individually unmanageable food matters. Regarding mosquitoes, it is assumed that the weight of the supplied *Culex* mosquito specimens were almost the same at individual level, and in that ground each mosquito individual of the said species was a manageable food element to adopt the individual transportation device. Naturally, each one of the supplied *Culex* species was transported individually by the ants of irrespective species. But participation of 2 to 3 ants to carry an *Anopheles* individual indicates that a specimen of this mosquito species is not manageable by a single ant individual to carry the same to the nest. This situation may have occurred because the legs of the said mosquito are longer than the lifting height of its main body when picked up by either of the two ant species. Thus, these appendages remain in contact of the ground and create obstruction and trouble in pushing and/or pulling. It is most likely that, being heavier in body weight compared to *Culex* individuals, transportation of an *Anopheles* mosquito species needs the cooperation of at least two ants during their transportation. This is justified from the fact of transportation of each of the 4 *Anopheles* mosquito specimens by pulling and pushing act of 2 ant individuals. Thus, to carry an *Anopheles* specimen there exist no problem when two ant individuals are involved. However, cooperation of 3 *P. roberti* workers in two occasions was recorded during transport of each of the 2 *Anopheles* specimens. Involvement of one more ant individual in ensuring transportation is certainly not due to weight but shape and/or size of the object concerned (Gross and Dorigo, 2004). Thus, it is concluded that evolution of cooperative transport in ants is induced by the characteristics of the food materials to be transported to the nest to ensure survival of the colony members.

Conclusion

Ants *P. roberti* and *P. longicornis* are equally apt to locate the food materials occurring in their foraging ground and to transport the same to their nest. These ants may carry the food particles individually either through pushing (lifting) or pulling (dragging), or cooperatively by synergistic pushing and pulling depending upon their ability to manage a food particle in their carrying act. Food transportation behaviours exhibited by *P. roberti* and *P. longicornis* during transportation of 6 different types of food materials offered, clearly indicate that, cooperative transportation strategy is evolved not only because of increasing weight of the food matters but also due to variations in shape and sizes of the same even if the object is easily manageable by an individual ant with respect to weight.

Acknowledgements

The authors are thankful to the Head of the Department of Zoology, University of Calcutta and to the Principal, Achhruram Memorial College, Purulia for the facilities provided. The ant specimens were identified by the Zoological Survey of India, Kolkata, India.

References

Campbell, R. C. 1989. Statistics for biologists. Cambridge University press, Cambridge.

Cerdá, X., Angulo, E., Boulay, R., Lenoir, A. 2009. Individual and collective foraging decisions : a field study of worker recruitment in the gypsy ant *Aphaenogaster senilis*. Behav Eco Sociobio 63, 551-562.

Czaczkes, T.J., Nouvellet, P., Ratnieks, F.L.W. 2010. Cooperative food transport in the Neotropical ant, *Pheidole oxyops* Insect. Soc. 58, 153-161.

Czaczkes, T.J., Ratnieks, F.L.W. 2013. Cooperative transport in ants (Hymenoptera: Formicidae) and elsewhere Myrmeco. news 18, 1-11.

Gross, R., Dorigo, M. 2004. Cooperative transport of objects of different shapes and sizes. In: Dorigo *et al.* (Eds) : Ants LNCS 3172, 106-117. Springer Verlag, Berlin.

Hölldobler, B., Stanton, R.C., Markl, H. 1978. Recruitment and food retrieving in Novomessor (Hymenoptera : Formicidiae): I. chemical signals. Behav. Eco. Sociobio. 4, 163-181.

McCreery, H. F., Breed, M. D. 2014. Cooperative transport in ants; a review of proximate mechanisms. Insect. Soc. 61, 99-110.

Naskar, K., S. K. Raut. 2014a. Food searching and collection by the ants *Pheidole roberti* Forel. Discovery 32, 6-11.

Naskar, K., S. K. Raut. 2014b. Judicious foraging by the ants *Pheidole roberti* Forel. Proc. Zoo. Society Kolkata 68, 131-138.

Naskar, K., S. K. Raut. 2014c. Ants forage haphazardly: a case study with *Pheidole roberti* Forel. Int. J. Sci. Nat. 5, 719-722.

Naskar, K., S. K. Raut. 2015a. Ants' foraging, a mystery. Int. J. Inno. Sci. Res. 4 (2), 064-067.

Naskar, K., S. K. Raut. 2015b. Foraging interactions between the Reddish brown ants *Pheidole roberti* and the Black ants *Paratrechina longicornis* Int. J. Res. Stud. Biosci. 3(3), 183-189.

Naskar, K., S. K. Raut. 2015c. Available food and ant's Response. Int. J. Eng. Sci. Res. Tech. 4 (4), 368-372.

Naskar, K., S. K. Raut. 2015d. Food-carrying strategy of the ants *Pheidole roberti.* Int. J. Tech. Res. App. 3 (3), 55-58.

Naskar, K., S. K. Raut. 2015e. Foraging behaviour following food contact in the ants *Pheidole roberti* Glo. J. Bio. Agri. Health Sci. 4 (2), 21- 24.

Naskar, K., S. K. Raut. 2015f. Cue for ant's trail development. Int. J. Res. Eng. App. Sci. 5 (5), 182-192.

Naskar, K., S. K. Raut. 2015g. Mysterious foraging of Pharaoh ant *Monomorium pharaonis.* Int. J. Res. Eng. App. Sci. 5 (7), 67-71.

Naskar, K., S. K. Raut. 2016a. Ants' food examination Proc. Zoo. Society Kolkata DOI 10.1007/s12595-016-0168-9.

Naskar, K., S. K. Raut. 2016b. Winter quarter-induced foraging in ants. Glo. J. Bio-Sci. Biotech. 5 (3), 318-323.

Robson, S.K., Traniello, J.F.A. 1998. Resource assessment, recruitment behaviour, and organization of cooperative prey retrieval in the ant *Formica schaufussi* (Hymenoptera: Formicidae). J. Insect Behav. 11, 1-22.

Traniello, J.F.A., Beshers, S.N. 1991. Maximization of foraging efficiency and resource defense by group retrieval in the ant *Formica schaufussi.* J. Insect Behav. 29, 283-289.

Traniello, J.F.A. 1983. Social organization and foraging success in *Lasius neoniger* (Hymenoptera: Formicidae) behavioural and ecological aspects of recruitment communication. Oeco. 59, 94-100.

– Segment 5 –

Medicinal Value of Insects

Chapter 8

Nutraceutical and Medicinal Insects: An Unexplored Research Field

☆ *Arias Barrera Ana Isabel,*
Pino Moreno José Manuel

ABSTRACT

In this research the chemical composition of some nutraceutical and medicinal insects is analyzed with emphasis on: royal jelly, honey, propolis, bee poison with specific emphasis on: alkaloids, amino acids, carotenoids, chitin, coumarins, crude fiber, fatty acids, flavonoids, glucosides, glycosides, irioids, peptides, phenolic compounds, phenoxazine, phytostanols, phytosterols, polyols, proteins, pterines, quinones, saponins, steroids, sterols, tannins, terpenes, triterpenes and tropolones. Besides, their function in the treatment and prevention of various types of diseases and their importance for the health of society are also discussed.

Introduction

In recent decades, a strong interest has been generated amongst people worldwide not only in promoting sustainable diets but also in knowing what kind of foodstuffs are being ingested, their composition and their effect on the human body; *i.e.* a correlation between food and health is sought. This new way of thinking has triggered the search for a deeper understanding of nutrients contained in different foodstuffs and hence renewed interest in what is commonly called "traditional medicine".

To some extent, we are rediscovering what in ancient times was already empirically applied. For example, Hippocrates said, "Let your food be your medicine and your medicine be your food" and something similar was uttered by Ludwig Feuerbach, German philosopher and anthropologist, "If you want to improve people, rather than speeches give them best food. Man is what he eats". From these aspects arises the concept of nutraceutical foodstuffs.

The term nutraceutical applies to some foodstuffs or components thereof that provide benefits to the health of humans or that are used for the prevention and treatment of patients suffering from certain diseases or ailments. The role played by active components of these foodstuffs has changed the role of diet on health. Nutraceuticals have changed the science of food and nutrition; they have been involved in the treatment of certain diseases and the reduction of others. Foodstuffs are evaluated not only in terms of macronutrients and micronutrients, but by their content of other "active components" and role in health promotion considering the concentration of the active components and the number of times they are exceeding the recommended daily intake.

The term functional foodstuff was first proposed in Japan in the early eighties (Alvídrez *et al.,* 2002). Since its inception there has been a lot of confusion with regard to the terminology used and the definition of these foodstuffs in the literature, thus being possible to find several definitions. The nutraceuticals are divided into three large groups listed below:

1. Nutrients: sugars, fats, amino acids, minerals and vitamins.
2 Chemical compounds: fiber, isoflavones, antioxidants, carotenes, lycopene, lutein, phenolic antioxidants, phospholipids, phytosterols.
3. Probiotics: beneficial microorganisms.

Some examples of substances with biological activity are described in Table 8.1 (Lopez *et al.,* 2012).

Table 8.1: Foodstuff Ingredients with Biological Activity: Nature and Properties

Bioactive Component	*Properties*	*Bioactive Component*	*Properties*
Fatty acids α-linolenic Omega 3	Anticancer, prevention of cardiovascular disease	Probiotic prebiotics	Anticancer, antimicrobial, Improve gastrointestinal disorders
Dietary fiber Soluble/Insoluble	Anticancer, Antihypertensive, Hypoglycemic, Hypocholesterolaemic	Proteins and peptides	Anticancer, Antioxidant, Hypocholesterolaemic, improvement of bone metabolism
Phytochemicals phytosterols, polyphenols, terpene thiols	Antihypertensive, anti-inflammatory, antioxidant, hypocholesterolaemic	Vitamins and minerals	Anticancer, hypercholesterolemic, Antihypertensive, Antioxidant, improved bone metabolism

Forerunners of Nutraceuticals and Medical Foodstuffs

In various ancient cultures the use of plants, fruits and animals for treatment or prevention of different diseases is widely known; some of these diseases are physical in character such as infections and others are more spiritual or emotional as the "scare". The product can be consumed as it is or as a mixture prepared in teas, ointments, *etc.* Unlike traditional modern medicine there is no specific dosage; even today in certain regions, people continue to carry out such practices either as a medical alternative or as an adjunct to allopathic medicine.

Traditional medicine, regardless of region or culture to which it belongs, is based on two key factors. The first is the test of trial and error, which implies a profound knowledge of the environment around the human group and the transmission of the use of natural resources to a select group of that population. The second factor is the belief and spirituality, *i.e.* religiosity awarding certain properties that can be beneficial, harmful, desirable or not to certain objects, animals or plants.

The use of insects and the products extracted from them have been part of the therapeutic resources in the medical systems of many cultures around the world (Costa Neto *et al.*, 2006). The use of these animals is widely described, for example, in Mexican prehispanic books like Chilam Balam of Ixil (Cahuich-Campos, 2013) and in many books of traditional Chinese medicine; for example in this country they are believed to have a remarkable action on spasms and seizures. Various species of medicinal insects have been reported around the world in countries such as Mexico (210), China (149), Brazil (48) and Cuba (6) (Boletin UNAM, 2009).

The therapeutic use of insects and various products derived from them is known as anthropoentomotherapy, both the knowledge and the various practices are transmitted through oral tradition from generation to generation. Although the use of insect species as medicinal resources is an ancient practice, anthropoentomotherapy is still relatively unknown academically (Costa Neto *et al.*, 2006).

Medicinal insects are used in various ways, many are used alone or mixed with various medicinal plants or other products such as honey, propolis, mud, bait, "pulque", alive to be ingested or smeared, crushed to form ointments, salves or poultices, mixed with their own products as in the case of the honey bee (honey, pollen or immature stages). They are boiled and both the body and the liquid in which they were boiled are consumed. Sometimes they are dried and ground to add a little water and apply as a poultice or ingested. When only one part of the body is used as the last femur of grasshoppers and crickets or the "horns" of the melolonthid beetles, they are boiled in water which is then sipped off and/or the powder obtained from scraping the horn is ingested as it is. (Ramos-Elorduy 2000; Pino *et al.*, 2009).

According to the dictionary of traditional Mexican medicine, for example, the jaws of worker ants "*Atta* spp." were used to close wounds after surgery, placing several ants on the injury, so that when they bit, they pierced skin on both sides; after being decapitated they formed a kind of antiseptic suture due to the secretions derived from salivary gland function.

Other references are found during the Crusades in which the honey of *Apis mellifera* (Linnaeus, 1758) (Hymenoptera, Apidae) was used to treat stomach, skin and eyes ailments; wax cured haemorrhoids, burns and wounds. The Ottomans used the bed bug *Cimex lectularius* (Linnaeus, 1758) (Hemiptera; Cimicidae), a kind of stink bug, to treat jaundice. The Ebers Papyrus, an Egyptian medical treatise dating from the XVI century B.C., contains several records of remedies obtained from insects and spiders. Pliny the Elder, in his Historiae Naturalis, recorded some remedies for various diseases using insects as main ingredient; in the Roman Empire in the first century BC, Dioscorides in the second book of Materia Medica, also mentions some remedies made from them (Costa Neto *et al.*, 2006).

Medicinal and Nutraceutical Insects

Food depends on various factors such as the place of origin of individuals, habitat, animal food resources and local vegetables, religion, ethnicity, education, social phenomena such as migration, colonization, trade, *etc.* The products consumed, how to cook them or keep them and store them give rise to ethnic markers; so it is that food is one of the elements that have contributed to determine identity among different ethnic groups (Acuña, 2010).

The anthropoenthomophagy, *i.e.* consumption of insects by humans, is an activity that has been practiced since the beginning of mankind, several countries such as China and Mexico have an ancient tradition in consumption of insects while in other countries, especially in Europe this practice has been recently adopted and valued (van Huis *et al.*, 2013).

Most insects feed on plants, particularly the foliage; for example, a caterpillar *Bombyx mori* (Linnaeus, 1758) (Lepidoptera: Bombycidae) consumes about forty thousand times its own weight in leaves to complete its development; that is why insects are considered accumulators of chemical constituents of plants, among which can be found pigments, sterols, toxic substances, *etc.* (Berenbaum, 1995). Insects are an important source of compounds for drug research because of their co-evolution with plants and the defensive products they produce, but historically they have received little attention in scientific research. For the reasons stated above we are interested in this research since this aspect is relatively recent and insects represent a source of "active principles" still very much underexplored.

Materials and Methods

This research was conducted through an extensive literature review of various scientific data bases, magazines and books about edible and medicinal insects worldwide; in these publications the compounds isolated are detailed. Based on the information obtained we determined which insects can be treated as nutraceuticals and so we could develop a view about them and discuss their importance in terms of health.

By consuming insects people are not only assimilating nutrients of good quality but also get an extra benefit for their "preventive healing" properties. There are even products that can be fully replaced by edible insects drastically reducing treatment costs.

Results

In the following paragraphs the nutraceutical and/or medicinal properties of the insects and their products are mentioned.

Medicinal or Nutraceutical Compounds

Royal Jelly

Royal jelly is secreted by the pharyngeal or frontal glands, in combination with others located in the upper chest, called cervical; these glands are well developed in bees from 5 to 12 days, commonly called nursemaids, and then they atrophy, which is what makes royal jelly expensive and scarce product (Espina and Ordetx, 1984). Because it contains minerals, vitamins and amino acids it is a highly beneficial health product, the chemical composition of royal jelly can be seen in Table 8.2.

Table 8.2: Chemical Analysis of Royal Jelly

Component	*Value*	*Component*	*Value*
Moisture by weight loss in drying at 100°F	24.15	Dextrose	11.70
Albuminoid (Factor 6.25)	38.62	Sucrose	3.35
Phosphorus	0.67	Fats	15.22
Sulfur	0.38	Other elements of unknown nature	3-4
Ashes	2.34		

Values are g/100g.

According to Haydack and Vivino (1950) 1 to 3 days old dry bee larvae have higher vitamin content than royal jelly (Table 8.3).

Table 8.3: Vitamin Composition of Royal Jelly Compared to the Value of Reported Dry Bee Larvae Weight

Age of Worker Larvae and Royal Jelly	*Thiamine mg/g*	*Riboflavin µg/g*	*Niacin mg/g*	*Pantothenic Acid µg/g*
		Royal jelly		
1-2 days	4.9	44.1	377	554
3-5 days	3.4	30.8	148	57
		Worker larvae		
1-3 days	29.9	161.3	393.6	916.1

However, chemical analyses of royal jelly are still insufficient so there are probably components in trace amounts that have a beneficial effect on humans. In Europe, mainly in Russia, Italy and France, it is officially recognized as a drug, while in the United States it has not been valued as such. It has been suggested that royal jelly has a therapeutic action in different disorders, but there are no studies to explain the mechanism of action. However, according to various authors cited by Reina (2010) royal jelly has the following properties:

- It stimulates metabolism and normalizes basal metabolic processes. Increases body weight and growth rate, improves growth in young children.
- Has enzymatic effects due to cholinesterase and phosphatase.
- Increases psychophysical performance.
- It stimulates the production of collagen type I, strengthens bones and promotes regeneration processes in fractures.
- Exerts tonic action on some centers of the hypothalamus, and as a result increases the secretion of adrenocorticotropic hormone in the pituitary.
- Acts favorably on disorders of the intestinal tract and strengthens the stomach and intestinal peristalsis.
- It stimulates cell metabolism and is an excellent epithelisation agent that regenerates tissue.
- Elevates the content of hemoglobin in the blood and leukocyte, glucose and red blood cells.
- Promotes endocrine system and helps normalize the autonomic functions.
- Is an excellent adjunct in gerontological treatment, slows the aging process and improves skin hydration and elasticity.
- It contains gamma globulin, a component that is able to slow senility and increase endurance and exerts antiviral, antimicrobial and antitoxic functions.
- Increases blood pressure in hypotensive states and has little effect in favoring hypertension and stimulates circulation.
- Three hours after ingestion of royal jelly, it was found to significantly curb blood sugar level up to 30 per cent in diabetic patients.

Honey

Honey is a natural product of stinging bees of the genus *Apis* (Hymenoptera, Apidae) and several genera of stingless bees (Hymenoptera, Apidae), which has medicinal properties recognized since antiquity. These effects have been variably attributed to the pH, the release of hydrogen peroxide, the high osmolarity created by its high sugar content, *etc.* Honey of genus *Apis* is the oldest substance recorded as being used for healing wounds and has been taken up today due to various advantages among which we can mention that it is easy to apply, it is not painful or harmful for tissues, it creates a useful moist environment for healing, has antibacterial properties and is a strong stimulator of granulation and epithelialization of wounds.

Possessing a great antibiotic and emollient profile, it has been used in the treatment of wounds, burns, ulcers, *etc.*; due to its high content of antimicrobial substances including the inhibin hormone that has a regulatory action on the pituitary, it has been used as an agent to promote fertility (Dold *et al.*, 1937). Vardi *et al.* (1998) showed that local application of honey on surgical wounds post infection in infants helped them to stay clean, sterile and closed after twenty-one days treatment. Bactericidal properties are attributed to honey because it interacts with pathogenic bacteria that include some types of bacilli, cocci, espirilos, and vibrios. Due to the diverse flowers they visit, honey has antibacterial properties against *Pseudomonas* (Migula, 1894), *Staphylococcus* (Rosenbach, 1884), *Escherichia coli* (Escherich, 1885), *Salmonella* (Lignieres, 1900), *Streptococcus* (Rosenbach, 1884), *etc.* It also has antifungal activity against some species of yeast and *Aspergillus* sp. (Micheli, 1729) and *Penicillium* sp. (Link, 1809) and against common dermatophytes; these are enhanced by its acid pH, viscosity, osmolarity and other phytochemical content. Antimicrobial, anti-inflammatory and stimulating rapid healing activity (Table 8.4) (Ayala, 2004) is then highlighted.

Table 8.4: Healing Properties of Honey in Wounds

Properties	*Clinical Results*	*Expected Mode of Action*
Antimicrobial activity	Sterilization of the wound. Inhibition of potential wound pathogens and digestive proteins, enzymes that destroy tissue. Deodorization of malodorous wounds. Wound protective barrier to prevent contamination of environmental pathogens	Production of hydrogen peroxide. Action of phytochemical components. Acidity (pH). Stimulation of the immune system: multiplication of B and T lymphocytes. Metabolization of glucose into lactic acid en lieu of serum amino acid metabolization. Physical barrier because of high viscosity.
Anti-inflammatory activity	Resolution of edema and exudates. Pain reduction. Reduction of keloid scarring	The high osmolarity allows fluid to create a layer of dilute plasma honey or lymph, resulting in wet conditions that enhance cure and no adhesion to the wound surface. Decrease inflammation associated with leukocytes. Suppression of inflammatory processes by scavenging free radicals by antioxidants. Preventing formation of serum exudates able to be colonized by bacteria.
Stimulation of rapidhealing	Phagocytic increase. Autolytic debridement increase. Increased angiogenesis. Cell proliferation. Collagen synthesis. Epithelialization, with less need for skin grafting	Stimulatory effect of glycosylated proteins of honey. Increased nutritional status in secondary tissues due to contact with lymph. Increased oxygen and acidity contact lymph. Production of hydrogen peroxide with antioxidant protection modifying proteins important for cell growth and debridement

Data from National Honey Board (2003).

Humans consume honey of different species of bees but the commercial form comes from the species *Apis mellifera* (Linnaeus, 1758) (Hymenoptera, Apidae). Bromatological analysis of honey produced by other species such as the *Melipona quadrifasciata* (Le Peletier, 1836) (Hymenoptera, Apidae) have been performed (Table 8.5). Comparing honey from the genders *Melipona*, *Scaptotrigona* and *Tetragonisca* with honey produced by *Apis mellifera*, it has been found that the former is more liquid, more acidic and its composition is not identical, this can make a significant difference in the effect they have on microorganisms

Table 8.5: Analysis of Honey from *Apis mellifera* and different Honeys from Meliponini

Components	*Melipona quadric-fasciata (Le Peletier, 1836)*	*Melipona beecheii (Bennet 1831)*	*Melipona solani (Cockerell, 1912)*	*Scaptotri-gona mexicana (Guérin-Meneville, 1845)*	*Tetragonisca angustula (Illiger, 1806)*	*Apis mellifera (Linnaeus, 1758)*
Water	34.68	17.32	19.66	18.74	26.08	17.70
Levulose	30.22					40.50
Glucose	28.28					34.02
Reducing sugars		68.77	75.7	57.22	54.83	
Sucrose	0.12	3.50	1.70	0.06	2.58	1.90
Dextrins	6.34					1.51
Ash	0.05	0.07	0.06	0.10	0.56	0.18
Diastase activity (DN)		21.29	8.31	18.62		
Without dosification	0.04					4.90
Free acidity (meq/kg)		23.23	4.97	12.68	12.33	
pH	4.1	3.67	3.8	4.04	4.03	3.6-4.2

Values are mostly g/100g. Different units are mentioned in parentheses.

Data from Espina and Ordetx, (1984); Dardón and Enriquez, (2008); Mendieta, (2002).

Currently various countries seek to rescue the cultivation of stingless bees for forest conservation and as a proposal for rural development: the idea is to market the products of the hive whose honey is highly valued because it is popularly used in the ancient Mesoamerican region for treatment of various respiratory, skin, eye and gastrointestinal conditions; this increases its value in relation to that of the honey of *Apis mellifera* L. (Dardón and Enriquez, 2008). Table 8.6 shows the antimicrobial activity of honey at different concentrations on microbial strains that cause common infections. Several studies have shown particular sensitivity of *S. aureus* (Rosenbach 1884) to honey, including antibiotic-resistant strains (Estrada *et al.*, 2004). Cooper *et al.* (2002) reported sensitivity to honey of 18 *S. aureus* methicillin resistant strains isolated from infected wounds and twenty strains of vancomycin resistant enterococci isolated from surfaces in hospital environment concentrations of 10 per cent (v/v).

Comparing the inhibitory capacity of stingless bee honey with stinger honey bee, we can see the greater inhibitory power of the first against *Staphylococcus aureus* and *S. epidermidis;* in the case of other microorganisms in Table 8.6, the inhibitory behavior is similar.

Table 8.6: Antibacterial Capacity of Meliponini Honey (a) and Honey of *Apis mellifera* (b) at Different Concentrations

Honey Concentration	*100 per cent (w/v)*		*75 per cent (w/v)*		*50 per cent (w/v)*		*25 per cent (w/v)*	
	a	b	a	b	a	b	a	b
Staphylococcus aureus	90	80	83	52	70	48	10	24
S. epidermidis	90	68	83	24	50	16	20	0
P. aeruginosa	24	76	43	36	30	24		8
E. coli	73	84	40	32	17	12		0
E. coli O157: H7	87		70		33		1	
S. enteritidis	83	92	63	44	27	20		8
L. monocytogenes	67	48	47	20	23	16	7	

Data from Zamora and Arias, (2011); Estrada et al. (2004).

Dardón and Enriquez (2008) determined Minimum Inhibitory Concentration (MIC) in bacterial (37 °C/24 hr) and mycotic (48 °C/24 hr) cultures reporting the averages of the different honeys of Meliponini (Table 8.7).

Table 8.7: *In vitro* Antimicrobial Activity of Honey from Stingless Bees. Solutions 10, 5, 2.5, 1.25 per cent (v/v)

Microorganisms	*Entomological Origin of Honey*							
	**Ga*	*Mb*	*My*	*Np*	*Pl*	*Sm*	*Sp*	Ta
	Solutions of honey (per cent v/v) with inhibition of microbial growth							
Bacillus subtilis	5	5	5	2.5	5	5	2.5	5
Candida albicans	(-)	10	5	5	10	5	5	10
Cryptococcus neoformans	5	5	5	2.5	5	5	2.5	5
Escherichia coli	5	5	5	5	5	5	5	5
Mycobacterium smegmatis	5	5	5	2.5	5	5	2.5	
Pseudomonas aeruginosa	5	5	5	5	5	5	2.5	10
Salmonella typhi	10	5	10	5	5	5	2.5	10
Staphylococcus aureus	10	5	5	5	5	5	2.5	10

* Ga: *Geotrigona acapulconis* (Strand, 1919) Mb: *Melipona beecheii.* (Bennet, 1831) My: *Melipona yucatanica.* (Camargo and Moure Roubik, 1988). Np: *Nannotrigona perilampoides.* (Cresson, 1878) Pl: *Plebeia* sp. (Schwarz, 1938) Sm: *Scaptotrigona mexicana* (Guérin-Meneville, 1845). Sp: *Scaptotrigona pectoralis* (Dalla Torre, 1896). Ta: *Tetragonisca angustula* (Illiger, 1806).

The microorganisms more susceptible to honey of stingless bees are *C. albicans* and *S. typhi*, however we know that both microorganisms reported high resistance to antibiotics. At the lower solution of 2.5 per cent, the honey from *S. pectoralis* proves effective to inhibit the growth of *S. aureus, S. typhi, M. smegmatis, B. subtilis, P. aeruginosa* and *C. neoformans*.

As already mentioned *S. aureus* often quickly develops resistance to antibiotics such as penicillin making it difficult to treat, so it is important to note that different honeys from stingless bees species reported in the above table inhibit growth at concentrations between 5- 10 per cent, except *S. pectoralis* inhibited at 2.5 per cent. Meanwhile *Mycobacterium* resistant to adverse environmental conditions, antibiotics, chemicals such as antiseptics and disinfectants such as *M. smegmatis* (Trevisan 1889) is inhibited in six cases by a 5 per cent (v/v) solution and the particular case is that of *Nannotrigona perilampoides* and *Scaptotrigona pectoralis* that are inhibited by a solution of 2.5 per cent (v/v), thus more studies on other strains of *Mycobacterium* are recommended (Dardón and Enriquez, 2008).

An important factor that allows us to explain the broad spectrum of antimicrobial activity and even the rest of the beneficial properties of honey is floral and geographical in origin, bees have different nutritional requirements and collect pollen along with other active compounds from various plants and therefore in some places its composition will be different even among honeys produced by the same species, therefore between the different types of honey there are detectable differences in their antimicrobial potential.

Other examples are investigations by Molan and Rhodes (2015) who showed that Manuka honey from New Zealand, which comes from the shrub *Leptospermum scoparium* (J.R. Forst. and G. Forst., 1775 Family Myrtaceae) has a great antibacterial activity that is due to hydrogen peroxide present and the component called methylglyoxal, a potent inhibitor of antibacterial growth (Lu *et al.*, 2013). Meanwhile, "Tualang" honey from Malaysia renders better results than the control and hydrofiber –hydrofiber of silver-on the reduction of bacterial growth of *Pseudomonas aeruginosa* in burn wound (Khoo *et al.*, 2010). Ahmed and Othman (2013) compared some properties and biochemical characteristics with Manuka honey and noted that their differences include higher amount of phenolics, flavonoids and HMF (hydroxymethylfurfural), which gives greater efficacy than Manuka honey against some strains of Gram negative bacteria in burn wounds (Schencke *et al.*, 2016).

The monofloral honey "Ulmo" (*Eucryphia cordifolia*) (Cav, 1798.) (Oxalidales; Cunoniaceae) from Chile, has also shown great bactericidal and efficiency in the treatment of wounds. Sherlock *et al.* (2010) stated that this honey has a stronger antimicrobial effect against *Staphylococcus aureus, Escherichia coli* and *Pseudomonas aeruginosa*, as compared with Manuka honey. Likewise Calderon *et al.* (2015) conducted clinical studies in patients with venous ulcers reporting highly effective treatment with tropical *Ulmo* honey associated with oral administration of ascorbic acid (Schencke *et al.*, 2016).

Besides its antibacterial, antioxidant and anti-inflammatory properties honey also creates a physical barrier and a moist local environment due to its high viscosity and high sugar concentration and constant fluid drainage by osmosis from the wound bed. It stimulates plasminogen thus increasing the plasmin activity of fibrin digesting enzyme (Molan, 2001, 2009; Schencke *et al.*, 2016). This promotes healing of burn wounds because these will heal faster in a moist environment rather than in a dry environment with crusting. A moist environment ensures the growth of epithelial cells, promotes the approach of fibroblasts to the wound margins and lack of adherence of the dressing to the wound, resulting in easy and painless changes, without the risk of breaking the newly formed epithelium (Molan, 2011). Furthermore, the characteristics of this locale allow autolytic debridement removing necrotic tissue and eschar (Schencke *et al.*, 2016).

Several studies have shown that the tensile strength in wounds also differs depending on the type of honey used since their action on collagen in the healing process shows differences. Moreover, Rozaini *et al.* (2004) observed that Manuka honeys, *Durio zibethinus* "Durian" (Murray, 1774) (Malvales; Malvaceae) and *Ananas comosus* "Nenas" (L.) (Merr, 1917) (Poales;. Bromeliaceae), have higher tensile strength as compared with wounds treated with two other honeys, *Melaleuca* sp. "Gelam" (Linnaeus, 1758) (Myrtales; Myrtaceae) and *Cocos nucifera* "Kelapa" (Linnaeus, 1758) (Arecales; Arecoideae) inferring that this is possible because the first group is able to increase collagen content and accelerate the healing process. Nisbet *et al.* (2010) showed that the level of hydroxyproline in collagen in groups treated with monofloral honey-Castaña- *Castanea sativa* (Mill, Gard Dict, ed 8., n 1, 1768.) (Fagales, Fagaceae), *Rhododendron luteum* -Azalea yellowing (Sweet, 1830) (Ericales, Ericaceae) and multiflora honey from the region of the Black Sea in Turkey, were significantly higher than in the untreated control group. In this study, no significant differences between the various types of honey were detected, but less areas were observed in the wounds unhealed group multiflora honey on days 7 and 14 post injury (Schencke *et al.*, 2016).

Propolis

Propolis is a viscous substance of reddish orange colour that comes from the resin of some trees that bees amalgamate with a secretion of saliva. As it is made up of many substances that also vary with the type of source, it does not have a specific chemical formula. It is used as cement or balsam welding honeycombs each other and the walls, to close the cracks and grooves of the hive. The human being generally uses it as a bactericide, dietary ingredient in candy, creams, soaps and as part of some medications. It is noted for possessing anti-inflammatory, antimicrobial, antioxidant and antineoplastic properties. While generally beneficial, some cases of allergies associated with some types of propolis have been reported. The antifungal properties of propolis, including their activity on various species of *Candida* (Berkh, 1923) and *Trichosporon* (Behrend, 1890), have been previously studied with positive effects. The antioxidant properties of propolis are also notable. Some antioxidant compounds include phenyl ester of caffeic acid (CAPE), quercetin, caffeic acid, galangine, ferulic acid, p-coumaric and CAPE (Peña, 2008). In general, according to Nagai *et al.* (2001) in its antioxidant properties propolis is more active than any other product of the hive.

Bee Poison

Bee venom is composed of molecules of melittin, mast cell degranulating peptide (MCD), apamin, hyaluronidase, and histamine. Melittin consists of twenty amino acids and has bactericidal, fungicidal, vasodilator, analgesic, anti-inflammatory activity; it also acts as a CNS inhibitor. The MCD peptide has twenty amino acids, has anti-inflammatory action against arthritis and is more potent than hydrocortisone. The apamine provides antigenic and anti-inflammatory qualities, besides having analgesic activity. Hyaluronidase improves circulation, increases the immune response and has antigenic properties. Histamine is a vasodilator and contains formic acid, hydrochloric acid, magnesium phosphate and some other microelements (Pardo, 2005). Apitoxin is fresh and purified bee venom, it is produced in the abdomen of worker bees, has anti-inflammatory and analgesic effects and stimulates blood circulation; it acts as a local anaesthetic and stimulates the adrenal glands that are responsible for the production of cortisone, which has antirheumatic properties. It is also used as a homeopathic remedy and is included in all homeopathic pharmacopoeias in the world (De Felice and Padin, 2012).

Chitin

The cuticle of insects is formed by sclerotin, chitin and protein. The thickness of cuticle varies according to gender and state of development. Chitin forms approximately 2-20 per cent of the insects on dry basis and varies between each order. For example in dry basis (g/100 g), the quantities of some orders are: Ephemeroptera 11.42, Odonata 12.53, Orthoptera 7.66, Blattodea 4.2, Isoptera 4.46, Hemiptera-Homoptera 12.26, Coleoptera 10.76, Lepidoptera 7.85 and Hymenoptera 6.14. Both chitin and chitosan can be considered as dietary fibre because the structure of chitin is similar to the plant cellulose and therefore may be included in this group, according to some researchers the fibre is of vegetable origin only. There are several nutritional studies that demonstrate that low consumption of this compound in the diet is associated with diseases such as colon cancer, diabetes, cardiovascular disease, diverticulitis, hypercholesterolemia, *etc.* (Biruete *et al.*, 2009). This fact is important because insects provide good quality protein and fiber.

Chitinase is an enzyme capable of digesting chitin reducing it to components that can be assimilated such as N-acetylglucosamine; in humans there are varying concentrations of chitinases, even in some cases they are not present, the function of these enzymes in the body during disease remains largely unknown. However chitotriosidase, enzyme produced by macrophages whose optimum activity is at pH 6 is found in large quantities in patients affected with Gaucher's disease and it has recently been found that this enzyme may be involved in the innate immune response, *e.g.* in diseases such as acute malaria, beta-thalassemia and other hemoglobinopathies its concentration is very high indicating that the macrophage activation is responsible for the expression of the chitotriosidase (Paoletti *et al.*, 2007).

Other enzymes of this group are the acid chitinase in mammals (AMCase) produced by the bronchial epithelium, that exhibits optimal activity at acidic pH and is involved in bronchial allergic asthma, but its function is not yet clear.

Pino *et al.* (2009) also describe that chitin has antibiotic, anticoagulant and hemostatic properties, and is as well a serum cholesterol reducer and non-allergenic drug transporter; in the industry as a protective insulating coating it serves to feed a thin edible film against pathogenic microorganisms. It also helps heal wounds and mild skin burns and repair tissue.

Other Molecules or Biologically Active Compounds

Fatty Acids

The fatty acids are obtained through the diet but there are certain denominated non-essential ones that can be synthesized in the body, they are two: α-linolenic and linoleic, both are polyunsaturated with double bonds in cis position. Although ω-3 fatty acids are important for the health of humans, both EPA and DHA can be synthesized through biochemical reactions from α-linolenic acid. As for the fatty acid arachidonic acid (20: 4) it is derived from linoleic acid, it will only be necessary if there is deficiency of its precursor linoleic acid.

Oleic acid slightly increases HDL and reduces LDL oxidation. It also produces a better profile of anticoagulant substances (antiplatelet therapy) and is a vasodilator which attenuates the thrombotic process. Overall, oleic acid has a more evident and positive cardiovascular effect than linoleic acid (Mataix, 2005). At the vascular level oleic acid decreases both systolic and diastolic blood pressure and also has been shown to exert a positive influence on the digestive functions.

Linoleic acid has cholesterol lowering effect by decreasing levels of total cholesterol and LDL but it slightly reduces HDL (antiatherogenic lipoprotein), *i.e.* having atheroma antiformation profile, which are lipid deposits in the arterial wall with production of yellowish induration masses and softening observed in atherosclerosis (Mataix, 2005).

In the body, linoleic acid and α-linolenic are responsible for training the eicosanoids, hormone-like substances of twenty carbons, these are necessary for normal immune function, for inflammation and clotting of blood. Omega 3s (EPA and DHA) may reduce the risk of cardiovascular disease by suppressing inflammatory factors. Omega-6 fats also reduce these diseases through various systems such as those affecting the availability of oxygen, the flow of blood, insulin resistance and blood pressure.

Arachidonic acid together with the cyclooxygenase enzymes originates what is known as the series 2 of prostaglandins (PGE2), prostacyclins (PGI2) and thromboxanes (TXA2) and series 4 of the leukotrienes (LTB4). Arachidonic acid may also suffer a spontaneous oxidation producing isoprostanes. On the other hand, omega-3eicosapentaenoic acid (EPA) gives rise to series 3 of the prostanoids (prostaglandin I3 (PGI 3) and the leukotrienes series 5(LTB 5) that are beneficial for health (Coronado *et al.*, 2006; Martínez and Rivas, 2005).

From enzymatic processes thromboxanes and prostaglandins are associated with homeostatic regulation and vasomotion, so a diet high in omega-3 has antihemostatic and antithrombotic effects by altering the balance between the various eicosanoids (Coronado *et al.*, 2006).

Thrombus formation is influenced by platelets and coagulation and fibrinolysis mechanisms. The intake of omega-3 fatty acids influences hemostasis through prolonged bleeding time and reduce platelet aggregation, but also exerting beneficial effects on erythrocyte deformability (Mataix and Gil, 2004). Moreover, consumption reduces the content of arachidonic acid that when found in high amounts in the cell membrane is transformed, through the hydrogenase enzyme, into thromboxane A2.

It is estimated that at least half of the deaths from coronary heart disease are the result of myocardial electrical instability leading to ventricular fibrillation. It has been shown that polyunsaturated omega 3 and omega 6 fatty acids have anti arrhythmogenic properties, omega-3 being the most potent (Mataix and Gil, 2004).

The omega-3 affect the number of leukocytes contained in the atheromatous plaque by reducing the expression of adhesion molecules in them, meaning they can stabilize it; they can also directly affect the expression of adhesion molecules in leukocytes themselves. Another point that influences the anti-inflammatory effect of omega-3 fatty acids is its molecular structure because different researchers have concluded that a double bond is the minimum necessary but sufficient for the fatty acids inhibit anti-inflammatory activity of the endothelium (Lopez and Macaya, 2006).

There have been changes in the composition of fatty acids in an organism by modifying the diet, for example the increase of polyunsaturated fatty acids in milk of American women of about 8 per cent in 1959, when the fat of animals was the main source of cooking fat, up nearly in 1977, when corn oil replaced much to these cooking fats (Castro, 2002; Neville and Picciano, 1997). Omega 6 has the effect of reducing the incidence in certain cancers such as breast and pancreas.

Types of fatty acids in insects are as follows:

The average content of lipids in dry basis in insects is approximately 30 per cent for larvae and 20 per cent for adults; the type of fatty acids is mostly polyunsaturated. Insects contain, however, significant amounts of short-chain fatty acids, which are rapidly absorbed in the large intestine and can be used as an energy source (Martí *et al.*, 2003).

In edible insects the fat fraction consists mainly of unsaturated fatty acids (UFA) (Table 8.8), that are important in preventing heart disease and decrease the high concentration of bad cholesterol in the blood. The profile of fatty acids in insects therefore is similar to white meat (chicken and fish).

Most insects with a high concentration of fat belong to the orders Lepidoptera, Coleoptera and Hymenoptera; this is due to the fact that exhibit a complete metamorphosis and to make that change an important energy investment is made, hence a high need for fat is needed in these organisms.

The predominant fatty acid in insects is the α-linolenic acid, which is one of the two essential fatty acids and predecessor of EPA and DHA; this is because they are accumulators of the chemical constituents of the plants where this fatty acid is present.

Table 8.8: Percentage of Edible Insects Present in some Insects

		ω-7	ω-9	ω-6		ω-3		
Food	Fat (g/100g)	(16: 1)	(18: 1)	(18: 2)	(20: 4)	(18: 3)	(20: 5)	(22: 6)
Bombyx mori (1) (Linnaeus, 1758)	30.00-35.00 (1)	1.50	29.64	5.30	-	33.34	-	-
Cirina forda (1) (Westwood, 1849)	12.24-14-30 (1)	0.20	13.90	8.10	0.1	45.30	-	-
Nudaurelia melanops (1) (Bouvier, 1930)	12.20 (2)	0.60	5.60	5.70	0.30	35.60	1.40	0.40
Samia ricinii (1) (Donovan, 1798)	26.00 (1)	1.82	17.39	5.49	-	42.23	-	-
Chondracis rosea (1) (De Geer, 1773)	18.49 (3)	1.00	20.20	12.30	-	40.10	-	-
Ruspolia differens (1) (Serville, 1869)	47.20 (1)	1.90	24.60	30.35	-	3.70	-	-
Aspongopus viduatus (1) (Fabricius, 1794)	54.20 (1)	10.62	45.53	4.90	-	0.43	-	-

Data from Rumpold and Schluter, (2013); Bukkens, (2005); Silistina and Jatin, (2015).

Insects can be used as protein and lipid source in animal feed, particularly poultry and pork, thus enriching meat products. Among the species most used for this purpose are the larvae of the black soldier fly *Hermetia illucens* (Linnaeus, 1758) (Diptera; Stratiomydae), the common housefly *Musca domestica* (Linnaeus, 1758) (Diptera, Muscidae) and the flour yellow worm *Tenebrio molito*r (Linnaeus, 1758) (Coleoptera, Tenebrionidae). In the special case of flies you can even modify their fatty acid profile by increasing the amount of omega-3 by supplying remains of fish, mainly blue, as food.

Iridoids

Its name comes from a genus of ants where it was first identified *Iridomyrmex* (F. Smith, 1858) (Hymenoptera, Formicidae); two of the first compounds identified were iridomirmecin and iridodial. In this group there are a series of bicyclic monoterpenes (C10) derived biosynthetically, the monoterpene geraniol which present as basic structure to iridan. The iridan core is frequently fused to a heterocycle comprising six atoms, one of which is oxygen, and this structural assembly is called iridoid (Lopez *et al.*, 2012; Boluda and Terrero, 2013; Jimenez *et al.*, 2006).

They can have open structures (secoiridoids) or closed structures (true iridoids) and generally appear as heteroside compounds, particularly as glucosides. They have beneficial effects on liver and biliary function. They have also been shown to have anti-inflammatory, antibacterial, anticancer, antitumor, choleretic and antiviral properties. In some cases they have even been taken as antidotes in poisoning by certain fungi such as those belonging to the genus *Amanita* (Pers, 1794).

Steroids, Sterols, Phytosterols and Phytostanols

All steroids contain a core composed of four saturated rings called gonane. Furthermore, as shown in cholestane, which is the basic structure of sterols (steroid alcohols), some have a side chain at the end of the ring. The four rings are not linear but present in angles. The three most important groups of steroids are sterols, bile acids and steroid hormones.

Sterols are steroid alcohols containing a hydroxyl group in β position in carbon three and one or more double bonds in ring B and the side chain; all of them derive from tetracyclic triterpenes. The composition of phytosterols and sterols in insects varies widely according to their diet; they have also important variations depending on whether they are carnivores, herbivores or omnivores. In (Table 8.9) sterol profile is presented in *Helicoverpa zea* (Boddie, 1850).

Table 8.9: Sterol Profile According to different Diets Applied to *Helicoverpa zea*

Dietsterols (Total per cent)	*Sterols in Tissues (per cent total)*
Cholesterol (100)	Cholesterol (100)
Sitosterol (100)	Cholesterol (80), sitosterol (20)
Stigmasterol (100)	Cholesterol (84), sitosterol (15)
Ergosterol (100)	Cholesta-5-7-dienol (41), ergosterol (36), cholesterol (23)
Clerosterol (100)	clerosterol (80), cholesterol (20)
Spinasterol (100)	lathosterol (63), spinasterol (34), cholesterol (5)
Lathosterol (100)	lathosterol (81), cholesterol (19)
Cholesterol (90), 24-dihydrolanosterol (10)	cholesterol (93), 24-dihydrolanosterol (7)
Cholesterol (70), 24-dihydrolanosterol (30)	cholesterol (88), 24-dihydrolanosterol (12)
Cholesterol (50), 24-dihydrolanosterol (50)	cholesterol (75), 24-dihydrolanosterol (25)
Cholesterol (30), 24-dihydrolanosterol (70)	cholesterol (50), 24-dihydrolanosterol (50)
Corn [sitosterol (51), campesterol + 22-dihydro-brassicasterol (27), isofucosterol (17), stigmasterol (6)]	cholesterol (80), campesterol + 22-dihydro-brassicasterol (11), sitosterol (10)
Lucerne [spinasterol (69), 22-dihydrospinasterol (17.3), avenasterol (13.3)]	lathosterol (54.5), 22-dihydrospinasterol (25.5), spinasterol (19.0), avenasterol (0.5)

Data from Behmer and Nes, (2003).

Among the physiological effects of sterols we have their anti-inflammatory, antitumor, antibacterial and fungicidal properties in humans; one of the functions of phytosterols and phytostanols is to inhibit cholesterol absorption facilitating excretion (Romero and Vázquez, 2012; Venezuela and Ronco, 2004; Ling and Jones, 1995). Some research has shown that phytosterols block the development of tumors in the colon, mammary glands and prostate, however the mechanisms by which this occurs are not clearly established, however it is known that phytosterols alter the transfer mechanisms through the cell membrane during the growth of tumors and reduce inflammation (Cuadro, 2004).

Both phytosterols and stanols have a hypocholesterolemic effect when ingested in a range of 1-3 g/day, so they are important in the prevention of cardiovascular disease. However, the administration of high doses of 20g/day occasionally causes diarrhea in humans (Venezuela and Ronco, 2004; Ling and Jones, 1995).

The joint action of sterols and/or stanols on these mechanisms produces a decrease in total plasma cholesterol and LDL cholesterol without changing HDL cholesterol levels (Venezuela and Ronco, 2004; and Heinemann *et al.*, 1986). Lowering of cholesterol plasma levels was detected two or three weeks after having been integrated in the diet (Romero and Vázquez, 2012).

Flavonoids

The term flavonoids denotes a very large group of polyphenolic compounds characterized by a benzo-Õ-pyran structure, which are widely distributed in vascular plants, they tend to have at least 3 phenolic hydroxyls and they are usually combined with sugars as glucosides, but also occur relatively frequently as free aglycones. Chemically, these substances are phenolic in nature and are characterized by two benzene aromatic rings linked by a bridge of three carbon atoms which can or cannot form a third ring (Cartaya and Reynaldo, 2001).

Humans cannot produce these protective chemicals, so they must be obtained through food or as supplements, among their features they stand out as antioxidants, cardiovascular disorders, anti-inflammatory, anti-cancer and antimutagenics. Flavonoids were discovered by the Nobel Prize winner Szent-Györgyi who isolated citrine from lemon peel in 1930; this substance acts as a regulator of capillary permeability (Martínez *et al.*, 2002) (Table 8.12).

Tannins

Tannins are sandwiched between the collagen fibers, establishing joints that allow them to create a high resistance against water and heat. They have the property of coagulating albumin mucous and tissues, creating an insulating and protective coating that reduces irritation and pain. Its astringent power makes them suitable for wound healing, especially given as poultices. They are useful as anti-inflammatory and antiseptic in cases of bronchitis, hemorrhoids, *etc.* They also have hypocholesterolemic and antioxidant activity, and are useful against diarrhea, intestinal cooling, vesicular diseases and as antidotes in case of poisoning by heavy metals and plant alkaloids (Cuadro 2004) (Table 8.12).

Quinones

Quinones or benzoquinones are one of the two isomers of cyclohexanedione. Its color can range from pale yellow to almost black, most being yellow to red.

Quinone is a common constituent of biologically relevant molecules such as vitamin K1 which is a phylloquinone. Others serve as electron acceptors in electron transport chains such as photosystems I and II of photosynthesis and aerobic respiration. They also have anti-cancer properties (Table 8.12).

One of the most important uses in the food industry is the use of crimson red quinones from cochineal *Dactylopius coccus* (Costa, 1835) (Hemiptera-Homoptera: Dactylopiidae), this color comes from carminic acid. The acid extract from cochineal is probably the best natural colorant with technological features, but is used less and less because of its high price. Its applications are diverse, an example is the food industry: in the preparation of jam, yogurt, ice cream and drinks with a nice red color; also it offers possibilities for its use in the cosmetic, textile and pharmaceutical industry.

Pterines

The pterines are heterocyclic compounds with a common double ring structure with 10 atoms, 4 of nitrogen and the others of carbon. Early works on pterines date from the late nineteenth century, there are published attempts to isolate yellow and white pigments called pterines from the wings of certain butterflies. However they could not be isolated with enough purity so as to characterize their structure, only in the mid-twentieth century it was proposed that they are derived from heterocyclic pyrazine [2,3-d] pyrimidine which was named pteridine (Lorente, 2003). The most important pterines are yellow xanthoperin (7-desoxileucopterina) (from the lemon butterfly) and leucopterin which is colorless (from white flowers butterfly) (Miale, 1985).

In 1936 it was found that xanthopterin cured experimental macrocytic anemia in rats and in 1941 in the silver salmon.

Antimicrobial Peptids

Antimicrobial peptides are effector molecules of the innate immune system, various types are found in almost all organisms, from bacteria to mammals. They are a group of versatile materials with complex mechanisms of action related to the interaction with the pathogen through its membrane, or affecting internal targets such as DNA replication and protein synthesis, and interacting with the host in immunomodulatory regulatory functions, the inflammatory process and healing (Tellez and Castaño 2010).

The first antimicrobial peptide isolated from an insect is the cecropine (1981) obtained from the pupa of *Hyalophora cecropia* (Linnaeus, 1758) (Lepidoptera, Saturniidae). Since this report over 170 antimicrobial peptides in insects have been discovered (Table 8.10). These compounds share some common characteristics such as low molecular weight below 5 kDa, a net positive charge at physiological pH and most have amphipathic α-helices or forks in folded β sheets or mixed structures (Bulet *et al.*, 1999; Ntwasa *et al.*, 2012).

Calmodulin

It is a thermo-stable intracellular acidic protein that is primarily located in the brain and heart; it is the prototype of the EF-hand family (Alvarado, 2009). It regulates the signal transduction of calcium into the cell, acting as a receiver by having four sites with high affinity for binding Ca^{2+} ion in a reversible manner (Table 8.12).

Table 8.10: Some Types of Antimicrobial Peptides: Functions and Properties

Antimicrobial Peptide	*Function and Properties*
Cecropine	It is also present in the genus *Drosophila*, *Bombyx* and *Sarcophaga*. It has a structure of two amphipathic helices, preserved trypsin, lysine and arginine residues. Has antimicrobial activity against Gram negative bacteria (*S. typhimurium*,/Ex Kauffmann and Edwards 1952, Leminor and Popoff 1987), *Acinetobacter calcoaceticus* (Beijerinck, 1911), *P. aeruginosa* (Schroetes 1872, Migula 1900) and Gram positive (*Bacillus megaterium* (Bary, 1884))
Defensin	It was first isolated in two independent sources, from cell culture *Sarcophaga peregrina* (Robineau-Desvoidy, 1830) (Diptera; Sarcophagidae) and *Phormia terraenovae* larvae (Robineau-Desvoidy, 1830) (Diptera, Calliphoridae), at present this peptide has been found in at least 40 species, especially in orders Diptera, Coleoptera, Hymenoptera, Trichoptera, Hemiptera-Homoptera and Odonata. They are active mainly against Gram positive bacteria and some Gram negative bacteria only; fungi and yeasts are affected by it. The lytic effect is more efficient at low salt concentrations
Drosomicin	The first antifungal peptide was isolated from the fruit fly *D. melanogaster*, Meigen 1830, it has forty-four residues including cysteine that are involved in the formation of four disulfide bridges. This peptide has many similarities with cysteine-rich peptides from plants with antifungal properties. It has a potent antifungal activity but is ineffective against yeasts and bacteria, has no hemolytic effect, is effective in amounts below 5mM against phytopathogenic fungi infecting humans. It also inhibits spore germination in high concentrations and slows the growth of hyphae at lower concentrations. This causes stunting evidenced as reduced hyphal elongation with a concomitant increase in branched hyphae.
Thanatin	Thanatin is the second peptide with antifungal properties, isolated from the spiny soldier bug *Podisus maculiventris* (Say, 1832) (Hemiptera-Homoptera: Pentatomidae), it has twenty-two cysteine residues including some involved in disulfide bond formation. This peptide has antimicrobial activity in filamentous bacteria (Gram positive and negative) and fungi, it exhibits its activity to below a 2.5 mM minimum inhibitory concentration and also has no hemolytic effect. Its activity and effectiveness is due to: 1. The C-terminal loop. 2. The three residues C-terminal. 3. The seven hydrophobic residues mostly linked to the N-terminal. 4. The three N-terminal residues that is required for antifungal activity but not antibacterial activity. D-thanatin has almost the same antifungal activity as L-thanatin but is virtually inactive against Gram negative and some Gram positive bacteria. This suggests that thanatin has more than one mechanism of action depending on the target microorganism.
Abaecin	Is a rich in proline peptide and was first isolated from *Apis mellifera*, but it also has been found in the bumblebee *B. pascuorum* (Scopoli, 1763) (Hymenoptera, Apidae) with 39 amino acids is the richest antibacterial proline peptide, the biggest isolated among insects. It has antimicrobial activity against Gram negative and positive bacteria.
Moricin	It was isolated for the first time from the silkworm *Bombyx mori*. It consists of a long alpha helix with eight turns of a sequence of 42 amino acids. The N-terminal amphipathic segment of the alpha helix is largely responsible for the increase in the permeability of the bacterial membrane.
Psacotheasin	Belongs to the "knottin-type" class is active against bacteria and fungi. The knottin fold topology is characterized by the type "abcabc" in disulfide bond formation.

Data from Bulet *et al.* (1999); Ntwasa *et al.* (2012).

Tropolones

They are a group of organic compounds with a central skeleton with the formula C_7H_5 (OH) O. They are of interest because of its unusual electronic structure and its role as a ligand precursor. Its structure was first described in 1954. Some of the most popular are hinokitiol, colchicine and purpurogallin. Many of them have pharmaceutical bioactivity (Table 8.12). They exhibit a broad spectrum of biological activities including antibacterial, antitumor and antifungal. Bioactivities of β-thujaplicin, for example, have been exploited since 1930; colchicine has been used since the antiquity as a remedy for the treatment of gout, pericarditis, Behcet's disease and atrial fibrillation (Al Fahad, 2014). 3-isopopenil-tropolone is a strong antimicrobial agent especially against *Staphylococcus aureus* (Nakano *et al.*, 2015).

Microsurgery

Lucilla sericata larvae (Meigen, 1826) (Diptera: Calliphoridae) remove necrotic or dead tissue, disinfect and stimulate granulation in tissues, secrete proteolytic enzymes (carboxypeptidases A and B, leucine aminopeptidase, collagenase and two serine proteases), degrading components of the extracellular matrix of dead tissue and in the case of being in contact with healthy tissue enzymes are denatured, and increase the degree of tissue oxygenation so that the necrotic tissue is removed (Ríos *et al.*, 2013).

It has been proposed that larvae eliminate infection by three mechanisms: 1) ingestion of bacteria, 2) irrigation of chemicals in the wound and 3) production of substances with antimicrobial properties (Ríos *et al.*, 2013).

1. Ingestion of bacteria: Given that any bacteria present are removed as they pass through the digestive tract of the larva, the antibacterial activity is due to different components among which figures the gut commensal *Proteus mirabilis* (Hauser, 1885) that produces two substances with antibacterial activity identified as phenylacetic acid and phenylacetaldehyde with particular action in the low pH of the intestine. Larvae inhibit the development of Gram-positive bacteria [including methicillin resistant *Staphilococcus aureus* (MRSA)] and anaerobic, with less effect on Gram negative and *Proteus* spp. (Hauser, 1885) or *P. aeruginosa* (Schroeter, 1872), in fact the latter is able to express virulence factors that kill larvae and limit their action on wounds intensely colonized by this bacteria. This is why larvae should not be used until previously treated against *P. aeruginosa*.
2 Irrigation chemicals: Excretion of ammonium bicarbonate and its derivatives on the wound neutralizes acid exudate produced by the wound inflammation, which contributes to raise the pH above seven and thus reduces the colonization of bacteria. Calcium carbonate was also found in the excretions of the larvae, it stimulates phagocytosis by calcium ions, facilitating the healing process in infected wounds.
3. Antimicrobial compounds: The lucifensine, which can be found in the salivary glands, fat body and hemolymph of *L. sericata* has proved effective against Gram positive microorganisms, specifically *Staphylococcus carnosus* (Schliefer and Fischer 1982), *Streptococcus pyogenes* (Rosenbach, 1884) and

Streptococcus pneumoniae (Klein, 1884), but also acts against MRSA and *S. aureus* (Rosenbach, 1884) with intermediate susceptibility to glycopeptides [glycopeptide intermediate *S. aureus* (GISA)].

Formation of granulation tissue is produced by fibroblast activation and endothelial cells forming a provisional matrix and therefore contributes to the development of this tissue. Some substances identified in the secretion of the larvae are: allantoin, urea, calcium carbonate, ammonium enzymes (trypsin, chymotrypsin, leucinaminopeptidases, carboxypeptidases A and B, serum proteases, collagenase, *etc.*) (Ríos *et al.*, 2013).

Larvae therapy is indicated primarily for cleaning and disinfection of chronic wounds with abundant necrotic fibronoid material or infected tissue. According to Ríos *et al.* (2013) there are many types of injuries upon they can be used, contraindications being rare (Table 8.11).

Table 8.11: Indications and Contra-indications of Therapy with Fly Larvae

Indications	Ulcers of diabetic patients, venous ulcers, non-diabetic neuropathic ulcers, arterial/ ischemic ulcers, pressure ulcers, thromboangiitis obliterans, post-traumatic wounds, necrotizing fasciitis, gangrenous pyoderma, pilonidal sinus, osteomyelitis, surgical site infections, surgical wounds that do not heal, wound infections projectile gun, burns, wounds infected by methicillin-resistant *Staphylococcus aureus* Rosenbach 1884, and ulcers of neoplastic origin.
Contra indications	Dry wounds, open wounds with communication with body cavities, wounds in the vicinity of blood vessels of large calibre and Patients allergic to soy, egg or fly larvae.

In the Table 8.12 we have shown a taxonomic list of nutraceutical insects and their "active ingredients".

Table 8.12: Nutraceutical Insects, and Chemicals they Contribute**

Order	*(General Chemicals)/Particular Chemicals*
Orthoptera	
Acrididae	(Sterols)/Cholesterol, sitosterol, two Δ5-sterol, cholestanol, Δ0-sterol
Acrididae/*Locusta migratoria* (Linnaeus, 1758)	(Sterols)/Δ7-sterol, Δ7,22-sterol, Δ5,22-sterol, Δ8-sterol, sitosterol, cholestanol, lactosterol, campesterol, sigmasterol, 9_,19-cyclopropylsterol
Acrididae/*Barytettix humphreysii* (Thomas, C. 1875)	(Sterols)/Cholesterol, sitosterol
Acrididae/*Trimerotropis pallidipennis* (Burmeister, 1838)	(Sterols)/Cholesterol, sitosterol
Acrididae/*Melanoplus differentialis* (Thomas, 1865)	(Sterols)/Cholesterol, sitosterol
Gryllidae/*Acheta domestica* (Linnaeus, 1758)	(Sterols)/Sitosterol, sigmasterol
	(Iridoids, coumarins, prostaglandin, triterpenoids and steroids)

Order	*(General Chemicals)/Particular Chemicals*
Pyrgomorphidae/*Schistocerca americana* (Drury, 1770)	(Steroid Hormone)/Cortisone
	(Sterols)/Cholestanol, sitosterol
Pyrgomorphidae/*Schistocerca gregaria* (Forsskål, 1775)	(Pterines)/
	(Isoprenoids)/Carotenoids
	(Pigments)/Ommocromos
	(Phytosterols)/7-dehidrocholesterol, desmosterol, (3β, 3α) colesta-8,14,24-trien-3-ol, 4,4-dimetil, (3β, 20R) colesta-5, 24-dien-3, 20-diol, fucosterol
Pyrgomorphidae/*Sphenarium purpurascens* (Charpentier, 1845)	(Isoprenoids)/Carotenoids
	(Flavonoids)/Catechins
	Iridoids, triterpenoids, tannins, coumarins, phenolic acid, polyols, free quinones, glycosides
Romaleidae (Brunner von Wattenwyl, 1893)/	(Sterols)/Cholesterol, sitosterol, two Δ5-sterol, cholestanol, Δ0-sterol
Romaleidae/*Romalea guttata* (Houttuyn, 1813)	(Sterols)/Cholesterol, sigmasterol
Blattodea	
Blattidae/*Periplaneta americana* (Linnaeus, 1758)	(Proteins)/Calmodulin
Blattellidae/*Blattella germánica* (Linnaeus, 1758)	(Sterols)/Desmosterol, ergoestanol
Hemiptera-Homoptera	
Coreidae/*Thasus gigas* (Klug, 1835)	(Alkaloids)/Fenoxazine
Dactylopiidae/*Dactylopius coccus* (Costa, 1835)	(Quinones)/Carminic acid
Pentatomidae/*Aspongopus viduatus* (Fabricius, 1794)	(Tocopherols, phospholipids)
Pentatomidae/*Edessa cordifera* (Walker, 1868)	(Coumarins alkaloids, iridoids "tropolones", phenolic acid, tannins)
Pentatomidae/*Edessa mexicana* (Stål, 1872)	(Iridoids, "tropolones", polyols, phenolic acid, coumarins and alkaloids)
Pentatomidae/*Edessa petersii* (Distant, 1881)	(Isoprenoids)/Carotenoids
	(Triterpenoids, steroids, condensed tannins, polyols, iridoides and alkaloids)
Pentatomidae/*Edessa* sp (Fabricius, 1803)	(Iridoids, «tropolones», phenolic acid, coumarins, free quinones, polyols alkaloids)
Pentatomidae/*Euschistus* (*Atizies*) *taxcoensis* (Ancona, 1933)	(Iodine, phenolic acid, coumarins alkaloids)

Order	*(General Chemicals)/Particular Chemicals*
Pentatomidae/*Euschistus crenator* (Fabricius, 1794)	(Tannins, triterpenoids, steroids,Iridoides, saponines, phenolic acid, coumarins, free quinones, alkaloids and polyols)
Pentatomidae/*Euschistus strennus* (Stål, 1862)	(Triterpenoids, steroids, iridoids coumarins and alkaloids)
Coleoptera	
Cerambycidae/*Acalolepta luxuriosa* (Bates, 1873)	(Peptids)/"Creopinas"
Cerambycidae/*Acrocinus longimanus* (Linnaeus, 1758)	(Peptids)/Alo-1, Alo-2, Alo-3 [Knottin-type]
Cerambycidae/*Psacothea hilaris* (Pascoe, 1857)	(Peptids)/Psacotheasin
Curculionidae/*Hypera postica* (Gyllenhal, 1813)	(Sterols)/Spinasterol, latosterol, avenasterol, 22-dihydrospinasterol
Scarabaeidae/*Allomyrina dichotomus* (Linnaeus, 1771)	(Peptids)/"Dicostatina", defensin
Tenebrionidae/*Tribolium* sp	(Peptids)/Defensins, "attacinas", "coleoptericina", creopinas
Tenebrionidae/*Tribolium castaneum* (Herbst, 1797)	(Sterols)/Cholesterol*, 7-dehydrocholesterol, demosterol, sitosterol, 5,7,24-cholestatrienol, 7-dehydrocholsterol
Tenebrionidae/*Tenebrio molitor* (Linnaeus. 1758)	(Sterols)/Cholesterol*, sitosterol, campesterol, sigmesterol, latoesterol
	(Peptids)/Tenicin 1
Tenebrionidae/*Ulomoides dermestoides* (Chevrolat, 1878)	(Antiinflammatories, limonene)
Tenebrionidae/*Zophobas atratus* (Fabricius, 1775)	(Peptids)/Defensin A, Defensin B
Lepidoptera	
Bombycidae/*Bombyx mori* (Linnaeus, 1758)	(Peptids)/"Attacina", "moricina", "drosocina", "cecropina", bombyxinas, lebocin 1, lebocin 2
Pieridae/*Catopsilia crocale* (Fabricius, 1775)	(Pterines)/Isoxanthopterine
Pyralidae/*Galleria mellonella* (Linnaeus, 1756)	(Sterols)/Cholesterol, sitosterol, campesterol
	(Acylglycerides)
Saturniidae/*Hyalophora cecropia* (Linnaeus, 1758)	(Peptids)/"Cecropina" A y B
Sphingidae/*Manduca sexta* (Linnaeus, 1763)	(Sterols)/Cholesterol, desmosterol, campesterol, sitosterol, cholest-5,7-dienol, ergosterol, stigmasterol, dihydrolanosterol, lanosterol, obtusifoliol, Δ7-obtusifoliol
Noctuidae/*Helicoverpa zea* (Boddie, 1850)	(Sterols)/Cholesterol, sitosterol, cholesta-5,7-dienol, ergosterol, latosterol, spinasterol, 24-dihydrolanosterol, 22-dihydrobrassicasterol, lathosterol, avenasterol.
Diptera	
Drosophilidae/*Drosophila* sp. (Fallén, 1823)	(Peptid)/"Andropinas", "drosocina", "drosomicina"

Order	*(General Chemicals)/Particular Chemicals*
Drosophilidae/*Drosophila-melanogaster* (Meigen, 1830)	(Peptid)/Defensin, "diptericina", "drosomicina", metchnikowin, moricin
Muscidae/*Musca domestica* (Linnaeus, 1758)	(Stanoles)/Cholesterol, campesterol, sitosterol
Hymenoptera	
Apidae/*Apis mellifica* (Linnaeus, 1758)	(Poison)/Apitoxin, "abaecina" and hyaluronidase.
	(Peptids [Honey])/Apidaecin "abaecina", "hymenoptaecina", "royalisina" and defensin
	(Protein)/Calmodulin
Eumenidae/*Anterhynchium flavomarginatum micado* (Kirsch, 1878)	(Peptids)/Eumenino mastoparan-AF
Formicidae/*Atta cephalotes* (Linnaeus, 17958)	(Sterols)/7-dehydro-24-methylenecholesterol, 22-dihydroergosterol, ergosterol
Formicidae/*Pogonomyrmex barbatus* (Smith, 1858)	(Triterpenoids, steroids)
Megachilidae/*Megachile rotundata* (Fabricius, 1787)	(Sterols)/Isofucosterol, 24-methylenecholesterol, sitosterol, campesterol
Vespidae/*Polybia occidentalis nigratella* (Buysson, 1905)	(Glucosides of steroid or of triterpenoids
	Steroids, triterpenoids, iridoids, alkaloids, polyols)/Saponines
Vespidae/*Vespula maculifrons* (Buysson, 1905)	(Peptids)/Apidaecin
Vespidae/*Vespula squamosa* (Drury, 1773)	(Peptids)/Apidaecin
	(Glucosides of steroid or of triterpenoids)/Saponines
	(Triterpenoids, steroids, Iridoides, coumarins alkaloids, polyols)

**References: Data from Costa Neto *et al.* (2006), Villarruel *et al.* (2004), Pino *et al.* (2009), Cheseto *et al.* (2015), Ntwasa *et al.* (2012), Mariod *et al.* (2011), Mustafa *et al.* (2008), Van der Horst, (1985), Behmer and Nes, (2003), Bulet *et al.* (1999), Mendoza and Saavedra, (2013). *Some cases have synthetic diets.

In the Table 8.13 we present the biological activity of some insects.

Conclusion

From the research papers found and reported it can be concluded that insects are an important niche of biologically active compounds. However, despite being a large group, there has not been enough scientific interest, as demonstrated by the few research papers that exist about the chemicals present in these animals taking into account the great diversity of them. It seems forgotten that they are one of the oldest groups on earth and have evolved with plants and, as such are bio-accumulators of the chemical constituents of these. Among the various compounds that can be found in insects are pigments, sterols, terpenes, antimicrobial peptides, alkaloids, coumarins, *etc.*

Table 8.13: Insect Compounds with Biological Activity

Order	*Scientific Name*	*(General Chemicals)/ Specific Chemicals*	*Activity*
Odonata			
Aeshnidae	*Aeshna cyanea* (O.F. Muller, 1764)	(Peptids)/Defensin	Antimicrobial
Dermaptera			
Forticulidae	*Forficula auricularia* (Linnaeus, 1758)	(Quinones)	
Hemiptera-Homoptera			
Notonectidae	*Notonecta glauca* (Linnaeus, 1758)	(Peptids)/Defensin	Antimicrobial
Pentatomidae	*Palomena prasina* (Linnaeus, 1761)	(Peptids)/Defensin, metalnikowin I, IIa, IIb, III	Antimicrobial
Pentatomidae	*Podisus maculiventris* (Say, 1832)	(Peptids)/Defensin 1, defensin 2	Antimicrobial
Pyrrhocoridae	*Pyrrhocoris apterus* (Linnaeus, 1758)	(Peptids)/Defensin	Antimicrobial
Neuroptera			
Chrysopidae	*Chrysopa perla* (Linnaeus, 1758)	(Peptids)/Defensin, pyrrhocoricin	Antimicrobial
Coleoptera			
Scarabaeidae	*Holotrichia diomphalia* (Bates, 1888)	(Péptids)/Holotricin 1, 1A,1B, 1C	Antimicrobial
Meloidae	*Lytta vesicatoria* (Linnaeus, 1758)	(Terpene)/Cantharidine	Vesicant
Trichoptera			
Limnephilidae	*Limmephilus stigma* (J. Curtis, 1834)	(Peptids)/Defensin	Antimicrobial
Lepidoptera			
Saturniidae	*Lonomia obliqua* (Walker 1855)		Antithrombotic
Pieridae	*Prioneris thestylis* (Doubleday, 1842)	Isoguanine	Anticancer
Diptera			
Tephritidae	*Cerratitis capitata* (Weidemann 1824)	(Peptids)/Ceratotoxins	Antimicrobial
Calliphoridae	*Calliphora vicina* (Robineau-Desvoidy, 1830)	(Peptids)/Defensin	Antimicrobial
Calliphoridae	*Lucilia sericata* (Meigen, 1826)	(Peptids)/(Allantoin	Antimicrobial
Calliphoridae	*Phormia terranovae* (Robineau-Desvoidy, 1830)	(Peptids) Defensin A, B	Antimicrobial
Culicidae	*Aedes aegypti* (Linnaeus, 1762)	(Peptids)/Defensin A,B,C	Antimicrobial
Muscidae	*Stomoxys calcitrans* (Linnaeus, 1758)	(Peptids)/SMD1, SMD2	Antimicrobial
Sarcophagidae	Sarcophaga peregrine (Linnaeus, 1758)	(Peptids)/Sarcotoxin IA, IB, IC, sapecina A,B,C	Antimicrobial
Syrphidae	*Eristalis tenax* (Linnaeus, 1758)	(Peptids)/Defensin	Antimicrobial
Hymenoptera			
Apidae	*Bombus pascuorum* (Scopoli, 1763)	(Peptids)/Defensin, apidaecina, abaecina	Antimicrobial
Apidae	*Bombus terrestres* (Linnaeus, 1758)	(Peptids)/Apidaecina	Antimicrobial

Order	*Scientific Name*	*(General Chemicals)/ Specific Chemicals*	*Activity*
Crabronidae	*Sphecius speciosus* (Drury, 1773)	(Peptids)/Apidaecina	Antimicrobial
Formicidae	*Formica rufa* (Linnaeus, 1761)	(Peptids)/Defensin	Antimicrobial
	—	(Phormic acid)	Homeophatic remedy against cystitis, rheumatism and gout
Formicidae	*Myrmecia gulosa* (Fabricius, 1775)	(Peptids)/Formaecin 1, formaecin 2	Antimicrobial
Ichneumonidae	*Coccygomimus disparis* (Viereck, 1911)	(Peptids)/Apidaecina	Antimicrobial
Pompilidae	*Anoplius samariensis* (Pallas, 1771)	(Neurotoxins Poison)/ α-pompilidotoxins, β-pompilidotoxins	Research and treatment of neurological diseases
Pompilidae	*Batozoonellus maculifrons* (Smith, 1873)	(Neurotoxins poison)/ α-pompilidotoxinas, β-pompilidotoxinas	Research neuroscientific and treatment of neurological diseases

*References: Data from Costa Neto *et al.* (2006), Villarruel *et al.* (2004), Pino *et al.* (2009), Cheseto *et al.* (2015).

One of the main problems worldwide is the resistance of bacteria to conventional antibiotics, so it is necessary to seek other alternatives to fight them. Insects and products derived from them have shown significant antimicrobial effects. In many cases they have no side effects unlike conventional drugs. Besides, they are an important source of nutraceutical foodstuffs that help prevent and decrease the risk of certain diseases or conditions apart from being a nutritious alternative that even has ecological and conservation benefits.

Traditional medicine is a starting point for the knowledge and development of beneficial chemicals for humans. Many of the pharmaceutical and nutritional compounds that are known today have been isolated from natural sources (mainly plants), and these have been part of oral tradition as healing remedies. Insects, like plants, are widely linked to traditional medicine in different cultures but have not been given the same attention. It is therefore important to emphasize in depth studies on them both to encourage consumption, in the case of edible insects, and to isolate active substances that can be used later for therapeutic or nutraceutical purposes.

Acknowledgements

The authors are grateful to Dr. Aurora Zlotnik Espinosa for her help in some aspects of the translation.

References

Acuña C.A.M. (2010). Etnoecología de insectos comestibles y su manejo tradicional por la comunidad indígena de los Reyes Metzontla, municipio de Zapotitlán Salinas, Puebla. Tesis para obtener el grado de Maestro en Ciencias.

Ahmed, S. and Othman, N.H. (2013). Review of the medicinal effects of Tualang honey and a comparison with manuka honey. Malaysian Journal Medical Sciences. 20 (3):6-13.

Al Fahad A. (2014). Tropolone and Sorbicillactone Biosynthesis in Fungi.University of Bristol. Degree of Doctor of Philosophy in the Faculty of Science, School of Chemistry.

Alvarado M.M.E. (2009). Estudio de la expresión de calmodulina y detección de posibles cambios en transducción de señales durante los dos procesos de diferenciación del parásito Giardia intestinalis. Trabajo de grado para el título de Doctorado en Química. Universidad Nacional de Colombia. Facultad de Ciencias. Bogotá.

Alvídrez M. A., González M. B. E. and Jiménez S. Z. (2002). Tendencias en la producción de alimentos: alimentos funcionales. Revista Salud Pública y Nutrición. 3 (3).

Ayala A.S.S. (2004). Efecto curativo de la miel de abeja en pacientes mexicanos con ulceras varicosas. Universidad Autónoma de Nuevo León. Requisito para el grado de maestría en metodología de las ciencias.

Behmer S.T and Nes D.W. (2003). Insect sterol nutrition and physiology.A global overview.Advances in insect physiology.31: 72.

Berenbaum M. R. (1995). Bugs in the system: insects and their impact on human affairs. (Helix Books collection) N.Y., Addison-Wesley.400 p.

Biruete G.A., Juárez H.E., Sieiro O.E., Romero V. R. and Silencio B.J.L. 2009. Los Nutracéuticos. Lo que es conveniente saber. Revista Mexicana de Pediatría. 79 (3): 136-145.

Boletín UNAM-DGCS-397. (2009). Los insectos recurso medicinal. UNAM.

Boluda J.C. and Terrero D.J. (2013). Iridoides y secoiridoides (1): Clasificación, biosíntesis, importancia ecológica, estrategias evolutivas y modificaciones semisintéticas. Revista de Fitoterapia. 13 (2): 153-161.

Bukkens S.G.F. (2005). Insects the human diet: Nutritional aspects.In Ecological Implications of Minilivestock: Potential of Insects, Rodents, Frogs and Snails. Ed. Paoletti M.G, 545-577 pp. Eneld, NH: Science Publ. 662 p.

Bulet P., Hetru C., Dimarcq JL. and Hoffmann D. (1999).Antimicrobial peptides in insects; structure and function.Developmetal and Comparative Inmmunology. 23: 329-344.

Cahuich-Campos D. (2013). Los artrópodos utilizados en la medicina tradicional maya mencionados en los libros de Chilam Balam de Chan Cah, Tekax y Nah e Ixil. Etnobiología. 11 (2): 16-23.

Calderón M.D., Figueroa C.S., Arias J.S., Sandoval A.H., and Torre F.O. (2015) Combinedtherapy of Ulmo honey (*Encryphia cordifolia*) and ascorbicacid to treatvenousulcers.Rev. Latino-Am. 23 (2) 259-266

Cartaya O. and Reynaldo I. (2001). Flavonoides: Características químicas y aplicaciones. Cultivos Tropicales. 22 (2):5.

Castro G.M.I. (2002). Ácidos grasos omega 3: Beneficios y fuentes. Interciencia. 27 (3):128-136.

Cheseto X., Kuate S.P., Tchouassi D.P., Ndung'u M., Teal P.E.A. and Torto B. (2015) Potential of the Desert Locust Schistocerca gregaria (Orthoptera: Acrididae) as an Unconventional Source of Dietary and Therapeutic Sterols". PLoS ONE. 10(5): e0127171.

Connor W.E. (1996).Omega-3 essential fatty acids in infant neurological development. Backgrounder. 1: 1-6.

Cooper R.A., Molan P.C., Harding K. (2002). The sensitivity to honey of gram-positive cocci of clinical significance isolated from wounds. Journal of Applied Microbiology. 93: 857-863.

Coronado H.M., Vega y León S., Gutiérrez T.R., García F.B. and Díaz G.G. (2006). Los ácidos grasos omega-3 y omega-6: nutrición, bioquímica y salud. REB. 25(3): 72-79.

Costa Neto E.M., Ramos-Elorduy J. and Pino M.J.M. (2006). Los insectos medicinales de Brasil: Primeros resultados. Boletín Sociedad Entomológica Aragonesa. 38: 395-414.

Cuadro M.L.F. (2004). Estudio bromatológico y fitoquímico de la jícama (*Smallanthus sonchifolia*) para determinar el tiempo óptimo de cosecha. Tesis de grado doctoral. Bioquímica y Farmacia. Escuela Superior Politécnica de Chimborazo. Riobamba, Ecuador.

Dardón M.J. and Enríquez E. 2008. Caracterización fisicoquímica y antimicrobiana de la miel de nueve especies de abejas sin aguijón (Meliponini) de Guatemala. Interciencia. 33 (12): 916-922.

De Felice J.L and Padin J. (2012). Apitoxina. Su preparado, especificaciones y farmacología. Ediciones argentinas and americanas. Primera edición.

Diccionario Enciclopédico de la Medicina Tradicional Mexicana. Biblioteca Digital de la Medicina Tradicional Mexicana. UNAM. Consulta Mayo 2016. http://www.medicinatradicionalmexicana.unam.mx/termino.php?l=1 and t=hormiga.

Dold H., Du D.H. and Dziao S.T. (1937). Nachweis antibakterieller hitzenden lichtempfindlicher Hemmungsstoffe (Inhibine) in Naturhonig (Blütenhoning). Z. Hyg. Infektkrank. 120: 155-167.

Espina P.D. and Ordetx G.S. (1984). "Apicultura Tropical". Tecnología de Costa Rica Editorial. 4ed, 506 p.

Estrada H., Gamboa M., Chaves C. and Arias M. (2004). Evaluación de la actividad antimicrobiana de la miel de abeja contra *Staphylococcus aureus, Staphylococcus epidermidis, Pseudomonas aeruginosa, Escherichia coli, Salmonella enteritidis, Listeria monocytogenes* y *Aspergillus niger*. ALAN. 55 (2):167-171.

Haydack M.H. and Vivino E. (1950).The changes in the tiamine, rivoflavin, niacin ad pantothenic acid content in the food of female honey bees during growth with a note vitamin activity of royal jelly and bee bread. Anales de la Sociedad Entomológica de América. 43 (3): 361-367.

Heinemann T., Leiss O. and Von Bergmann K. (1986). Effect of low-dose of sitostanol on serum cholesterol in patients with hypercholesterolemia.Atherosclerosis. 61:219-223.

Jiménez E.M., Saad V.I., Sánchez C.A, Reyes C.R. (2006) Estudiosobre monoterpenos e iridoides. Ed. Romo de Vivar R.A. En Química de la flora mexicana. Investigaciones en el Instituto de Química. UNAM 227 p.

Khoo Y.T., Halim A.S., Singh K.K. and Mohamad N.A. (2010). Wound contraction effects and antibacterial properties of Tualang honey on full-thickness burn wounds in rats in comparison to hydrofiber. BMC Complementary Alternative Medicine. 10:48.

Ling W.H. and Jones P.J. (1995). Dietary Phytosterols: A review of metabolism, benefits and side effects. Life Sciences. 57:195-206.

López F.A. and Macaya C. (2006) Efectos antitrombóticos y antiinflamatorios de los ácidos grasos omega 3. Revista Española de Cardiología 6 (Supl. D): 31-37.

López C.N., Miguel M. and Aleixandre A. (2012) Propiedades beneficiosas de los terpenos iridoides sobre la salud. Nutrición Clínica y Dietética Hospitalaria 32 (3): 81-91.

Lorente C. (2003). Fotofísica y propiedades fotosensibilizadoras de pterinas en solución acuosa. Tesis de doctorado. Facultad de Ciencias Exactas. UNLP.

Lu J., Carter D. A., Turnbull L., Rosendale D., Hedderley D., Stephens J., Gannabathula, S., Steinhorn G., Schlothauer R. C., Whitchurch C. B. and Harry E. J. (2013). The effect of New Zealand kanuka, manuka and clover honeys on bacterial growth dynamics and cellular morphology varies according to the species. PLoS ONE, 8(2):e55898.

Mariod A., Matthaus B. and Ibrahim A.S. (2011). Fatty acid, tocopherols of Aspongubus viduatus (melón bug) oil during different maturity stages". International Journal of Natural Product and Pharmaceutical Sciences. 2 (1): 20-27.

Martí A., Moreno M.J and Martínez J.A. 2003. Alimentos probióticos, prebióticos y simbióticos. 61-77 pp. In: Astiasarán A.I., Lasheras A. B., Ariño P. A. H. (Eds). Alimentos y Nutrición en la Práctica Sanitaria. Diaz de Santos. Madrid, España.

Martínez F.S., González G.J., Culebras J.M and Tuñón M.J. (2002). Los flavonoides: Propiedades y acciones antioxidantes. Nutrición Hospitalaria. 17 (6): 271-278.

Martínez C.A. and Rivas A.S. (2005). Funciones de las prostaglandinas en el sistema nervioso central. Revista de la Facultad de Medicina. UNAM. 48 (5):210-216.

Mataix V. J. (2005). Nutrición para educadores. Ediciones Díaz de Santos, S.A. Madrid, España. 72p.

Mataix V. J. and Gil A. (2004). Libro blanco de los Omegas-3: Los ácidos grasos poliinsaturados Omega-3 y monoinsaturados tipo oleico y su papel en la salud. Editorial Médica Panamericana. Madrid, España. 153 p.

Mendieta C.R.J. (2002). Comparación de la composición química de la miel de tres especies de abejas (*Apis mellifera, Tetragonisca angustula* y *Melipona beecheii*) de El Paraíso, Honduras". Trabajo de graduación presentado como requisito parcial para el título de Ingeniero en Agroidustria.

Mendoza D.L. and Saavedra A.S. (2013). Chemical composition and anti-irritant capacity of whole body extracts of *Ulomoides dermestoides* (Coleoptera, Tenebrionidae). Revista de la Facultad de Química Farmacéutica. Colombia. 20 (1): 41

Miale J.B. (1985). Hematología: Medicina de laboratorio. Editorial Reverté S.A. Barcelona, España.

Molan P. (2001). Why honey is effective as a medicine. 2. The scientific explanation of its effects. Bee World. 82(1): 2240.

Molan P. (2009).Debridement of wounds with honey.Journal of Wound Technology. 5: 12-17.

Molan, P. (2011).The evidence and the rationale for the use of honey as a wound dressing. Wound Practice and Research. 19 (4): 204-220.

Molan P. and Rhodes, T. (2015). Honey: A biologic wound dressing. Wounds. 27(6):141-51.

Mustafa N.E.M., Mariod A.A. and Matthaus B. 2008.Antibacterial activity of *Aspongopus viduatus* (melon bug) oil.Journal of Food Safety. 28: 577-586.

Nagai, T., Sakai M., Inoue R., Inoue H., and Suzuki N. (2001).Antioxidative activities of some commercially honeys, royal jelly, and propolis. Food Chemistry.75: 237-240.

Nakano K., Chigira T., Miyafusa T., Nagatoishi S., Caaveiro M.M.J. and Tsumoto K. (2015). Discovery and characterization of natural tropolones as inhibitors of the antibacterial target Cap F from *Staphylococcus aureus*. Scientific Reports. 5(http://www.nature.com/articles/srep15337)

National Honey Board (2003). Honey Health and Therapeutic Qualities Associated Marketing, Chicago II: 1-23.

Neville M. and Picciano M.F. (1997). Regulation of milk lipid secretion and composition. Ann. Rev. Nutr.17:159-184.

Nisbet H.O., Nisbet C., Yarim M., Guler A. and Ozak A. (2010).Effects of three types of honey on cutaneous wound healing.Wounds. 22 (11): 275-283.

Ntwasa M., Goto A. and Kurata S. 2012. Coleopteran antimicrobial peptides: Prospects for Clinical applications. International Journal of Microbiology. 2012: 1-8.

Paoletti G.M, Norberto L., Damini R. and Musumeci S. (2007). Human Gastric Juice Contains Chitinase that can degrade chitin. Annals of nutrition and Metabolism. 51: 244-251.

Pardo G. A. 2005. Descubra el poder de la miel. Colección natural. Grupo Imaginador de Ediciones. Argentina, Buenos Aires. 122 pp.

Peña C.R. 2008. Estandarización en propóleos: antecedentes químicos y biológicos. Ciencia e Investigación Agraria. 35 (I):17-26.

Pino M.J.M, Ángeles C.S., García A. (2009). Substancias curativas encontradas en insectos nutracéuticos y medicinales. Entomología Mexicana. 8: 256-261.

Ramos-Elorduy, J. (2000). La etnoentomología actual en México en la alimentación humana, en la medicina tradicional y en el reciclaje y alimentación animal. Mem. XXXV Cong. Nac. de Ent. 3-46.

Reina P.E.T. (2010). Producción y análisis financiero de la obtención de jalea real de abejas (*Apis mellifera*) por el método doolittle. Escuela Politécnica Nacional. Facultad de Ingeniería Química y Agroindustria. Proyecto previo a la obtención del título de ingeniera agroindustrial.

Ríos Y.J.M., Mercadillo P.P., Yuil de Ríos E. and Ríos C.M. 2013. "Terapia con larvas de mosca para heridas crónicas: alternativa de una época de creciente resistencia a los antimicrobianos. Dermatología Cosmética, Médica y Quirúrgica. 11 (2): 134-141.

Romero P.J. and Vázquez T.E.M. (2012). Fitoesteroles y Fitoestanoles: eficaces para la disminución de lípidos plasmáticos. Revista CES Salud Pública. 3 (2): 165-173.

Rozaini M.Z., Zuki, A.B.Z., Noordin M., Norimah Y. and Hakim A.N. (2004). The effects of different types of honey on tensile strength evaluation of burn wound tissue healing. International Journal of Applied Research in Veterinary Medicine. 2: 290-296.

Rumpold B.A. and Schlüter O.K. (2013). Nutritional composition and safety aspects of edible insects.Molecular Nutrition and Food Research. 57: 802-823.

Schencke C., Vásquez B., Sandoval C. and Del Sol M. (2016). El rol de la miel en los procesos morfofisiológicos de la reparación de heridas. International Journal Morphology. 34 (1): 385-395.

Silistina N. and Jatin S. (2015).Proximate composition of wild edible insects consumed by the Bodo tribe of Assam, India.International Journal of Bioassays. 4 (07): 4050-4054.

Sherlock O., Dolan A., Athman R., Power A., Gethin G., Cowman S. and Humphreys H. (2010). Comparison of the antimicrobial activity of Ulmo honey from Chile and Manuka honey against methicillin-resistant Staphylococcus aureus, *Escherichia coli* and *Pseudomonas aeruginosa*.BMC Complementary and Alternative Medicine. 10 (1): 47.

Téllez A.G. and Castaño C.J. (2010).Péptidos antimicrobianos. Infectio. 14 (1): 55-67.

Van der Horst D.J. (1985). Insect lipids and lipoproteins, and their role in physiological processes. Progress in Lipid Research. 24: 19-67.

Van Huis A., Van Itterbeeck J., Klunder H., Mertens E., Hollaran A., Muir G.M. and Vantomme P. (2013). Edible insects: future prospects for food and feed security. FAO Forestry paper 171. 182 p.

Vardi A., Barzilay Z., Linder N., Cohen H.A. and Garet G. (1998). Local application of honey for the treatment of neonatal post-operative wound infection. Acta Paediatrica 87 (4): 429-432.

Venezuela A. and Ronco M. A.M. (2004). Fitoesteroles y fitoestanoles: Aliados naturales para la protección de la salud cardiovascular. Revista Chilena de Nutrición. 21 (1): 161-169.

Villarruel R., Huizar R., Corrales M., Sánchez T. and Islas A. (2004). Péptidos naturales antimicrobianos: escudo esencial de la respuesta inmune. Investigación en Salud. VI (3): 170-179.

Zamora L.G. and Arias L.M. (2011). Calidad microbiológica y actividad antimicrobiana de la miel de abeja sin aguijón. Revista Biomédica. 22 (2):59-66.

– Segment 6 –

Insects as Food for Human and Livestock

Chapter 9

Edible Insects in some Latin American Countries: Advantages and Prospects

☆ *Pino Moreno José Manuel, Reyes Prado Humberto and García Flores Alejandro*

ABSTRACT

In many countries, the consumption of insects as food dates back to ancient times. In Latin America, anthropo-entomophagy has been primarily restricted in ethnic groups despite the introduction of other eating habits; and this food resource is being used in specific ways. A significant number of insect species provide an unlimited source of animal protein to the Latin American ethnic groups, which is lacking in food staples of the so called economically higher class. Additionally, most of these species can be obtained directly from their natural environments. In countries such as Argentina, Bolivia, Brazil, Colombia, Ecuador, Paraguay, Peru, and Venezuela, approximately 100 genera of insects are consumed as food; this number is comparatively less than half of the genera reported from Mexico, where 240 genera and over 500 species of edible insects have been recorded. In Latin America, the various species of edible insects constitute a potential and actual natural renewable food resource with a high degree of acceptance. It could be affirmed that records, awareness, and consumption of insects will increase as further research is performed on this eating habit, and in the near future insects will play an important role in nutritionally enriching humans, the animals in our diet, and even our pets.

Introduction

Anthropo-entomophagy is the consumption of insects or their products by man and is an ancient custom. Despite the abundant insect resources that ecosystems provide to society, malnutrition continues to prevail in Latin America and the Caribbean. Malnutrition affects people's health and physical integrity and impacts school performance. The latter consequence causes underachievement and early school dropouts that subsequently affect the future employment of individuals and condemn new generations to perpetual poverty. Malnutrition is also among the leading causes of child mortality (UNICEF, 2016).

Socioeconomic differences are conspicuous in Latin America, and problems of marginalization, lack of access to mass media, and educational deficiencies are faced daily. These phenomena cause problems, such as nutritional misinformation and the blending of different cultures and traditions that lead to poor nutrition. The evident problems of hunger and malnutrition in Latin America have been noted by several authors and institutions and have been identified as priority issues. Malnutrition is the most important factor contributing to child mortality in developing countries (Guardiola and Gónzalez-Gómez, 2010).

A low nutrient intake has long been one of the main reasons that native people use numerous insect species, which provide nutrients that are lacking in basic foods (Bodenheimer, 1951; Meyer, 2010).

Mexico is one of the Latin American countries where most of the research on anthropo-entomophagy has been performed. Studies have reported taxonomic aspects (Ramos-Elorduy *et al.*, 2008), biogeographic aspects (Ramos-Elorduy and Pino, 1992), world biodiversity (Ramos-Elorduy and Conconi, 1994), sustainability (Ramos-Elorduy, 1997a), importance in the diet of rural communities (Ramos-Elorduy, 1997b), and nutritional value (Ramos and Pino, 2001; Ramos-Elorduy *et al.*, 2012). Insects directly and/or indirectly are an excellent nutritional option for human because they are also a part of the animal food webs in various land and water environments (DeFoliart, 1999; Makkar *et al.*, 2014).

Edible Insects and Ethnicity

Over 100 million Latin Americans suffer from hunger or malnutrition, and the figures are even higher in Africa and Asia. Between 1500 and 2000 insect species are consumed globally by approximately 3,000 ethnic groups in 113 countries (Yen, 2009), including some Latin American countries, such as Argentina, Bolivia, Brazil, Venezuela, and Mexico. Most insects consumed belong to the orders Coleoptera, Orthoptera, Blattodea, Hymenoptera, and Diptera (Makkar *et al.*, 2014; van Huis, 2016).

In this context, we considered it pertinent to conduct a retrospective taxonomic analysis of the edible insects in Latin America with an emphasis on Mexico. We report the advantages and prospects of this food resource in the following sections.

Materials and Methods

Various sources were consulted for the present investigation, including Paoletti (2005) and various publications concerning edible insects in Mexico produced under the direction of Dr. Julieta Ramos-Elorduy Blázquez.

Results and Discussion

Table 9.1 presents some edible insects from several Latin American countries, including the Order, Family, scientific name, and ethnicities that consume insects.

Table 9.1: Taxonomic Relationship of Edible Insectsin Latin America

Order	*Family*	*Scientific name*	*Ethnic Group*
		ARGENTINA	
Hymenoptera	Vespidae	*Brachygastra lecheguana* L.	Lechiguana, qa'tek, nowa.'lhek
		Polistes versicolor O.	kotag'ma, chyos'tax, kyos'tax
		Polistes cavapyta S.	Bala-puka, Ko:'yeta Gañima# asi´tax
		Polybia ignobilis H.	Karán negro, ye#e´lax, ñi#e 'hala
		Polybia ruficeps S.	bala, waGa´to, wu#na
		Polybia sericea O.	karán, pe´gela tik#lha
		Agelaia multipicta H.	we' naq, ma 'sa
	Apidae	*Apis mellifera* L.	qona'yaq, 'poleo, naku:'tax
		Scaptotrigona jujuyensis S.	Yana, ' ñie#e, ma#age, pi:'ni, ko:'yik
		Tetragonisca angustula fiebrigi S.	Señorita, ha'ma, wu:'sa
		Lestrimelitta limao S.	penaGa'di, pem'cax
		Melipona favosa orbignyi G.	moro moro Qo# na'yaq, Na:'kwu
		Plebeia catamarcensis H.	mestizo, supfwe'tax
		Plebeia molesta P.	pinoGo'daq, chiyoGo'hek
		Plebeia sp.	Pusquilla, se:h'mat, ce:'mat, qonole 'ci
		Xylocopa ordinaria S.	'ho 'poleo, qaca:wu'tax
Data from Arenas, 2003			
		BOLIVIA	
Isoptera	Termitidae	*Nasusitermes corniger* M.	Lecos
Hymenoptera	Formicidae	*Atta* sp.	Lecos
Data from Paoletti y Dufour (2005)			
		BRAZIL	
Isoptera	Termitidae	*Labiotermes labralis* H.	Kayapo
		Syntermes sp.	Maku, Cobia
Coleoptera	Bruchidae	*Caryobruchus* sp.	Surui
		Pachymerus cardo F.	Surui
		P. nucleorum F.	Timbira
	Buprestidae	*Euchroma gigantea* L.	Kayapo
	Curculionidae	*Rhinostomus barbirostris* F.	Surui
		Rhynchohphorus palmarum L.	Bororo, Caingua, Maku, Surui
	Scarabaeidae	*Megasoma acteon* L.	Kayapo
		Strateagus sp.	Kayapo
Hymenoptera	Formicidae	*Atta cephalotes* L.	Kayapo
		A. sexdens L.	Kayapo
		Atta sp.	Kayapo, Surui

Order	*Family*	*Scientific name*	*Ethnic Group*
	Meliponiidae	*Melipona grandis* G.	Surui
		Melipona schwarzi M.	Surui
		Nannotrigona bipunctata polystica M.	Surui
		Nannotrigona sp.	Surui
		Nannotrigona xanthotricha M.	Surui
		Plebeia sp.	Surui
		Ptilotrigona lurida S.	Surui
		Tetragonisca angustula L.	Surui
		Tetragonisca branneri C.	Surui
		T. dorsalis S.	Surui
Data from Posey, (1979, 1987); Milton, (1984); Coimbra, (1984); Coimbra, (1985)			
		COLOMBIA	
Orthoptera	Tettigoniidae	*Neoconocephalus* sp.	Tukanoans
Isoptera	Termitidae	*Labiotermes labralis* H.	Tukanoans
		Syntermes aculeosus E.	Tukanoans
		S. spinosus E.	Tukanoans
		S. tanygnathus C.	Tukanoans
Anoplura	Pediculidae	*Pediculus* sp.	Maku
Hemiptera-Homoptera	Membracidae	*Umbonia spinosa* F.	Tukanoans
Coleoptera	Buprestidae	*Euchroma gigantea* L.	Tukanoans
	Cerambycidae	*Acrocinus longimanus* L.	Tukanoans
	Curculionidae	*Rhynchophorus palmarum* L.	Maku, Tukanoans
	Passalidae	*Veturius sinuosus* D.	Tukanoans
	Scarabaeidae	*Dynastes hercules* L.	Tukanoans
Lepidoptera	Hesperidae	*Hesperidae* sp.	Tukanoans
	Mimallonidae	*Outunima* sp.	Tukanoans
	Noctuidae	*Batya* sp.	Tukanoans
	Saturniidae	*Automeris* sp.	Tukanoans
		Diorphia sp.	Tukanoans
Hymenoptera	Formicidae	*Atta cephalotes* L.	Tukanoans
		Atta laevigata S.	Tukanoans
		Atta sexdens L.	Tukanoans
	Meliponidae	*?*	Tukanoans
	Vespidae	*Apoica toracica* R. de B.	Tukanoans
		Polybia rejecta F.	Tukanoans
		Stelopolybia angulata F.	Tukanoans
Data from Milton, 1984; Dufour, 1987			

Order	*Family*	*Scientific name*	*Ethnic Group*
		ECUADOR	
Odonata	Aeschnidae	*Aeschna brevifrons* H.	Quichuas
		A. marchali R.	Quichuas
		A. peralta R.	Quichuas
		Coryphaeschna adnexa H.	Otavalos, Quichuas
Orthoptera	Acrididae	*Schistocerca* sp.	Quichuas
Hemiptera-Homoptera	Membracidae	*Umbonia spinosa* F.	Quichuas
Coleoptera	Cerambycidae	*Oncideres* sp.	Awa, Saraguros
		Psalidognathus atys W.	Saraguros
		P. cacicus W.	Saraguros
		P. erithrocerus R.	?
		P. modestus T.	?
		Macrodontia cervicornis L.	?
	Curculionidae	*Cosmopolites sordida* G.	Ashuaras, Shuaras
		Dynamis nitidula G.	Ashuaras, Cofanes, Huaoranis, Quichuas, Secoyas, Shuaras
		Dynamis perryi W.	Ashuaras, Cofanes, Huaoranis, Quichuas, Secoyas, Shuaras.
		Metamasius cinnamominus P.	Ashuara, Shuaras
		M. dimidiatipennis J.	Ashuaras, Shuaras
		M. hemipterus L.	Ashuaras, Shuaras
		M. sericeus O.	Ashuaras, Shuaras
		Rhinostomus barbirostris F.	Ashuaras, Awa, Cofanes, Huaoranis, Negreg. Esmerald, Quichuas, Shuaras, Sionas
		Dynamis nitidula G.	?
		D. perryi W.	?
		Rhynchophorus palmarum L.	Ashuaras, Awa, Cofanes, Huaoranis, Negreg Esmerald, Quichuas, Secoyas, Shuaras, Sionas, Tsachilas
	Lucanidae	*Sphaenognathus feisthameli* G-M.	Quichuas, Saraguros
		S. lindenii M.	Quichuas
		S. metallifer B. & L.	Canaris
	Scarabaeidae	*Ancognatha castanea* E.	Canaris, Otavalos, Pilahuines, Quichuas, Salazacas, Saraguros
		A. jamesoni M.	Quichuas
		A. vulgaris A.	Canaris, Otavalos, Quichuas
		Clavipalpus antisanae B.	Quichuas

Order	Family	Scientific name	Ethnic Group
		Coelosis biloba L.	Awa, Tsachilas
		Democrates burmeisteri R.	Salazacas, Quichuas
		Dynastes hercules L.	Huaoranis
		Golopha aeacus B.	Otavalos, Quichuas
		G. aegeon D.	Otavalos, Quichuas,
		Heterogomphus bourcieri G.	Quichuas
		Leucopelaea albescens B.	Otavalos, Quichuas, Salazacas
		Pelidnota nigricauda B.	Quichuas
		Platycoelia forcipalis O.	?
		P. parva K.	?
		P. rufosignata O.	?
		P. Lutescens B.	?
		Praogolofa unicolor B.	Otavalos, Pilahuines, Quichuas, Saraguros
Lepidoptera	Brassolidae	*Brassolis astyra* G. and S.	Huaoranis, Quichuas
		B. sophorae L.	Quichuas
	Castnidae	*Castnia daedalus* C.	Quichuas
		C. licoides B.	Huaoranis, Quichuas
		C. licus D.	Huaoranis, Quichuas
		Eupalamides cyparissias F.	?
	Hepialidae	*Hepialus* sp.	Negreg. Esmerald, Quichuas
	Nymphalidae	*Panacea prola* D.	Napo district
Hymenoptera	Apidae	*Apis mellifera* L.	Negr. Esmerald
		Bombus atratus F.	Canaris, Otavalos
		B. ecuadorius M.	?
		B. funebris S.	Canaris, Otavalos
		B. robustus S.	?
	Meliponiidae	*Tetragonisca angustula* L.	Canaris
	Formicidae	*Atta cephalotes* L.	Ashuaras, Awa, Cofanes, Huaoranis, Negr.es, Quichuas, Secoyas, Shuaras, Iona smerald, Quichuas, Secoyas, Sionas
		Atta sexdens L.	Ashuaras, Awa, Cofanes, Huaoranis, Negr. esmerald
	Megachilidae	*Megachile* sp.	Otavalos, Pilahuines, Quichuas, Salazacas, Saraguros
	Vespidae	*Angiopolybia paraensis* S.	Ashuaras, Shuaras
		Apoica pallens F.	Ashuaras, Shuaras
		Apoica pallida O.	Ashuaras, Shuaras
		A. strigata R.	Ashuaras, Shuaras
		A. toracica R. de B.	Ashuaras, Shuaras

Order	Family	Scientific name	Ethnic Group
		Brachygastra lecheguana L.	Colonos de la Costa
		Brachymenes wagnerianus S.	Ashuaras, Shuaras
		Mischocyttarus rotundiocollis C.	?
		M. tomentosus Z.	?
		Montezumia dimidiata S.	Ashuaras, Shuaras
		Polistes bicolor L.	Ashuaras, Shuaras
		P. deceptor S.	Ashuaras, Shuaras
		P. occipitalis D.	Ashuaras, Shuaras
		P. testaceicolor B.	?
		P. testaceicornis B.	Ashuaras
		Polybia aequatoriales Z.	Shuaras
		Polybia dimidiata F.	Shuaras
		P. emaciata L.	Shuaras
		P. flavifrons S.	Shuaras
		P. testeicornis B.	Shuaras
		Stelopolybia baezae R.	Canaris
		S. corneliana R.	?
		S. lobipleura R.	?
		S. ornata D.	?
		S. virginea F.	?
Data from Onore, 1997; Onore, 2004			
PARAGUAY			
Coleoptera	Cerambycidae	*Macrodantia cervicornis* L.	Ache
	Curculionidae	*Rhynchophorus palmarum* L.	Ache, Guayaki
	Passalidae	*Passalus interruptus* L.	
Hymenoptera	Apidae	*Apis mellifera* L.	Ache
Data from Clastres, 1972; Hawkes *et al.*, 1982; Hurtado *et al.*, 1985; Hill *et al.*, 1987			
PERU			
Megaloptera	Corydalidae	*Corydalus armatus* H.	Cordillera
		C. peruvianus K. Davis	Cordillera
Coleoptera	Bruchidae	*Pachymerus nucleorum* F.	Mashco.-Piro
Hymenoptera	Formicidae	*Atta* sp.	Campa, Gr. Pajonal
Data from Denevan, 1971; Hill and Kaplan, 1983; Menzel and D´ Alusio, 1998			
VENEZUELA			
Odonata	Gomphidae	*Zonophora* sp.	Ye kuana
	Libellulidae	*Dasythemis* sp.	Yekuana
Orthoptera	Acrididae	*Aidemona azteca* S.	Yukpa
		Orphulella sp.	Yukpa

Order	*Family*	*Scientific name*	*Ethnic Group*
		Osmilia flavolineata De G.	Yukpa
		Osmilia sp.	Yukpa
		Schistocerca sp.	Yukpa
		Tropidacris latreillei P.	Yukpa
		T. cristata L.	Guajibo
		Rhammatocerus sp.	Guajibo
	Tettigoniidae	*Conocephalus angustifrons* R.	Yukpa
Isoptera	Termitidae	*Labiotermes* sp. *cf- labralis* H.	Yanomamo
		Nasusitermes sp.	Yanomamo
		N. ephrateae H.	Yanomamo
		Syntermes sp.	Guajibo, Piaroa, Cuao river, Yanomamo
		S. tterritus E.	Yanomamo
		S. aculeosus E.	Yekuana
Anoplura	Pediculidae	*Pediculus* sp.	Yanomamo, Yukpa
Hemiptera-Homoptera	Belostomatidae	*Belostoma* sp.	Ye´kuana
	Naucoridae	*Ambrysus* sp.	Ye´kuana
	Cicadidae	?	Piaroa, Cuao River
Megaloptera	Corydalidae	*Corydalus* sp.	Yukpa
Coleoptera	Bruchidae	*Caryobruchus* sp.	Yukpa
	Curculionidae	*Anthonomus* sp.	Yukpa
		Dynames borasi F.	Guajibo, Piaroa
		Metamasius cinnamominus P.	Piaroa
		Rhinostomus bartbirostris F.	Guajibo, Hoti, Piaroa
		Rhynchophorus palmarum L.	Baniwa, Bari, Hoti, Piaroa, Yanomamo, Ye´kuana
	Scarabaeidae	*Pelidnota* sp.	Yanomamo
		Podischnus agenor O.	Yukpa
Trichoptera	Hydropsychidae	*Leptonema* sp.	Yukpa
Lepidoptera	Noctuidae	*Laphygma frugiperda* S.	Yukpa
		Mocis repanda F.	Yukpa
	Sphingidae	*Erinnyis ello*	Yanomamo
Diptera	Stratiomyidae	*Chryschlorina* sp.	Yukpa, Yanomamo
Hymenoptera	Eumenidae	*Eumenes canaliculata*	Yukpa
	Formicidae	*Atta cephalotes* L.	Ye, kuana
		Atta sp.	Guajibo, Piaroa, Cuao river, Yanomamo, Ye kuana

Order	*Family*	*Scientific name*	*Ethnic Group*
	Meliponidae	*45 e.n.*	Curripaco, Guajibo, Piaroa, Yanomamo
		Frieseomelitta varia L.	Yanomamo
		Partamona testaceae K.	Yanomamo
	Meliponidae	*Scaprotrigona xanthotricha*	Yanomamo
		Trigona (Tetragona) clavipes F.	Yukpa
		Trigona amulthea O.	Yukpa
	Vespidae	*15 e.n.*	Curripaco, Guajibo, Piaroa, Cuao river, Yanomamo
		5 e.n	Piaroa, Cuao river
		Copina sp.	Yanomamo, Ye kuana
	Vespidae	*Mamisiciacorin* sp.	Yanomamo
		Mischocyttarus sp.	Yukpa
		Polistes canadensis L.	Yukpa
		Polistes pacificus F.	Yukpa
		Polistes versicolor O.	Yukpa
		Polybia ignobilis H.	Yukpa
Data from Ruddle, 1973; Smole, 1976; Beckerman, 1977; Zent, 1992; Govoni, 1998; Paoletti and Bukkens, 1997; Onore, 1997;Paoletti and Duffour, 2005, e.n= Ethnic names.			

In the case of Mexico (Table 9.2), the number of species in each genus is also indicated.

Table 9.2: Registered Edible Insects of Mexico

Order	*Family*	*Genus*	*Number of Species*
Ephemeroptera	Ephemeridae	*Ephemera*	1
	Leptophlebiidae	*Thraulodes*	2
	Baetidae	*Baetis*	1
Odonata	Aeschnidae	*Aeschna*	2
		Anax	1
	Libellulidae	*Erytrodiplax*	1
	Coenagrionidae	*Enallagma*	1
		Ischnura	1
Orthoptera	Acrididae	*Orphula*	1
		Schistocerca	4
		Boopedon	5
		Silvittetix	1
		Opeia	2
		Orphulella	4
		Rhamatocerus	2
		Aidemona	1

Order	*Family*	*Genus*	*Number of Species*
		Pedies	1
		Melanoplus	7
		Arphia	3
		Encoptolophus	1
		Xanthippus	1
		Locusta	1
		Spharagemon	2
		Trimerotropis	4
		Abracris (*Osmilia*)	1
		Plectrottetia	1
		Homocoriphus	1
	Romaleidae	*Chromacris*	2
		Taeniopoda	2
		Xyleus	1
	Pyrgomorphidae	*Sphenarium*	4
	Tettigoniidae	*Pyrgocorypha*	1
		Microcentrum	2
		Petaloptera	2
		Scudderia	1
		Stilpnochlora	2
		Idiarthron	1
		Liparoscelis	1
Phasmatodea	Phasmatidae	*Bacteria*	1
Blattodea	Blaberidae	*Blaberus*	2
		Epilampra	1
	Bláttidae	*Blattaria*	1
		Periplaneta	2
	Blatellidae	*Chorisoneura*	2
		Pseudomops	1
		Blatella	1
		Paratropes	1
Isoptera	Termitidae	*Macrotermes*	1
Psocoptera	Psocidae	*Methylophorus*	1
Anoplura	Pediculidae	*Pediculus*	1
Hemiptera-Heteroptera	Belostomatidae	*Abedus*	3
		Belostoma	2
		Lethocerus	2
	Coreidae	*Acantocephala*	4
		Anasa	2
		Colophoserus	1

Order	Family	Genus	Number of Species
		Mamurius	2
		Pentascelis	2
		Piezogaster	2
		Sephina	1
		Thasus	4
	Corixidae	*Buenoa*	1
		Hesperocorixa	1
		Trichocorixa	1
		Corisella	5
		Graptocorixa	2
		Krizousacorixa	3
	Lygaidae	*Neocoriphus*	1
	Notonectidae	*Notonecta*	2
	Pentatomidae	*Brochymena*	2
		Chlorocoris	4
		Edessa	13
		Euschistus	17
		Mormidea	1
		Monomorpha	1
		Nezara	2
		Banasa	1
		Pellaea	1
		Oebalus	2
		Padaeus	2
		Pharypia	1
		Proxys	2
	Rhopalidae	*Jadera*	1
	Fulgoridae	*Fulgora*	3
	Cicadidae	*Cicada*	11
		Dundubia	1
		Odopoea	2
		Proarna	2
		Tibicen	2
		Quesada	1
	Membracidae	*Aetalion*	5
		Anthiante	1
		Hoplophorion	1
		Umbonia	3
	Dactylopiidae	*Dactylopius*	4

Order	*Family*	*Genus*	*Number of Species*
Megaloptera	Corydalidae	*Corydalus*	2
Coleoptera	Gyrinidae	*Gyrinus*	2
	Haliplidae	*Haliplus*	2
		Peltodytes	2
	Noteridae	*Suphiselus*	1
	Dytiscidae	*Laccophilus*	2
		Rhantus	3
		Dytiscus	3
		Thermonectus	3
		Cybister	5
		Megadytes	2
	Carabidae	*Cicindela*	2
	Hydrophilidae	*Berosus*	1
		Dilobocelus	1
		Tropisternus	4
		Hololepta	2
	Staphylinidae	*Oxytelus*	1
	Lucanidae	*Lucanus*	2
	Passalidae	*Passalus*	5
		Paxillus	1
		Heliscus	1
		Popilius	3
		Oileus	1
		Verres	1
	Scarabaeidae	*Cyclocephala*	1
		Dynastes	2
		Enema	1
		Megasoma	2
		Strategus	3
		Xylorictes	4
		Melolontha	1
		Phyllophaga	2
		Chrysina	1
	Buprestidae	*Chalcophora*	1
		Euchroma	1
	Elateridae	*Chalcolepius*	2
		Pyrophorus	2
	Erotylidae	*Dichomorpha*	1
	Zopheridae	*Zopherus*	2
	Tenebrionidae	*Eleodes*	3

Order	*Family*	*Genus*	*Number of Species*
		Stophagus	1
		Tenebrio	2
		Zophobas	1
	Meloidae	*Meloe*	4
	Cerambycidae	*Aplagiognathus*	2
		Callipogon	1
		Derobrachus	2
		Stenodontes	5
		Trichoderes	1
		Arhopalus	2
		Cerambyx	1
		Eburia	1
		Megacyllene	1
		Acrocynus	1
		Lagocheirus	1
		Ornithia	1
		Polyrhaphis	1
		Prosopocera	1
		Cisa	1
	Chrysomelidae	*Blepharida*	2
		Leptinotarsa	1
		Lactica	1
	Curculionidae	*Rhynchophorus*	1
		Metamasius	1
		Scyphophorus	1
Trichoptera	Hydropsychidae	*Leptonema*	2
	Leptoceridae	*Oecetis*	1
	Hydrobiosidae	*Atopsyche*	2
Diptera	Syrphydae	*Campylostoma*	1
		Copestylum	2
		Eristalis	1
	Ephydridae	*Ephydra*	2
		Mosillus	1
	Muscidae	*Musca*	1
	Calliphoridae	*Macellaria*	1
	Stratiomydae	*Hermetia*	2
Lepidoptera	Hepialidae	*Phassus*	3
	Cossidae	*Comadia*	1
	Sessidae	*Synanthedon*	1
	Castniidae	*Synpalamides*	1

Order	Family	Genus	Number of Species
	Pyralidae	*Laniifera*	1
	Hesperidae	*Aegiale*	1
	Papilionidae	*Neographium*	1
		Papilio	1
	Pieridae	*Catasticta*	2
		Phoebis	1
		Eucheira	2
	Nymphalidae	*Danaus*	2
		Vanessa	2
		Chlosyne	1
		Cisa	1
		Hamadryas	1
	Geometridae	*Synopsia*	1
	Bomycidae	*Bombyx*	1
	Lasyocampidae	*Eutachyptera*	1
	Saturniidae	*Arsenura*	2
		Caio	1
		Eacles	1
		Hemileuca	1
		Paradirphia	1
		Hylesia	3
		Antherea	1
		Actias	2
	Sphingidae	*Clanis*	1
		Manduca	1
		Cocytius	1
	Noctuidae	*Ascalapha*	2
		Thysania	1
		Latebraria	1
		Guerra	1
		Helicoverpa	1
		Spodoptera	3
Hymenoptera	Dioprinidae	*Neodiprion*	1
		Zadiprion	1
	Formicidae	*Eciton*	1
		Atta	3
		Acromyrmex	2
		Pogonomyrmex	2
		Azteca	2

Order	Family	Genus	Number of Species
		Dolichoderus	1
		Liometopum	3
		Tapinoma	2
		Camponotus	2
		Myrmecosistus	4
	Vespidae	*Apoica*	1
		Brachygastra	4
		Parachartegus	3
		Polybia	6
		Synoeca	1
		Mischocyttarus	4
		Polistes	7
		Vespula	5
		Anmophila	1
	Apidae	*Apis*	4
		Eulaema	1
		Cephalotrigona	2
		Lestrimelitta	3
		Melipona	8
		Nannotrigona	2
		Partamona	3
		Plebeia	2
		Scaptotrigona	4
		Trigona	11
		Trigonisca	6
		Bombus	7

Data from Ramos-Eloduy *et al.* (2008).

According to Table 9.1, eleven insect orders have been reported (Anoplura, Coleoptera, Diptera, Hemiptera-Homoptera, Hymenoptera, Isoptera, Lepidoptera, Megaloptera, Odonata, Orthoptera and Trichoptera), with 35 families and 93 genera.

Table 9.2 summarizes the edible insects recorded to date in Mexico, indicating the order, family, genus, and number of species. Overall a total of 15 orders, 75 families, 240 genera, and 507 species that are consumed by different ethnic groups (*i.e.*, Amuchas, Amuzgo, Cora, Cuicateco, Chinanteco, Chochos, Chol, Chontal, Huasteco, Huave, Huichol, Kehia, Jonaz, Lacandon, Mame, Matlazincas, Maya, Mazahua, Mazatec, Motozintleco, Nahuatl, Ñähñus, Ocuilteco, Otomi, Otopamean, Popolaca, Tarahumara, Tarasco, Tepehua, Tepehuano, Tlapaneco, Tojolabal, Totonaco, Triques, Tzeltal, Tzotzil, Yutoazteca, Zapotec, and Zoque). These species are distributed in different states of the Mexican Republic, including Campeche,

Chiapas, Chihuahua, Federal District, Durango, State of Mexico, Guerrero, Hidalgo, Jalisco, Michoacán, Morelos, Nayarit, Nuevo Leon, Oaxaca, Puebla, Queretaro, Quintana Roo, San Luis Potosi, Tabasco, Tamaulipas, Tlaxcala, Veracruz, Yucatan, and Zacatecas.

Preparation, Consumption, and Storage of Edible Insects

Indigenous and rural communities consume insects more extensively and in larger varieties; nevertheless, it is the ethnic groups that have preserved anthropo-entomophagy in Latin America. As an example we can refer to Ramos-Elorduy and Pino (1989), who conducted an investigation regarding anthropo-entomophagy among indigenous people and some other groups practicing anthropo-entomophagy in several states of Mexico.

The consumption of edible insects occurs to varying degrees in rural areas, where it constitutes a complementary and daily part of the diet of the population and in some cases the local economy because the insects are sold in commerce. Anthropo-entomophagy is more common in rural areas with nuclei of small farmer populations due to various factors, including conditions of poverty, the ease of obtaining insects in the same environment, knowledge of their habits, distribution and seasonality, and the weather. In Brazil, insects are an important food source for 39 indigenous groups and urban communities. Some authors have reported that approximately 500 species of insects are distributed in more than 260 genera and 70 families are largely used in Latin America (Costa and Ramos-Elorduy, 2006).

Edible insects are consumed in several stages of development (*e.g.*, eggs and/or larvae, larvae and pupae, or adults for different species). This phenomenon is related to the life cycles and geographical distributions of the species and depends on both the abiotic and biotic conditions and whether a species is terrestrial or aquatic. In the latter case, water bodies can be stable sources, such as lakes, artificial ponds, and springs, or vary according to the season (*i.e.*, dams and rivers) (Ramos-Elorduy and Pino, 1989).

Various species are collected and dry-stored for later consumption, especially during months of scarcity, such as ahuahutle, axayacatl, jumiles, madroño worms, chicatana ants, and escamoles. Unfortunately, insect consumption is often seen as an "unhealthy" habit that is associated with a high level of primitivism. However, anthropo-entomophagy dates back to ancient times in cultures that exploited the environment efficiently and rationally and then skilfully integrated insects into a varied diet. Today, these eating habits have been displaced by other food sources and by ethnocentrism, which has disrupted the process of domestication of new species (van Huis, 2016).

Nutritional Benefits of Edible Insects

Edible insects present several advantages as a food source, such as being a numerically dominant group of animals with high reproductive potential, short life cycles, various food habits, high food conversion efficiency, wide geographical distributions, and great adaptability to all types of habitats. Insects also have been demonstrated to have high nutritional value. Therefore, their systematic and rational use currently provides a food source with high biological value to the people in Africa, Asia, and America (Ramos-Elorduy and Conconi, 1994; Ramos-Elorduy and Pino, 1989: Ramos-Elorduy, 2004; Yen, 2009; van Huis, *et al.*, 2016.).

Insects are an unlimited source of animal protein that is largely wasted. This resource would ensure sufficient input for acceptable nutrition. To achieve this state, it is important to implement mass rearing or incentivize feeding and breeding of animals using insects as a significant ingredient in feed. Studies conducted on the quantity and quality of protein, fatty acids, minerals, and vitamins in insects show that they have a high nutritional value and that, if exploited systematically, constitute a food source that fulfils two crucial characteristics: they are sufficiently numerous and acceptably edible. When one realizes that insects are our main competitors for food, their importance becomes obvious. According to some authors, insects consume approximately one-third of our food during the growing season and while in storage (Rumpold and Schluter, 2013; Payne *et al.*, 2016; Nowak *et al.*, 2016).

In Latin America, the awareness and consumption of edible insects will increase as further studies are performed. Most of these known species are collected directly from their natural surroundings. However, available data on the quantities of insects consumed are scarce. According to Rumpold and Schluter (2013) and Nowak *et al.* (2016), the most consumed types of insects are beetles (Coleoptera) (31 per cent), caterpillars (Lepidoptera) (18 per cent), bees, wasps, ants (Hymenoptera) (14 per cent), grasshoppers, locusts, crickets (Orthoptera) (13 per cent), cicadas, planthoppers and leafhoppers, scale insects and bugs (Hemiptera) (10 per cent), termites (Isoptera) (3 per cent), dragonflies (Odonata) (3 per cent), flies (Diptera) (2 per cent) and other orders (5 per cent).

Sustainability of the Insect Resource as Food

The following exploration and management considerations should be taken into account when considering wild insects to be used as food and for the protection of insect populations in their natural environments: observe the diet and lifestyles of local people in the management and conservation of natural insect habitats, allow sustainable harvesting of edible insects by the local population in protected areas, regulate the use of pesticides to prevent the bioaccumulation of contaminants in the food chain, develop methods to control harvest levels and prevent populations of beneficial insects from becoming endangered, integrate systems as much as possible for the total or partial domestication of insects to supplement the insects captured by collection in the wild and to provide a continuous supply when wild populations fluctuate due to seasonal factors, and prevent the release of non-native domesticated insect species into natural environments (Rumpold and Schluter, 2013; van Huis *et al.*, 2013; Looy *et al.*, 2014).

Conclusion

We have shown that many cultural groups in Latin America use insects in their diet and that the taxonomy of the organisms consumed is likewise diverse. For these reasons, insects have been the food of the past, the present and certainly will play an important role in the future in human nutrition, the nutrition of meat resources such as cows, sheep, poultry, and fish and the nutrition of companion animals. Therefore, we must increase studies on their biology and rearing under controlled conditions and monitor the various microbiological and safety aspects of these traditional foods.

References

Arenas P. 2003 Etnografía y alimentación entre los Toba-Nchilamole#ek y Michilhuku´tas del chaco central Argentina Ed. Consejo Nacional de Investigación Científica y Técnica CONICET, Centro de Estudios Farmacológicos y Botánicos CEFIBO y Instituto de Botánica Darwinion IBOD. 1ª Ed. 562 p.

Beckerman, S., 1977.The use of the palm by the Bari Indians of the Maracaibo Basin, Principes.22,143-154.

Bodenheimer, F. S., 1951., Insects as Human Food: A Chapter of the Ecology of Man. Dordrecht: Springer Netherlands. pp. 7-38.

Clastres, P., 1972. The Guayaki, in: Bicchieri, M.G. (Ed.) Hunters and Gatherers Today. A Socio-economic study of Eleven Such Cultures in the Twentieth Century. Holt, Reinhart and Winston, Inc., New York pp. 139-159.

Coimbra, C. E. A. jr., 1984.Estúdios de ecologia humana entre os Surui do Parque Indígena Aripuana, Rondónia 1. O. Uso de larvas de Coleopteros (Bruchidae e Curculionidae) na alimentaçao. Rev. Bras. Zool. 2, 35-47.

Coimbra, C. E. A., 1985. Estúdios de ecologia humana entre os Surui do Parque Indígena Aripuana, Rondónia. Aspectes alimentares, Bol. Museu Paranense Emílio Goeldi. Antropol. 2, 57-87.

Costa-Neto, E., Ramos-Elorduy, J., 2006. Los insectos comestibles de Brasil: etnicidad, diversidad e importancia en la alimentación. Bol. SEA. 38, 423-442.

DeFoliart, G., 1999. Insects as food: Why the Western Attitude Is Important. Ann. Rev. Entomol. 44, 21–50.

Dufour, D. L., 1987 Insects as food: a case of study from the Northwest Amazon Amer. Anthropol.89, 383-39.7

Guardiola, J., Gónzález-Gómez, F., 2010.The influence of inequality on undernutrition in Latin America: an economic perspective. Nutr. Hosp., 25, 38-43.

Govoni, G., 1998. The Paca (*Agouti paca*): General characteristics and farming opportunities. Bull. BEDIM.7, 16.

Hawkes K., Hill., K., O´Connell, J.F., 1982 Why hunters gather: optimal foraging and the Achè of the eastern Paraguay. Amer. Ethnol. 9, 379-398.

Hill, K., Kaplan, H., 1983. Field report and research summary. The Mashco-Piro of Manu National Park, Peru, Manuscript Dept. Anthropology, Univ. Utah, Salt Lake City, UT. USA.

Hill, k., Kaplan, H., Hurtado, A.M., 1987 Foraging decision among Achè hunter-gatherers.New data and implication for optimal foraging models.Etnol. Sociobiol. 8, 1-36.

Hurtado, A.M., Hawkes, K., Hill, K., Kaplan, H., 1985. Female subsistence strategies among Achè hunter.Gatherers in Eastern Paraguay.Human Ecol. 13, 1-28.

http://www.unicef.org/lac/Fecha de consulta Mayo 2016.

Makkar, H. P. S., Tran G., Heuzé, V., Ankers, P., 2014. State-of-the-art on use of insects as animal feed.Anim. Feed Sci. Tech.197, 1-33.

Meyer, R. V., 2010. Entomophagy and its impact on world cultures: the need for a multidisciplinary approach, in: Durst, P., Johnson, D., Leslie, R.N., Shono, K. (Eds.), Forest insects as food: humans bite back. Food and Agriculture Organization of the United Nations. Bangkok, Thailand,pp. 23-32.

Menzel, P., Aluisio, D´., 1998. Man eating bugs. Ten Speed Press, Berkeley, CA. 192p.

Milton, K., 1984 Protein and carbohydrate resources ok Maku Indians of north-western Amazonia. Amer. Anthropol. 86, 7-27.

Nowak, V., Persijn, D., Rittenschober, D., Charrondiere, U. R., 2016.Review of food composition data for edible insects.Food Chem. 193, 39-46.

Looy, H., Dunkel, F. V., Wood, J. R., 2014. How then shall we eat? Insect-eating attitudes and sustainable foodways.Agric. Human Values, 31, 131-141

Onore, G., 1997. A brief note on edible insects in Ecuador, Ecol. Food Nutr.36, 277-285.

Onore, G., 2004. Edible insects in Ecuador, in: Paoletti, G.M. (Ed.), Ecological implications of minilivestock (Potential of insects, rodents, frogs and Snails) Science Publisher, Inc, Enfield (NH),USA, pp. 343-352.

Paoletti, M.G., Bukkens, S.G.F., 1997. Minilivestock.Ecol. Food Nutr.36, 95-346.

Paoletti, G.M., 2005. Ecological implications of minilivestock (Potential of insects, rodents, frogs and Snails) Ed. Science Publisher, Inc, Enfield (NH),USA. 648p.

Paoletti, G.M., Dufour, D.L., 2005 Edible invertebrates among Amazonian Indians. A critical review of disappearing knowledge, in: Paoletti, G.M. (Ed), Ecological implications of minilivestock (Potential of insects, rodents, frogs and Snails) Science Publisher, Inc, Enfield (NH),USA. pp. 293-342.

Payne, C. L. R., Scarborough, P., Rayner, M., Nonaka, K., 2016.A systematic review of nutrient composition data available for twelve commercially available edible insects, and comparison with reference values.Trends Food Sci. Tech. 47, 69-77.

Posey, D.A., 1979. Ethnoentomology of the gorotire kayapo of central Brazil.PHD. Univ. Georgia, Athens, 177p.

Posey, D.A. 1987. Ethnoentomological survey of brazilian indians. Entomol. Gener.12, 191-202.

Ramos-Elorduy, J., 1997a. Insects: a sustainable source of food?.Ecol. Food Nutr. 36, 247-276.

Ramos-Elorduy, J., 1997b.The importante of edible insects in the nutrition and economy of people of the rural areas of Mexico.Ecol.Food Nutr. 36, 349-366.

Ramos-Elorduy, J., 2004., La etnoentomología en la alimentación, la medicina y el reciclaje, in: Llorente B.,J.E., Morrone, J.J., Yañez, O.O., Vargas, F.I. (Eds), Biodiversidad taxonomía y biogeografía de artrópodos de México. Hacia una síntesis de su conocimiento CONABIO, Facultad de Ciencias UNAM, Instituto de Biología UNAM. México, pp.329-413.

Ramos-Elorduy, Pino Moreno, J.N., 1989. Los insectos comestibles en el México antiguo estudio etno-entomologico. A.G.T., México, 108 pp.

Ramos-Elorduy, J., Pino-Moreno, J., 1992. Biogeographical aspects of some edible insects from mexico. Abstracts of the III International Congress of Ethnobiology, UNAM, México D.F. 143 p.

Ramos-Elorduy, J., Conconi, M., 1994., Edible insects of the world Fourth International Congress of Ethnobiology Abstracts, Lucknow, India, p. 311.

Ramos-Elorduy, J., Pino-Moreno, J. M., 2001 Contenido de vitaminas en algunos insectos comestibles de México, Rev. Soc. Quím. de Méx. 45, 66-76.

Ramos-Elorduy, J., Pino-Moreno, J.M., Martínez, C.V.H., 2008. Base de datos de la Colección Nacional de Insectos Comestibles de México. UNIBIO-IBUNAM. 68 p.

Ramos-Elorduy, J., Pino-Moreno, J. M., Martínez, C.V.H., 2012. Could grasshoppers be a nutritive meal? Food Nutr. Sci. 3, 164-175.

Rumpold, B. A., Schluter, O. K., 2013. Nutritional composition and safety aspects of edible insects. Mol.Nutr. Food Res.57, 802-823.

Ruddle, K., 1973. The human use of the insects: examples from the Yukpa. Biotropica. 5, 94-101.

Smole, W.J., 1976. The Yanoama Indians.A cultural Geography, Univ. Texas Press, Austin Texas, 272 p.

van Huis, A., 2013. Potential of insects as food and feed in assuring food security. Annu Rev Entomol. 58, 563-583.

van Huis, A. 2016. Edible insects are the future?.Proc. Nutr. Soc. 24, 1-12.

Yen, A. L., 2009. Edible insects: Traditional knowledge or western phobia. Entomol. Res. 39:289-298.

Zent, S., 1992. Historical and Ethnographic Ecology of the Upper Cuao River Votiha: for an Interpretation of Native Guianese Social Organisation, PhD, Columbia University, New York.

Chapter 10

The "Jumiles" (Hemiptera: Pentatomidae and Coreidae) in the State of Morelos, Mexico: Taxonomy, Distribution and Commercialization

☆ *Pino Moreno José Manuel, García Flores Alejandro, Monroy Martínez Rafael and Barreto Sánchez Sandra Denisse*

ABSTRACT

In Mexico and throughout the world, anthropo-entomophagy has been practiced for centuries, and in the Mexican republic this tradition that was started even before the arrival of the Spaniards still exists. In this ethno-entomologic investigation, we have registered 30 species of "Jumiles" that belong to the families Pentatomidae and Coreidae, the main genera being Edessa, Euschistus and Acantocephala. These insects are an alimentary resource that is natural, renewable, with high nutrient potential, and are chiefly exploited by the rural low income population due to their high abundance and high degree of approval in the different localities. The cooking and preservation strategies range from rudimentary to the complex and some are traditional since precolonial times.

The "Jumiles" are used for self-consumption or are widely commercialized by diverse types of vendors both in established markets and itinerating markets (tianguis); for these vendors they are a source of economic income and by this socioeconomic activity, collectors, distributors and restaurants are benefited.

We conclude the study with a note that the consumption of "Jumiles" persists in the state of Morelos thanks to the rural indigenous population that still maintains the anthropoentomophagic tradition, despite the influence of miscegenation and the introduction of different alimentary habits.

Introduction

The Class Insecta is the animal group that is numerically dominant on earth; it is one of the most important and diversified groups of arthropods in the entire animal kingdom with almost a million species described, even when in reality many of them are yet to be classified.

Insects possess numerous forms, structures and adaptations that derive from their evolutionary diversity; they have been able to populate broadly both aquatic and terrestrial environments; which means they have colonized all the habitats available (Coronado and Márquez, 1976, http://www.natureduca.com/zoo_inverteb_insectos1.php.)

Insect Classification

Due to their diversity, the characteristics of insects are enormously varied and this makes the process of identification really difficult, which is in fact very necessary for their taxonomy. Nevertheless, international nomenclatural rules have been adopted for classifying insects, taxonomical procedures must adjust to them (Coronado and Márquez, 1976). The Class Insecta is divided into orders on account of wing structure, mouth parts, metamorphosis and other diverse characteristics from which the names of the orders derive.

In general, as has already been mentioned, between entomologists there exist differing points of view as to the limits of certain of the orders. There are cases in which two groups are considered by certain authors in a single order, while other authorities place them in two or more separate orders; and cases in which two groups considered by some entomologists as separate orders, are integrated by other scientists into a single one; some entomologists even consider as orders those groups that have been reported as separate classes within the insects by others (Domínguez, 1979, Vázquez, 2007).

Grimaldi and Engels (2005) consider the Class Insecta is made up by 40 orders, one of these is the Order Hemiptera. We discuss briefly about this order below.

General Characteristics of the Order Hemiptera

This order includes almost 23,000 species, their size is small to large; body is cylindrical, elongated, oval, flattened or shield shaped; the head has an oral apparatus of the sucking type, they have well developed compound eyes and, when eyespots (ocelli) are present, there are two of them; the antennae are short or long and have 4 to 5 segments; their legs are normal or prehensile; they generally have two pairs of well-developed wings, the first pair has the anterior part hardened and the posterior part membranous, this is why they are called hemelytra, the second pair is membranous; the abdomen is frequently made up of ten segments and carrying frequently a well-developed ovipositor; and finally, they undergo incomplete metamorphosis (Coronado and Márquez, 1976).

General Characteristics of the Families Pentatomidae and Coreidae

Family Pentatomidae is comprised of about 317 recognized species in Mexico that are commonly known as stink bugs. Their body is oval or shield shaped, small triangular head, with big compound eyes and two eyespots, the rostrum or beak made up of 4 segments. Their antennae have 5 segments. The prothorax is triangular in form, the wing tips extend beyond the end of the abdomen, which has four pairs of odoriferous glands in the nymphs of certain species (Coronado and Márquez 1976, Reyes, 2007).

Ancona (1932) and Ramos-Elorduy (2003) have carried out some studies on the biology of Mexican stink bugs, and the first author describes them partially as follows: the eggs are yellowish, cylindrical, they measure approximately 2 mm in length and remain glued to the under surface of leaves from the end of April or the beginning of May until the last days of August, when the small pale green insects emerge. Afterwards they change color to yellowish green as they grow up gradually. Ancona (1932) also points out that the time in which they are more abundant is between November and February and that the last "Jumiles" disappear with the first rains. Ortega and Zurita-García (2013) pointed out in a study on the genus *Edessa* that these insects are widely diverse in size, shape and color, which makes them difficult to study.

The body of Coreidae is more or less elongated; the head is narrow and shorter than the prothorax, the beak is relatively long; compound eyes and eyespots present; the antennae have four segments, in the legs the femora and tibiae sometimes show modifications in the form of a leaf, making a wide abdomen that is generally concave (Coronado and Márquez, 1976).

General Characteristics of the "Jumiles"

"Jumil" is the common name given in Mexico to some species of edible hemipteran insects of the family Pentatomidae, that mainly belong to the genera *Euschistus, Edessa, Monomorpha, Brochymena, Oebaleus, Padaeus, Proxys, Mormidea* and *Disderia*; and of the family Coreidae the genera, *Acantocephala, Thasus, Piezogaster, Mamurius, Sephina* and *Anasa*.

The common name derives from the Nahuatl *Xomilli* (from *Xotlmilli*), the term *Xotl* (foot) and *milli* (hotbed) indicate that these insects live in crop fields or amongst fallen oak leaves.

These insects are amply used as foodstuff and are very abundant in the states of the Mexican republic that are situated in the central plateau, they measure less than a centimeter (the females are bigger than the males), and they are consumed mainly in the states of Chiapas, Chihuahua, Guerrero, Hidalgo, Jalisco, México, Michoacán, Morelos, Oaxaca, Puebla, Querétaro and Veracruz (Ramos-Elorduy, 2004). They have a characteristic cinnamon flavor that is derived from the stems and leaves of the oaks upon which they feed. In a chemical study of the "Jumiles" done by Esparza (1966), an abundant unsaturated alcohol was obtained; when oxidized to its aldehyde form it rendered the characteristic odor of these insects.

In Taxco, Guerrero and in numerous municipalities of the state of Morelos such as Amatlán, Ayala, Cuautla, Jojutla, Tepoztlán, Tilzapotla, they are even eaten alive. They are very much appreciated in certain indigenous communities because of the liquid they exude that has a very special odor and flavor which is used and appreciated as condiment; it is also valued due to its medicinal properties as it is ascribed healing, and even aphrodisiac qualities (Ramos-Elorduy, 1987, Ramos-Elorduy and Pino 1988). Different research papers confirm that certain "Jumiles" have analgesic and anesthetic properties, and it is known they have a high iodine content (Ramos-Elorduy, 2003, Ramos-Elorduy *et al.*, 2000).

According to certain local beliefs, the "Jumiles" are considered the souls of the dead that return to be once more with their significant others and bring them good fortune; eating them is the way to recognize they are the materialized presence of the beloved departed ones (Ramos-Elorduy and Pino, 1989). In a similar way, the inhabitants of Taxco account for their presence each year at the site.

In precolonial times Mexican people used to pilgrimage to the Cerro del Huizteco, in the municipality of Taxco, Guerrero, to go up to a temple dedicated to the "Jumil" (Castillo, 2011); and these organisms were also collected for the feast in honor of the dead.

In the state of Guerrero, the Fiesta del "Jumil" ("Jumil" Feast) is celebrated annually since 1943 on the first Monday after the Day of the Dead ("Dia de Muertos"), specifically in the Cerro El Huizteco, located in the municipalities of Taxco de Alarcón and Tetipac. Likewise, Ramos-Elorduy (2003) describes *Euschistus taxcoensis* as the representative species of the "Jumiles" of Taxco, the mystical, magical, religious and nutritional status of the insect being such that a temple was built to honor it. The temple was excavated on a great steep rock at the top of the Cerro del Huizteco; it is bordered by a channel in whose central part is a bridge that contacts to a circular stone, very possibly used for ritual purposes, that ends up in a precipice. In the "Díadel Jumil" thousands of locals and visitors socialize among eateries, regional dancing, music and contests, and "Jumiles" are collected lavishly. During the festivity another one of the most visited places is "La Cueva", where mass is held; according to Ruíz (1999) and Ramos-Elorduy (2003), this place constitutes a precolonial site of interest where a religious ceremony is celebrated for the "Jumil" which is considered a sacred insect. The practice of "Jumil" pilgrimage is even held to date, on the first Monday after the Day of the Dead.

Santa María (1978) mentions that these insects smell like bugs and produce good oil. Ramos-Elorduy (2003) points out that in the region of Taxco and its vicinity their abundance period ranges from November to February and that the last "Jumiles" disappear with the start of the first rains; this is the reason why, year after year, from the first days of October the "Jumiles" season begins.

To assess the nutritional value of some of the most consumed edible insects of Mexico, such as the "Jumil", Ramos and Bourges (1977, 1982) analyzed the protein content of the adults of the Taxco "Jumil" *Euschistus taxcoensis* (Hemiptera: Pentatomidae). They found that the protein content is 70.3 per cent on dry matter basis. This establishes "Jumil" as the insect with the highest protein content among edible insects. Nutrition experts explain that this proportion is highly compared with that of meat which supplies only 40 per cent.They are considered a exceedingly nutritious foodstuff, better than poultry when in 1982 they made a comparison with the FAO/OMS guidelines on the composition of indispensable amino-acids. They have equally been mentioned as the edible insects that are richer in fats, as in them the following fatty acids have been found: caproic, caprilic, lauric, miristic, palmitic, palmitooleic, stearic, oleic, linoleic and linolenic (Pino and Ganguly, 2016).

Anthropo-entomophagy

The diet of the Mesoamerican cultures was varied, for example, the Aztec, Mexica, Maya, Mixtec, Zapotec people, among others, had diets that were well balanced and diversified. They combined corn, beans and amaranth with animal proteins from diverse sources which included different types of insects such as crickets, mosquitoes used as bird feed, "Jumiles", maguey worms, ants and wasps, as well as greens, flowers, algae and a great variety of fruits.

This is why we say that in Mexico anthropo-entomophagy has been practiced and recorded since precolonial times, as this habit intrigued the first writers that chronicled the conquest and colonization. For example, Fray Bernardino de Sahagún (1980,1988), mentions in the Florentine Codexthe edible insects consumed by the indigenous people: locusts, corn worms and some aquatic insects like the "ocuiliztac", "atelepitz", "atopinan" and "ahuihuilla". Even today, this activity is still carried on in many states of the Mexican republic like Aguascalientes, Campeche, Chiapas, Chihuahua, Durango, Guanajuato, Guerrero, Hidalgo, Jalisco, México, Michoacán, Morelos, Nayarit, Nuevo León, Oaxaca, Puebla, Querétaro, Quintana Roo, San Luis Potosí, Tabasco, Tamaulipas, Tlaxcala, Veracruz, Yucatán and Zacatecas (Ramos-Elorduy, 2004).

Edible insects have a set of advantages that make them of great importance both in human and animal diets, these are: high digestibility, vitamin, mineral and caloric richness, and their nutritional value with respect to protein and amino-acid content. It is observed that the insects are rich in nutrients, this is why they are a source of macro and micronutrients in the diets of various indigenous groups throughout Mexico (Ramos-Elorduy, 1997, Ramos-Elorduy and Pino, 1989, 2001), as well as in other countries being an important source of animal protein for mankind (Bodenheimer, 1951, Bergier, 1941, Van Huis *et al.*, 2013). These insects have short life cycles, a high metabolic efficiency and wide geographical distribution, their reproductive potential is enormous, and they are easy to collect, cook, preserve and store, so that their systematic and rational use provides people a type of food that has a high biological value (Ramos-Elorduy and Pino, 1989).

Recently, the FAO (Van Huis, 2013) has reported that throughout the world different communities of indigenous people collect edible insects as a source of nutrients, and they are considered a healthy, clean and nutritious food that can be cooked, presented and preserved in countless ways both in Mexico and other parts of the world (Ramos-Elorduy and Conconi, 1994, Ramos *et al.*, 2013, Van Huis *et al.*, 2013).

The lack of nutrients in the diet of the native tropical human groups is one of the main reasons that have made them use numerous insect species, which have a high percentage of proteins which are absent in the basic staple foods (Bodenheimer, 1951).

In Mexico, today there are 545 recorded species of edible insects, among which are dragonflies, lice, crickets, "ahuahutle", "axayacatl", "Jumil", various worms like maguey (white and red, from sticks, arbutus, corn, "jarilla" and "mesquite"), "escamol", melliferous ants and"chicatanas", stingless bees, wasps, *etc.* are a few to be mentioned (Ramos Elorduy *et al.*, 2012).

Ethnology and Nutrition Status of the State of Morelos

a) Geographic Location

The state of Morelos is located in the central part of the country, in the southern slope of the Ajusco mountain range, and within the basin of the Balsas River. The coordinates in which it is localized are: 19°08′ north, 18°20′ south, 98°38′ east, 99°30′ west; it limits to the north with the state of Mexico and with Mexico City; to the east and southeast with Puebla; to the south and southwest with Guerrero and to the west with the state of Mexico (INEGI, 2007b). Map 10.1 shows the geographic location of the state of Morelos.

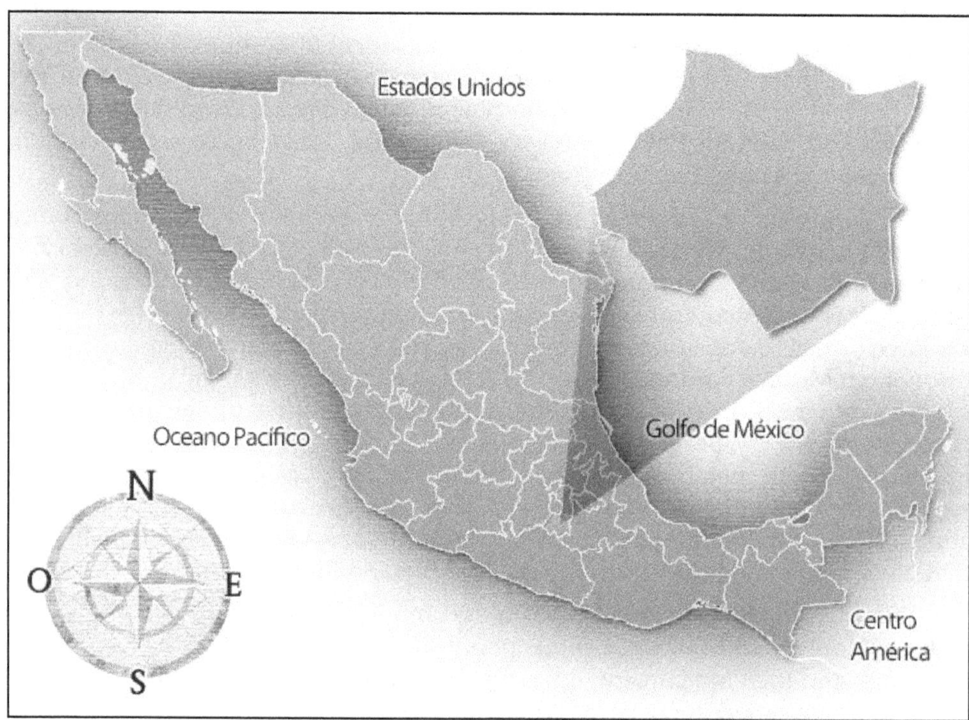

Map 10.1
***Source*: http://www.travelbymexico.com/estados/morelos**

Its altitude is varied; it ranges from 3,000 MASL, in the limits with Mexico City, to 850 MASL in the region of Huixtla. The state's surface is 4,958 square kilometers, which represent 0.25 per cent of the national total (INEGI, 2007b).

b) Ethnology

The indigenous population of Morelos is distributed in the 33 municipalities that are part of the state, however, in 15 of these is where the higher proportion is concentrated, these inhabitants are found mainly in the communities of Tetelcingo (Nahua; Mosehualli), Xoxocotla (Nahua), Cuentepec (Nahua, Tlapaneco), Coatetelco (Nahua, Zapoteco), Hueyapan (Nahua), these are all poor and neglected communities (CDI, 2007).

According to the number of speakers of indigenous languages recorded by the INEGI (2007b), in some of the municipalities the migrant indigenous population of Mixtec, Tlapanec and Zapotec origin that has arrived from the states of Puebla, Guerrero and Oaxaca include more representatives than the native Nahua population whose origin is in Morelos.

c) Malnutrition and Hunger

In the state of Morelos malnutrition and hunger prevail mainly among the inhabitants of rural zones, mostly due to their subsistence economy. This nutritional situation, as well as the socioeconomic importance of the "Jumiles", were our motives to carry out the present ethno-entomological investigation. The objectives of the present chapter are to track, collect and identify taxonomically the edible insects known as "Jumiles", systematizing some aspects of their distribution by municipalities, as well as their commercialization and preservation.

Materials and Methods

The applied methodology can be divided in the following aspects:

a) Literature Analysis

We initially carried out a bibliographical and cartographic analysis of the different municipalities of the state of Morelos with the aim of learning their general characteristics with respect to location, municipal distribution, topography, climate, fauna and flora, nutritional situation, and as well to establish study and sampling zones.

b) Field Work

Temporary visits to the selected localities were carried out to collect "Jumiles" and to go to the markets and tianguis where they are commercialized.

Informal interviews were applied among the inhabitants of each locality as well as semistructured surveys with an ethical focus in different markets and tianguis (Costa Neto, 2012) by means of a basic questionnaire where we had included open, free-answer questions with the aim of gaining knowledge about the species of "Jumiles", their collection sites, the places they come from, the places where they are sold, their costs, as well as the ways they are cooked and preserved, *etc.* All these taking into account diverse factors that include the length of the interview and the accessibility and disposition of each person.

According to its regionalization, we selected 51 localities that correspond to municipal heads, small villages as well as cities, markets and tianguis (from nahua *tianquiztli*; it is a market on a public place that does not have a permanent space, it is installed in designated days to sell, buy or interchange goods and services), because in these places people from all the neighboring towns gather to sell or buy various foodstuffs. Thus we made a systematic search of the species of edible insects amply distributed in the state during the months of November and December 2014 and January, February, March 2015, that correspond to the season of abundance and commercialization as well as of collection/purchase of "Jumiles" in some markets and tianguis. Subsequently, the organisms were mounted, labeled, identified and catalogued. We also searched for the ways in which they are cooked, preserved and commercialized. The present study took place in 51 localities, 33 of them are municipalities of the state of Morelos. Table 10.1 shows the localities we studied.

Table 10.1: List of Sampling Locatities in the State of Morelos

Municipality	*Locality*
1 Amacuzac	Amacuzac
2Atlahuacan	Atlahuacan
3Axochiapan	Axochiapan
4 Ayala (come from Atlixco Puebla and Jantetelco).	5Xalostoc 6 San Juan Chinameca
7 Chalcanzingo	Chalcanzingo
8 Coatlan del Río	Coatlán del río
9 Cuautla	10Tetelcingo
11 Cuernavaca	Cuernavaca
12 Emiliano Zapata	
13 Huitzilac	
14Jantetelco	Hill of Jantetelco,15 Amayuca
16Jiutepec	
17 Jojutla de Juárez	Market
18 Jonacatepec	
19 Miacatlán	
20 Ocuituco	21 Jumiltepec (cerro)
22Puente de Ixtla	23 La Tigra y 24 El zapote)
	25 Tilzapotla
	26 Xoxocotla (market)
27 Temixco	28 Cuentepec
29 Temoac	30 Amilcingo
	31Huazulco
32 Tepalcingo	

Municipality	Locality
33Tepoztlán	34 Amatlán
	Hill Mazatepec
35 Tetecala	
36 Tetela del Volcán	
37 Tlalnepantla	
38 Tlaquilenango	39 Tianguis, 40 Cerro Negro,Cerro Frío
	41San José de Pala
42Tlatizapan	
43 Tlayacapan	
44 Totolapan	
45 Xochitepec	46 Atlacholoaya
47 Yautepec	48 Oaxtepec
49 Yecapixtla	
50 Zacatepec de Hidalgo	Market
51 Zacualpán de Amilpas	

c) Laboratory Activities

The insect species we collected and bought from the markets and tianguis during their abundance and sale season, were preserved in ethanol 70 per cent inside glass jars (Pastrana, 1985). These samples were labeled with data such as: collection and/or purchase date, site of collection and/or purchase, cost and location of the collection site.

All the organisms were mounted by using entomological pins as described by Triplehorn and Johnson (2005), labeled and identified by means of the pertinent taxonomical keys (Brailovsky, 1985, 1988, 1992, Brailovsky and Barrera, 1994, Brailovsky and Márquez, 1973, Brailovsky and Sanchez, 1983, Brailovsky *et al.*, 1992, Rolston, 1974), at the Biology Institute UNAM (IBUNAM), specifically at the Department of Zoology, Entomology Laboratory. Later the species were catalogued and incorporated into the national collection of edible insects.

Results and Discussion

Collection and Purchase

The collection and/or purchase of the "Jumiles" begins in the month of October and finishes in the month of February, one can obtain nymphs and adults. Whole families participate in this activity, but in a higher proportion the men, as in certain localities like Jantetelco, one has to climb perilous mountain trails and this is the reason why women and children do not participate. The insects are picked by hand and placed in plastic or plastic mesh bags that are fastened tightly so that they do not escape; when most abundant, each family can collect between 2 and 5 kilograms per day.

In the locality of Jumiltepec we were told that in the season when the jumiles are abundant the locals search in the sky for clouds of "Jumiles" and observe closely to identify the site where the jumiles "let themselves fall" so that afterwards they can go and collect the insects.

Taxonomic Analysis

In Table 10.2 we present the species that were identified for the different localities that were sampled. In it we also indicate families, genera, species, edible stages of development and their distribution by municipality and/or locality.

Table 10.2: Taxonomic Analysis of the "Jumiles" (Hemiptera: Pentatomidae and Coreidae) of the State of Morelos, Mexico

Genus	*Species*	*Edible Stage of Development*	*Distribution*
	Family	***Pentatomidae***	
1 *Euschistus*	*comptus* Walker 1868.	Nymphs and adults	Cuautla*
2 *Euschistus*	*egglestoni* Rolston 1974	Nymphs and adults	Market : Amayuca and Jojutla, Tianguis de Tlaquiltenango, Temoac Yecapixtla,Cerro de Tilzapotla
3 *Euschistus*	*biformis* Stäl 1862	Nymphs and adults	Tepoztlán *
4 *Euschistus*	*crenetor orbiculator* Rolston 1974	Nymphs and adults	Cuautla*
5 *Euschistus*	*strenuus= E. zopilotensis* Stäl 1862	Nymphs and adults	Hill of Jantetelco, Jonacatepec, Zacualpán de Amilpas, Tilzapotla, Puente de Ixtla, MarketCuautla, Jojutla,Cuernavaca, San José de Pala**, Cerro de Mazatepec
6 *Euschistus*	*sulcacitus* Rolston 1971	Nymphs and adults	Cuautla*
7 *Euschistus (Atizies)*	*taxcoensis* Ancona 1933	Nymphs and adults	Cuautla**
8 *Euschistus*	sp.	Nymphs and adults	Cuautla,Jumiltepec, Ocuituco, Amayuca, Zacualpan de Amilpas, Tetelcingo, Jonacatepec, Axochiapan, Tepalcingo, Amilcingo, Huazulco, San Juan Chinameca, Puente de Ixtla**.
9 *Edessa*	*cordifera* Walker 1868	Adult	Tilzapotla*, El Zapote, Cuernavaca (Market Adolfo López Mateos), Market Zacatepec, New market Cuautla, Xoxocotla, Puente de Ixtla, Jumiltepec
10 *Edessa*	*mexicana* Stäl 1872	Adult	Cuautla, Yautepec*
11 *Edessa*	*fuscidorsata* Distant 1881	Adult	Yautepec*

Genus	Species	Edible Stage of Development	Distribution
12 *Edessa*	*conspersa* Stal1872	Adult	El Zapote, Market Zacatepec, Market Xococotla, Jumiltepec, New market Cuautla, Cuernavaca (Markety Adolfo López Mateos), Market of Zacatepec, Puente de Ixtla
13 *Edessa*	sp.	Nymphs and adults	Jumiltepec, Puente de Ixtla, Cuernavaca*, Tilzapotla* (El Zapote*, Cerro Frío*), Market : Xoxocotla, Market Zacatepec,Yautepec, Miacatlán, Tetecala, Cuentepec, Jojutla de Juárez, El Zapote, Cuernavaca** New market Cuautla.
14 *Monomorpha*	*tetra* Walker 1868	Adult	Tepoztlán*
15 *Brochymena*	*arcana tenebrosa* Walker 1867.	Nymphs and adults	Cuernavaca*,
16 *Oebalus*	*mexicanus = solubea mexicana* Sailer 1944	Adult	Tepoztlán, Oaxtepec* Xoxocotla, New market of Cuautla
17 *Padaeus*	*trivitattus* Stäl 1872	Nymphs and adults	Cuernavaca, Tetela del Volcán*
18 *Padaeus*	*viduus* Vollenhoven 1968	Nymphs and adults	Oaxtepec*
19 *Proxys*	*punctulatus* Palisot de Beauvois 1818	Adult	Atlacholoaya, Hill of Jumiltepec, La Tigra, Puente de Ixtla*, Market Jojutla, Tlaquiltenango
20 *Mormidea*	sp.	Nymphs and adults	Yautepec*
21 *Disderia*	*inornata* Ruckes 1959	Adult	Zacatepec, New market of Cuautla
	Family	***Coreidae***	
22 *Acantocephala*	*declivis* Say 1832	Adult	Amatlán, Tilzapotla, Cuautla*
23 *Acantocephala*	*luctuosa* Stäl 1855	Adult	Tilzapotla*
24 *Acantocephala*	*femorata* (F.) 1775	Adult	Cuernavaca (Market Adolfo López Mateos), New marketof Cuautla.
25 *Acantocephala*	sp.	Adult	Amatlán*
26 *Thasus*	sp.	Adult	Xalostoc*
27 *Piezogaster*	*Indecorus* Walker 1871	Nymphs and adults	Tepoztlán y Tétela del Volcán*
28 *Mamurius*	sp.	Adult	Tepoztlán*
29 *Sephina*	*Vinula* Stäl 1862.	Nymphs and adults	Xalostoc*
30 *Anasa*	*maculipes* Stäl 1862	Adult	Zacatepec, New marketof Cuautla.

*Ramos-Elorduy *et al.*, 2012. ** Reyes P.H., 2007.

When we analyze Table 10.2 we can observe that in the different localities that we studied, 30 species of "Jumiles" were found, that belong to two families: a) Pentatomidae: in this we report 21 species (70 per cent), among them the genus

Euschistus is the best represented with 8 species which are *E. egglestoni, E. strennus=zopilotensis, E, biformis, E.comptus, E. crenator orbiculator, E. sulcacitus, E. taxcoensis* and *E. sp.* it is followed by the genus *Edessa* with five species: *E. conspersa, E. mexicana, E. cordifera, E. fuscidorsata* and *E.* sp., then we have the genus *Padaeus* with two species *P. viduus* and *P. trivittatus* and the rest of the genera: *Monomorpha tetra, Brochymena arcana tenebrosa, Oebaleus mexicanus=Solubea mexicana, Proxys punctulatus, Mormidea* sp., and *Disderia inornata* each encompass a single species.

For the Family b) Coreidae we report 9 species which correspond to 30 per cent of the total, the genus *Acanthocephala* is the best represented with four species: *A. femorata, A. declivis, A. luctuosa* and *A.* sp., and the other genera encompass only one species each: *Thasus* sp., *Piezogaster indecorus, Mamurius* sp., *Sephina vinula* and *Anasa maculipes.*

In the following localities we find that these species of "Jumiles" are sold: Cuautla (*Acantocephala femorata, Edessa conspersa, Edessa cordifera, Edessa sp., Disderia inornata, Anasa maculipes* and *Oebalus mexicanus*), Cuernavaca: Market Adolfo López Mateos: (*Edessa conspersa, Edessa cordifera, Acantocephala femorata*), El Zapote (*Edessa conspersa, Edessa cordifera*), Jojutla (*Proxys punctulatus, Euschistus egglestoni*), Jumiltepec (*Edessa conspersa, Edessa cordifera, Edessa sp.*), Puente de Ixtla (*Edessa conspersa, Edessa cordifera, Edessa sp.*), Tlaquiltenango (*Proxys punctulatus, Euschistus egglestoni*), Xoxocotla (*Edessa conspersa, Edessa cordifera, Edessa sp., Oebalus mexicanus*) *and* Zacatepec (*Edessa conspersa, Edessa cordifera, Edessa sp., Disderia inornata, Anasa maculipes*).

However, *Euschistus egglestoni* was collected or sold/purchased in the following localities: Amacuzac, Amayuca, Atlahuacan, Ayala, Cerro de Tilzapotla, Emiliano Zapata, Huitzilac, Jantetelco, Jiutepec, Jojutla, Jonacatepec, Puente de Ixtla, Ocuituco, Temoac, Tepoztlán, Tlalnepantla, Tlaquiltenango, Tlatizapán, Tlayacapan, Totolapan, Xoxocotla, Yautepec, Yecapixtla and Zacualpán de Amilpas, thus we can conclude that this is the species most amply distributed in the state. Quite the contrary happened for *Acantocephala femorata,* which was only collected in Cuautla, and it is the species with the most limited distribution.

In Table 10.3 we report the number of species of "Jumiles" recorded for each locality, Cuautla is the municipality in which we determined the highest number of species, with 15, this is probably because of being the second most important city of the state. It is followed in decreasing order by Puente de Ixtla (6 species), Tepoztlán, Tilzapotla, Xococotla, Jumiltepec and Zacatepec (5 species), Jojutla,Yautepec and Cuernavaca (4 species), El Zapote (3 species), Jonacatepec, Zacualpán de Amilpas, Oaxtepec, Tetela del Volcán and Amatlán (2 species), and finally with one species for each locality we have: Tlaquiltenango, Yecapixtla, Jantetelco, Cerro de Mazatepec, San José de Pala, Ocuituco, Amayuca, Tetelcingo. Axochiapán, Tepalcingo, Amilcingo, Huazulco, San Juan Chinameca, Cerro Frío, Miacatlán, Tetecala, Cuentepec, Atlacholoaya, la Tigra, and Tlaquiltenango. This distribution is shown in Table 10.3.

Table 10.3: Number of Species of "Jumiles" Reported by Locality

Amatlán	1. *Acantocephala declivis*, 2. *Acantocephala* sp.	2
Axochiapan	1. *Euschistus* sp.	1
Ayala		
a) San juan Chinameca	1. *Euschistus* sp.	1
b) Xalostoc	1. *Thasus* sp., 2. *Sephina vinula*	2
Cuautla	1. *Euschistus comptus*, 2. *E. crenator orbiculator*, 3. *E. strennus* = *E. zopilotensis*, 4. *Euschistus sulcacitus*, 5. *Euschistus* (*Atizies*) *taxcoensis*, 6. *Euschistus* sp., 7. *Edessa mexicana* 8. *Edessa* sp. 9. *Edessa cordifera*, 10. *Edessa conspersa*, 11. *Oebalus mexicanus* =*Solubea mexicana*, 12. *Disderia inornata*, 13 *Acantocephala declivis*, 14. *Acantocephala femorata*, 15. *Anasa maculipes*	15
Tetelcingo	1. *Euschistus* sp.	1
Cuernavaca	1. *Euschistus strennuus*= *E: zopilotensis*, 2. *Edessa* sp., 3. *Edessa conspersa*, 4. *Padaeus trivittatus*	4
a) Adolfo López Mateos (Market)	1. *Edessa cordifera*, 2. *Acantocephala femorata*	2
Jantetelco		
a) Hill ofJantetelco	1. *E. strenuus* = *E. zopilotensis*	1
b) Amayuca	1. *Euschistus* sp.	1
Jojutla (Market)	1. *E. egglestoni*, 2. *E. strennus* = *E. zopilotensis*, 3. *Edessa* sp., 4. *Proxys punctulatus*	4
Jonacatepec	1. *E. strennus* = *E. zopilotensis*, 2.*Euschistus* sp.	2
Miacatlán	1. *Edessa* sp.	1
Ocuituco	1. *Euschistus* sp.	1
a) Jumiltepec	1. *Euschistus* sp., 2. *Edessa* sp., 3. *Edessa cordifera*, 4. *Edessa conspersa*, 5. *Proxys punctulatus*	5
Puente de Ixtla	1. *E. strennus* = *E. zopilotensis*, 2. *Euschistus* sp., 3. *Edessa* sp., 4 *Edessa cordifera*, 5. *Edessa conspersa* 6. *Proxys punctulatus*	6
El Zapote	1.*Edessa cordifera*, 2. *Edessa conspersa*, 3 *Edessa* sp.	3
b) Market of Xoxocotla	1. *E. egglestoni*, 2. *Edessa* sp., 3. *Edessa cordifera*, 4. *Edessa conspersa*, 5. *Oebalus mexicanus* =*Solubea mexicana*.	5
c) Hill of Tilzapotla	1. *Euschistus egglestoni*	1
d) Tilzapotla	1. *E. strenuus* = *E. zopilotensis*, 2. *Edessa cordifera* 3. *Edessa* sp., 4. *Acantocephala declivis*, 5. *Acantocephala luctuosa*	5
e) La Tigra	1.*Proxys punctulatus*	1
Temixco Morelos		
a) Cuentepec	1. *Edessa* sp.	1
Tepalcingo	1. *Euschistus* sp.	1
Temoac	1 *Euschistus egglestoni*	1
a) Amilcingo	1. *Euschistus* sp.	1
b) Huazulco	1. *Euschistus* sp.	1

Tepoztlán	1. *E. biformis*, 2. *Monomorpha tetra*, 3. *Oebalus mexicanus=Solubea mexicana*, 4. *Piezogaster indecorus*, 5. *Mamurius* sp.	5
Amatlán	1. *Acantocephala declivis*, 2. *Acantocephala* sp.	2
Hill of Mazatepec	1. *Euschistus strenuus = E. zopilotensis*	1
Tetecala	1. *Edessa* sp.	1
Tetela del Volcán	1. *Padaeus trivitattus*, 2. *Piezogaster indecorus*	2
Tlaquiltenango ("Tianguis")	1. *E. egglestoni*	1
Jantetelco		
a) Hill of Jantetelco	1. *E. strenuus = E. zopilotensis*	1
b) Amayuca	1. *Euschistus* sp.	1
Tilzapotla	1. *E. strenuus = E. zopilotensis*, 2. *Edessa cordifera*, 3. *Edessa* sp., 4. *Acantocephala declivis*, 5. *Acantocephala luctuosa*	5
Tlaquiltenango	1. *Proxys punctulatus*	1
San José de Pala	1. *E. strenuus = E. zopilotensis*	1
Cerro Frío	1. *Edessa* sp.	1
Xochitepec		
a) Atlacholoaya	1. *Proxys punctulatus*	1
Yautepec	1. *Edessa mexicana*, 2. *Edessa fuscidorsata*, 3. *Edessa* sp., 4. *Mormidea* sp.	4
a) Oaxtepec	1. *Oebaleus mexicanus= Solubea mexicana*, 2. *Padaeus viduus*	2
Yecapixtla	1. *E. egglestoni*	1
Zacualpán de Amilpas	1. *E. strennus = E. zo*pilotensis, 2. *Euschistus* sp.	2
Zacatepec de Hidalgo	1. *Edessa* sp., 2. *Edessa cordifera*, 3. *Edessa conspersa*, 4. *Disderia inornata*, 5. *Anasa maculipes.*	5

As can be seen these edible insects have an ample distribution which is a sign of their acceptance and wide use in the diet, as well as of its commercialization by the inhabitants of the different localities we sampled. Cuautla is the municipality in which we find the highest number of species consumed.

According to the data obtained during our field work, in Table 10.4 we show the sites of collection and commercialization or site of sale of the "Jumiles" recorded for the different municipalities of the state of Morelos; as can be seen, the most traded species is *Euschistus egglestoni* (Table 10.4).

Also in Table 10.4 we show that certain species of "Jumiles" such as *Euschistus egglestoni* come from different municipalities and localities of the state of Morelos such as Cuernavaca, Tlaquiltenango, Chalcanzingo, Jantetelco, Amacuzac as well as from Iguala and Taxco in Guerrero and from Atlixco in Puebla, and are commercialized in: Amacuzac. Atlahuacan, Ayala, Hill of Tilzapotla, Emiliano Zapata, Huitzilac, Jiutepec, Jonacatepec, New market of Cuautla, Ocuituco, Temoac, Tepoztlán, Tlaltizapán, Tlalnepantla, Tlayacapan, Totolapan, Yautepec, Yecapixtla, and Zacualpan de Amilpas.

Edessa conspersa, E. cordifera and *Acantocephala femorata* come from Taxco, Guerrero and are sold at the Adolfo López Mateos market; and *Acantocephala femorata* comes from Iguala, Guerrero and is sold at the Newmarket of Cuautla; *Edessa conspersa, Edessa cordifera, Edessa* sp., *Oebaleus mexicanus* and *Anasa maculipes* come from Iguala, Guerrero and are sold at the New market of Cuautla. Places like Jantetelco and Amacuzac in the state of Morelos, Atlixco, Puebla, Iguala and Taxco, Guerrero can be considered as collection centers of the "Jumiles" that send them to the state of Morelos for their sale and consumption (Table 10.4, Map 10.2).

Table 10.4: Route Marketing of some Species of "Jumiles" in Morelos

Genus and Specie	*Place of Origin*	*Sales Place*
Euschistus egglestoni	Cuernavaca	Yecapixtla
E. egglestoni	Tlaquiltenango	Hill of Tilzapotla
E. egglestoni	Iguala Gro.	Market Nuevo de Cuautla
E. egglestoni	Chalcanzingo and Jantetelco	Zacualpán de Amilpas, Jonacatepec
E. egglestoni	Jantetelco	Ocuituco, Yecapixtla and Temoac
E. egglestoni	Atlixco Puebla and Jantetelco	Ayala
E. egglestoni	Jantetelco and Puebla	Atlahuacan
E. egglestoni	Jantetelco, Amacuzac and Puebla	Totolapan
E. egglestoni	Taxco Guerrero and Amacuzac	Tlayacapan, Jiutepec, Tlaltizapán
E. egglestoni	Taxco Guerrero.	Tlalnepantla
E. egglestoni	Taxco Guerrero and Atlixco Puebla.	Tepoztlán, Huitzilac
E. egglestoni	Amacuzac Morelos	Yautepec
E. egglestoni	Taxco Guerrero.	Amacuzac
E. egglestoni	Atlixco Puebla and Taxco Guerrero.	Emiliano Zapata
E. egglestoni	Jantetelco	Jonacatepec
Edessa conspersa. E. cordifera, Acantocephala femorata	Taxco Guerrero.	Market Adolfo López Mateos
Acantocephala femorata	Iguala Guerrero.	New market Cuautla
Edessa conspersa, Edessa cordifera, Edessa sp., *Oebalus mexicanus, Anasa maculipes*	Iguala Guerrero.	New market Cuautla

We even observed a number of vendors that are native to the states of Guerrero and Puebla, which may be due to the cited species having acceptance in the localities of Morelos where they are commercialized, as is the case of the "Jumiles" of Taxco, Guerrero. In addition, the "Jumiles" are recognized as an autochthonous foodstuff thatcharacterizes the state of Morelos and that has a high nutritional value (Ramos-Elorduy and Pino, 1990).

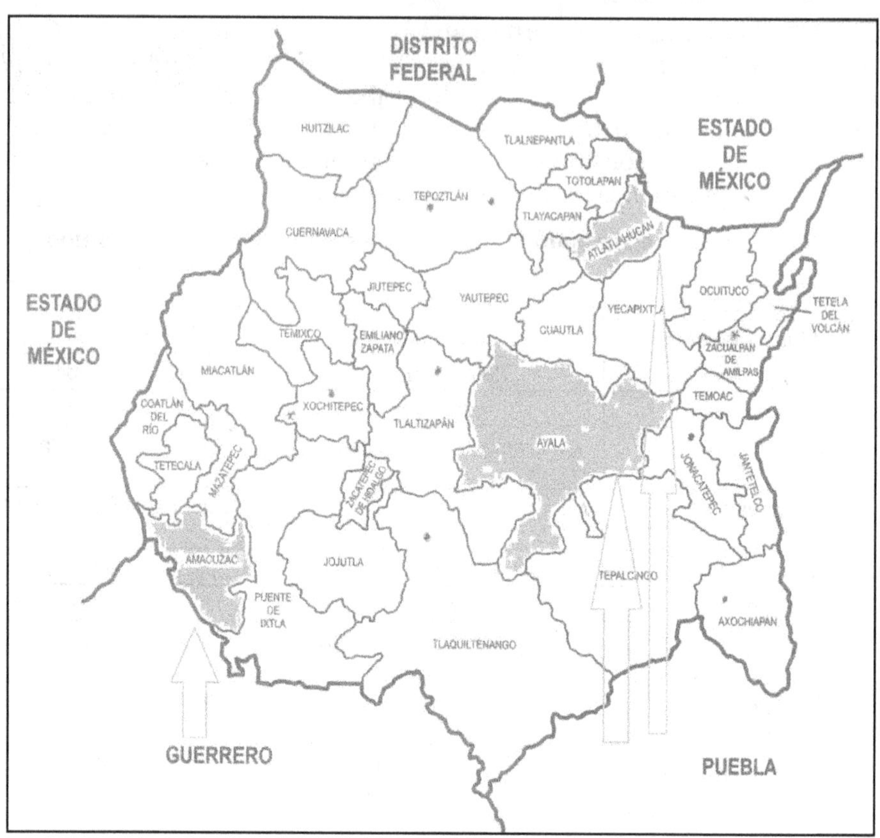

Map 10.2: Centres of Origin and Marketing Places of "Jumiles". (http://museoscomunitariosmor.blogspot.mx/2011/04/los-museos-son-en-axochiapan-esta-el.html)

Commercialization

The "Jumiles" are generally commercialized alive, they are offered in a series of small paper cones ("cucuruchos"), placed on the middle of a pewter bowl, so that the customers can have a look at them; they can also be found in plastic bags of different sizes. In market or tianguis days one can find up to three vendors that sell exclusively "Jumiles", and five vendors that offer "Jumiles" and other foodstuffs. According to Reyes (2007), in Table 10.5 we show the prices for different measure units in some of the localities in which they are commercialized. The way in which they are sold varies with the locality, for example, in smaller villages the cost is lower in comparison with the bigger cities of the state where a kilogram can cost up to $ 667.00 pesos.

Table 10.5: Prices of some Species of "Jumiles" in the State of Morelos

Localities	*Paper Cone 15 grs*	*Paper Cone 25 grs*	*Plastic Bag 30 grs*	*Per Kilogram (1000 grs)*
Jumiltepec Ocuituco Amayuca Zacualpan de Amilpas Amilcingo Huazulco San Juan Chinameca	$5.00	$ 10.00	$ 15.00	From $ 333.00 to $ 500.00
Cuautla, Cuernavaca Puente de Ixtla Jojutla Yautepec	$ 10.00	$ 15.0	$ 20.00	From $ 600.00 to $ 666.00
Chalcanzingo	$ 10.00	10 organisms		

In these localities the prices goes from $ 333.00 to $ 666.00.

Other prices recorded during this investigation are shown in Table 10.6.

Table 10.6: Price of some Insects in different Localities Visited

Scientific Name	*Weight grs.*	*# of Organisms*	*Price*	*Price per kilogram ($)*	*Observations*
E. comptus	15	123	$ 10.00	666.0	
E. egglestoni	2.5	81	$ 10.0	4000.00	
E. egglestoni	10.62	344	$ 10.00	941.00	
E. egglestoni	4.94	160	$ 15.00	3036.00	
E. egglestoni	4.44	144	$ 10.0	2252.00	
E. egglestoni	6.1	198	$ 10.00	1639.00	
E. egglestoni	5.18	168	$ 10.0	1990.00	
E. egglestoni	5.74	186	$ 10.00	1742.00	
E. egglestoni	4.50	146	$10.00	2222.00	
E. egglestoni	4.63	150	$ 10.0	2159.00	
E. egglestoni	5.15	167	$ 15.00	2912.00	
E. egglestoni	2.40	78	$ 15.00	6250.00	
E. egglestoni	5.18	168	$ 10.000	1930.5	
E. egglestoni	5.27	171	$ 15.00	2846.2	
E. egglestoni	5.34	173	$ 15.00	2808.9	
E. egglestoni	5.0	162	$ 15.00	3000.00	
E. egglestoni	5.21	169	$ 10.0	1919.38	
E. egglestoni	5.24	170	$ 10.0	1908.39	
E. egglestoni	4.60	149	$ 15.0	3260.86	
E. egglestoni	5.46	177	$ 10.0	1836.50	

Scientific Name	Weight grs.	# of Organisms	Price	Price per kilogram ($)	Observations
E. egglestoni	6.7	217	$ 10.	1492.53	
E. egglestoni	2.62	85	$ 5.00	1908.39	
E. conspersa, E. cordifera, A. femorata	9.7	315	$ 10.00	1030.92	The can of sardines costs $ 50.00
Edessa conspersa, E. cordifera, E. sp. *Oebaleus mexicanus* and *Anasa maculipes*	5.8	81	$ 10.0	1724.13	
Edessa and *Euschistus*	*Edessa* 4.3 *Euschistus* 0.3 total 4.6 g	37 10	Of both $10.00	1123.59	They are from Iguala Guerrero, were purchased in the New market Cuautla
Edessa	5.8	81	$ 10.00	1,724.00	They are Iguala Guerrero, were purchased in the New market Cuautla
Euschistus egglestoni	6.7	217	$ 10.00	1,492.53	They are from Amayuca market.
Euschistus egglestoni	2.3	85	$ 5.00	2,173.9	They are of Jonacatepec marketcome from Jantetelco.
Average	6.0	164.64	$ 12.31	2152.64	

The weight of the "Jumiles" purchased in retail generally in "cucuruchos" or plastic bags varies (4.3, 5.8, 6.7 and 2.3 g) and their prices fluctuate from $ 5.00 to $ 10.00 and extrapolating these data we have that the cost per kilogram ranges from $ 666.00 to $ 6250.00, the average is $ 2152.64. That is, the amount of money that the vendors obtain in markets and tianguis due to the sale of "Jumiles" is highly significant, especially if we consider the widespread poverty and marginalization that prevail in the state.

In relation to the commercialization of the "Jumiles", we confirmed that there is an intermunicipal distribution (I.M.D.) and an interstate distribution (I.E.D.), this can be observed in Table 10.7.

The "Jumiles" I.E.D. mainly involves the state of Puebla, Atlixco in particular, and in the case of Guerrero, the city of Taxco and they are sold in Amacuzac, Tlalnepantla and Atlahuacan, that is, in the region of the state of Morelos that limits with the state of Guerrero; however, the highest proportion of commercialized "Jumiles" come from Guerrero.

The "Jumiles" vendors from Guerrero also travel to the main cities of the state of Morelos, like Cuernavaca and Cuautla to sell them, we must point out that the insects from Guerrero are larger, they are called "chumiles"and correspond to the genus *Edessa* spp.

Table 10.7: Intermunicipal and Interstate Distribution of "Jumiles"

Place of Origin	*Sales Places*	
Taxco and Guerrero	Amacuzac, Tlalnepantla	*I.E.*D
Jantetelco, Morelos and Puebla	Atlahuacan	*I.E.*D., I.M.D.
Taxco, Guerrero, Atlixco and Puebla	Emiliano Zapata and Huitzilac, Tepoztlán	*I.E.*D.
Taxco, Guerrero, Amacuzac and Morelos	Jiutepec, Tlaltizapán, Tlayacapán	I.M.D.
Chalcanzingo and Jantetelco and Morelos	Jonacatepec, Zacualpán de Amilpas	I.M.D.
Jantetelco and Morelos	Ocuituco, Temoac, Yecapixtla	I.M.D.
Jantetelco, Amacuzac, Morelos and Puebla	Totolapan	*I.E.*D., I.M.D.

In a similar manner in the I.M.D. the localities of Jantetelco, Amacuzac and Chalcanzingo function as providers of insects that are commercialized in the following places: Atlahuacan, Jiutepec, Tlatizapán, Tlayacapán, Jonacatepec, Zacualpán de Amilpas, Ocuituco, Temoac, Yecapixtla and Totolapan. This interstate and intermunicipal distribution network shows us the commercial importance of these insects.

Cooking/Preparation Methods

The "Jumiles" are the chief ingredient essential for the preparation of delicious recipes, either ground in different sauces or as condiment (Ruíz, 1999). Of the edible insects few are eaten alive as in the case of the "Jumiles", most often they are consumed fried or roasted; it is also common for them to be boiled and then fried with onion and chili and then prepared in tacos (Ramos-Elorduy, 1984). In the localities of Amilcingo and Amayuca, the "Jumiles" are consumed alive wrapped in a tortilla, or they are simply eaten as any other nut. In other localities, such as Chalcanzingo they are prepared as a sauce and for this they are ground in a "molcajete" together with tomatoes and chili; once the sauce is ready it is accompanied with cheese, pork or simply by a tortilla, they are equally used as a condiment.

Those that consume them alive believe they have medicinal properties, for example, for the treatment of scrofula, tuberculosis, as well as for kidney, liver and gastrointestinal conditions. Their use as an analgesic and anesthetic, to relieve rheumatic and arthritic pains has also been reported (Ramos-Elorduy y Pino, 1988). Their taste is reportedly similar to that of peppermint.

Preservation

Some families of the community of Huazulco preserve them, by boiling them and then letting them dry; in this way they do not deteriorate, and they store them in small fabric sacks in dry places.

Particularly in Cuautla, Morelos, Carbajal (2008) reported the socioeconomic and cultural aspects associated with the sale of the "chumiles" (*Euschistus sulcacitus* R.), and he concluded that the collectors of this species have an organized and holistic knowledge that lets them carry out a sustainable management of the species. In that same study, a matriarchate associated with the sale of the species was reported; this is of great importance as the female leader takes the short and long term decisions all along the process that involves every step from collecting the insects to their sale.

Conclusions

In the state of Morelos anthropo-entomophagy is widely acknowledged and it implies the collection, preparation, distribution, preservation and commercialization of the "Jumiles". This activity is beneficial for the people involved as it generates temporary jobs and maintains and promotes this habit in the state, that is, by means of the retail sale of "Jumiles" they obtain economic resources that help them meet their clothing, footwear and alimentary expenses.

Also, as we have already mentioned, edible insects possess advantages that give them a special importance in the human diet, not only of the state but on a worldwide basis because, as Van Huis *et al.* (2013) have emphasized, insects form part of the diet of 2000 million people in many countries around the world.

Acknowledgements

The authors thank M. Cristina Mayorga Martínez for her help in the determination of some species, Dra. Aurora Zlotnik Espinosa for her help in some aspects of the translation and to Dr. Oscar Francke Ballvé for his invaluable help in style correction.

References

Ancona. L. H. 1932. Los "Jumiles" de Taxco, (Guerrero.). *Anales del Instituto de Biología.UNAM. México*.**III.** : 149-162.

Bergier, E. 1941. Peuples entomophages et insectes comestibles: étude sur les mœurs de l'homme et de l'insecte. Avignon, Imprimerie Rulliere Freres.238 pp.

Bodenheimer, F.S. 1951. Insects as human food; a chapter of the ecology of man. The Hague, Dr. W. Junk Publishers, 352 p.

Brailovsky, H. 1985. Revisión del género *Anasa* Amyot-Serville (Hemiptera-Heteroptera-Coreidae-Coreini). *Monografía. Inst. Biol. Univ. Nal. Autón. México* **2**: 266 p.

Brailovsky, H. 1988. Hemiptera-Heteroptera de México XXXVIII. Los Pentatomini de la Estación de Biología Tropical "Los Tuxtlas" Veracruz (Pentatomidae). *An. Inst. Biol. UNAM.Ser. Zool.* **58** (1):69-154.

Brailovsky, H. 1992. Un género y tres especies nuevas de Coreidos Neotropicales (Hemiptera-Heteroptera-Coreidae-Coreinae-Coreini). *An. Inst. Biol. Univ. Nac. Autón. México. Ser. Zool.* **63** (2): 185-199.

Brailovsky, H. and E. Barrera. 1994. Descripción de cuatro especies y una subespecie nuevas de la Tribu Anisoscelidini (Hemiptera-Heteroptera-Coreidae). *An. Inst. Biol. Univ. Nal. Autón. México. Ser. Zool.* **65** (1): 45-62.

Brailovsky, H. and C. Márquez M. 1973. Notas sobre algunos Hemípteros de Cuautla, Morelos. *An. Inst. Biol. Nal. Autón. México, Ser. Zoología* **44**(1): 67-76.

Brailovsky, H. and C. Sánchez. 1983. Hemiptera-Heteroptera de México XXVI. revisión de la Familia Coreidae Leach. Parte 3. Tribu Spartocerini Amyot-Serville. *An. Inst. Biol. Univ. Nal. Autón. MéxicoSer. Zool.* 53(1): 181-203.

Brailovsky, H. Cervantes and C. Mayorga. 1992. Hemiptera-Heteroptera de México XLIV. Biología, Estadios Ninfales y Fenología De la Tribu Pentatomini (Pentatomidae) en la Estación de Biología Tropical "Los Tuxtlas", Veracruz *Publ. Espec. Inst. Biol. UNAM*. 8: 204 p.

Carbajal V, L. A. 2008. Aspectos socioeconómicos y culturales asociados al manejo tradicional del germoplasma de Chapulines y chumiles de los mercados ladinos de Cuautla, Morelos. Tesis Profesional. Facultad de Estudios Superiores Iztacala, UNAM.101p.

Castillo G., X. 2011. Impacto ambiental de la fiesta tradicional del "Jumil", insecto comestible, *Edessa* spp. (Hemiptera: Pentatomidae) en el Cerro El Huizteco, Taxco de Alarcón y Tetipac, Guerrero. Tesis Prof. (Biología) Facultad de Estudios Superiores Iztacala UNAM.

CDI, 2007. *Nahuas de Morelos*. Comisión Nacional para el Desarrollo de los Pueblos Indígenas (CDI). 67 p.

Coronado P.R. and A. Márquez D.1976. Introducción a la Entomología (Morfología y Taxonomía de insectos). Ed. Limusa México2ª reimpresión, 282 p.

Costa Neto, M.E. 2012. Manual de Etnoentomología. Manuales y Tesis. *Soc. Ent. Aragonesa* **4**: 104 p.

Esparza C. M. de J. 1966. Estudio químico de los insectos *Atizies taxcoensis* ("Jumiles"). Universidad Nacional Autónoma de México. Tesis ProfesionalFacultad de Química. 70 p.

Grimaldi D. and M. S. Engels. 2005. Evolution of Insects. Cambridge University Press, New York USA. 755 p.

INEGI, 2005. Anuario Estadístico del Estado de Morelos. Gobierno del Estado de Morelos. INEGI, Cuernavaca.

INEGI, 2007a. Carta de Climas, Escala 1:1'000,000.

INEGI, 2007a. Censo general de población y vivienda 2005. Resultados definitivos del ll conteo de Población y Vivienda 2005 para el Estado de Morelos, México.

INEGI, 2007b. Información Geográfica, Datos Generales. Morelos, México.

Ortega L.G. and Zurita García M.L., 2013 Descripción de los estadios ninfales de *Edessa reticulata* y *Edessa jugata* (Heteroptera: Pentatomidae: Edessinae) para Oaxaca y Veracruz, *Rev. Mex. de Biodiversidad* **84** (3) Versión On-Line 2007-8706.

Pastrana J.A. 1985 Caza, preparación y conservación de insectos. Ed. El Ateneo.234 pp.

Pino, M. J. M. and A. Ganguly 2016. Determination of fatty acid contents in some edible insects of México. *J. of Insects as Food and Feed* **2** (1):37-42.

Ramos-Elorduy J. 1984. Los insectos como recurso actual y potencial. Seminario de la alimentación en México. Instituto de Geografía. UNAM. p. 120 - 139.

Ramos-Elorduy J. 1987. Los insectos como fuente de proteínas en el futuro. Limusa. 2ª Edición. México. 149 p.

Ramos-Elorduy, J. 1997. Insects: A sustainablesource of food? *Ecol. of Food and Nut.***36**: 247-276.

Ramos-Elorduy J. 2003. Les "Jumiles"punaisessacrées au Mexique. In: E. Motte-Florac and J. M. C. Thomas (eds.) Les "insectes" dans la tradition orale, Ethnosciences**11** : 325-353. Eds. Peeters Louvain- Paris- Dudley, MA.

Ramos-Elorduy, J. 2004. La etnoentomología en la alimentación, la medicina y el reciclaje [pp. 329-413]. In: J. E. Llorente B., J. J. Morrone, O. Yánez O, I. Vargas F. (Eds). *Biodiversidad, Taxonomía y Biogeografía de Artrópodos de México. Hacia una síntesis de su Conocimiento*Vol. IV. ED. CONABIO, Facultadde Ciencias, UNAM, Instituto de Biología, UNAM, México, D. F.

Ramos-Elorduy J. and H. Bourges R.1977.Valor nutritivo de ciertos insectos comestibles de México y lista de algunos insectos comestibles del mundo. *Anales del Instituto de Biología. UNAM. México.* Serie Zoología **48**:165 - 186.

Ramos-Elorduy J. andH. Bourges R.1982. Valor nutritivo y calidad de la proteína de algunos insectos comestibles de México. *Folia Entomológica Mexicana.* **53**: 111-118.

Ramos-Elorduy, J. and M. Conconi.1994. Edible insects of the world *Fourth International Congress of Ethnobiology* Abstracts, Lucknow, India, p 311

Ramos-Elorduy, J. and J. M Pino M. 1988. The utilization of the insects in the empirical medicine of ancient Mexicans. *J. of Ethnobiology.* **8**(2): 195-202.

Ramos-Elorduy J. andJ.M. Pino. M.1989. Los insectos comestibles en el México antiguo. AGT editor. México. 1ª Ed.108 p.

Ramos-Elorduy J. and J.M. Pino M. 1990. Contenido calórico de algunos insectos comestibles de México (Caloric content of some edible insects of Mexico), *Rev. Soc. Mex. de Quím.* **34** (2): 56-68.

Ramos-Elorduy, J. and J. M. Pino M. 2001. Contribución de la entomofauna silvestre en la alimentación de las etnias de México. *Resúmenes del IV Congreso Nacional de Etnobiología,* ITA Núm. 6, Huejutla, Hidalgo. p. 72.

Ramos-Elorduy, J., E. Motte-Florac, J. M. Pino M. and C. Andary. 2000. Les insects utilisés en Médecine Traditionnelle au Mexique: perspectives. In: Healing, yesterday and today, tomorrow? Vol. III. Ethnopharmacology (pp. 271-290). Erga Edizioni, Italia.

Ramos-Elorduy J., J. M. Pino Moreno and V.H. Martínez C. 2012. Checklist de insectos comestibles. UNIBIO, Instituto de Biología.

Ramos, R.B., J. Ramos-Elorduy, J. M. Pino M., S. Ángeles C. and A. García P. 2013. Insectos comestibles. Gastronomía y turismo en la zona arqueológica de San Juan Teotihuacán, Estado de México. *Entomología Mexicana* **12** (1) :563-569.

Reyes P.H. 2007. Los insectos como alimento humano en algunas localidades del Estado de Morelos, Tesis Prof. (Biología) Escuela Nacional de Ciencias Biológicas. del Instituto Politécnico Nacional. 75 p.

Ruiz O. J. 1999. Cuadernos de Taxco. Día del "Jumil". Año 1.No.1. México.

Rolston H. R. 1974 Revision of the genus *Euschistus* in middle America (Hemiptera. Pentatomidae, Pentatomini), *New York Entomological Society* **48** (1) 1-102.

Sahagún, F. B. de. 1980. *Códice Florentino.* Ed. Archivo General de la Nación México, Reproducción Facsimilar Libro IIIp. 221, 247-260.

Sahagún, F. B. de. 1988. *Historia General de las cosas de la Nueva España.* Tomo 2, Cap. XIII, Editorial Mexicana Consejo Nacional para la Cultura y las Artes. Gobierno de la República Mexicana. 514 p.

Santa María J. 1978 Diccionario de Mexicanismos, Ed. Porrúa,México,1207 p.

Triplehorn C. A. and N. F. Johnson 2005. Borror and Delong's Introduction to the Study of Insects, Seven Ed. Thompson, Australia, 881 p.

van Huis A., J.Van Itterbeck, H. Klunder, E. Mertens, A. Halloran, G. Muir and P. Vantomme 2013. Edible Insects: Future prospects for food and feed security. FAO, Roma, 187p.

Referemces from Electronic Media

Domínguez 1979 http://www.natureduca.com/zoo_inverteb_insectos. (https://es.wikipedia.org/wiki/Morelos) Wikipedia 2014

Vázquez, a. 2007. http://entomologiajalapa.wordpress.com/2007/12/29/muy-interesante/

Chapter 11

Nutrients and Anti-nutrients Present in *Oxya fuscovittata* (Orthoptera: Acrididae): Looking beyond Proximate Composition

☆ *Arijit Ganguly and Parimalendu Haldar*

ABSTRACT

Oxya fuscovittata has been lately emerged as a potential insect protein source in India. Various authors described its nutritional value in terms of proximate composition minerals and energy. Recently this insect has been proved to be a potential alternative nutrient supplement for fish and poultry. Despite having these information it cannot be established as a proper nutrient supplement unless and until we have sufficient knowledge on the presence of other nutrients and anti-nutritional factors. With this view in the present study we have aimed to evaluate nutrients such as fatty acids, amino acids, vitamins, minerals, as well as anti-nutrents like, tannin, oxalate, and phytin present in the body tissues of O. fuscovittata. The results were very much encouraging because the insects were found to be rich in amino acids including all the indispensible ones, and above 1945 KJ/100g of energy was also obtained. A total of six fatty acids have been detected in the present study showing a pretty high amount of linoleic and linolenic acid. Quite good amount of vitamin contents were also perceived in this insect, especially Retinol, Ascorbic acid and Niacin. Although a trace of various anti-nutritional factors were obviously detected, they were all under tolerance limit because the values were either similar or lower in content as found in other edible insects and edible plant materials. We have concluded the study with the statement that this insect definitely is a key nutrient resource that can be utilised as a supplementary ingredient in the diets of various livestock.

Introduction

Insects are an important protein source for human and livestock, such as poultry, fish and pigs (Yen, 2015). Many of the food insect researchers have claimed that edible insects hold a great potential to replace conventional protein sources from formulated diets of livestock (Das and Mandal, 2014). Among insects, the order Orthoptera (*i.e.* crickets and grasshoppers) has been documented as a very rich protein resource in the literature (Zhou and Han, 2006). According to a study conducted by Ramos-Elorduy *et al.* (2012), the amount of crude protein of grasshoppers varies from about 43-77 per cent that could be utilised to formulate good quality feed for livestock. The efforts of Anand *et al.* (2008a) and Das *et al.* (2012) to estimate the potential of annual biomass production of some grasshoppers have put forward some encouraging results when the insects were reared in captive condition. In both the cases the genus *Oxya* was found to have the ability to produce greater biomass compared to its contemporaries. In another report Anand *et al.* (2008b) proposed that the grasshopper *Oxya fuscovittata* (Marschall) is also nutritionally rich in terms of proximate composition and minerals.

Observing the great potential of *O. fuscovittata* to be a promising food insect for mass scale production to feed the livestock, Ganguly *et al.* (2010) made an attempt to find out the appropriate food plant that could be used for feeding these insects. The research outcome sorted out the seedlings of *Sorghum halepense* as the most suitable one. Later, Das *et al.* (2012) and Ghosh *et al.* (2014) also reported about using different plants for mass production of another grasshopper species under the genus *Oxya*. Recently, Ganguly *et al.* (2014) ushered on the prospective of *O. fuscovittata* as a potential nutrient supplement in the formulated diets that can compete with the fish meal supplemented diets for the aquarium fish *Poecilia sphenops*. In the same report proximate composition of *O. fuscovittata* again confirmed that the insect is a good nutrient resource in terms of crude protein (about 65 per cent) that could be exploited by the livestock.

However, merely a good proximate value cannot instate this grasshopper species as a good nutrient supplement, because only a high crude protein value cannot establish the quality of the protein unless and until we have the idea of its amino acid contents (Siriamornpun and Thammapat, 2008). Likewise the amount of fatty acids, vitamins, minerals and anti-nutritional factors are also needed to be evaluated. Although the insect *O. fuscovittata* has been found to be nutrient-rich in terms of proximate composition, minerals and energy, still sufficient information is lacking about the other nutrients and anti-nutritional factors that should also be present in the insect's body tissue. Keeping this in mind, in the present study we have estimated the energy content, fatty acids, amino acids, vitamins, minerals and anti-nutritional factors to add up more information to our knowledge about the nutritional value of *O. fuscovittata*. This information would be of immense importance to have ample idea whether this species of insect could really be utilised as an alternative nutrient supplement for fishes as well as other livestock. This information will also help us to point out what are the abundant and limiting macro and micro nutrients present, thus enabling to formulate quality feed in future.

Methodology

Collection and Preparation of Sample Species

Eight hundred and fifty five (855) adult *O. fuscovittata* were collected by sweeping method with standard insect nets from the grasslands of Santiniketan (23° 39′N, 87° 42′E), West Bengal, India. They were then taken to the insect museum of Entomology Research Unit, Department of Zoology of Visva-Bharati University, for taxonomic confirmation where locally available insect species were preserved and identified by the Zoological Survey of India, Kolkata. After that the insects were first frozen in a refrigerator (Godrej, Mumbai, India) and then oven dried (Indian instrument manufacturing company, Kolkata, India) at 60°C till a constant body weight was obtained consecutively for three times (Akinnawo and Ketiku, 2000). Legs and wings were then removed and finally crushed to powder form prior to chemical analyses.

Estimation of Nutrients and Anti-nutrients in the Body Tissues of *O. fuscovittata*

Energy content in the tissues of *O. fuscovittata* was estimated using oxygen bomb calorimeter (IKON instruments, New Delhi, India). Atomic absorption spectrophotometry (Varian Techtron,Victoria, Australia) was employed to determine the contents of Ca, Fe, Zn, Mg, Cu and Mn using standard reference chemicals as previously reported by Ganguly *et al.* (2013). Fatty acid extraction from the insect samples and their methyl ester preparation were conducted according to Bettelheim and Landesberg (2001). Gas chromatographic (Agilent 6890N,Palo Alto, USA) analysis was then performed with the purified methyl esters of the fatty acids. Afterwards, the percent fatty acid compositions were obtained from the GC peak. For amino acids, the samples were first hydrolysed with 6 N HCl containing 1 per cent phenol for 22 h at 105°C, finally quantification was obtained at 38°C by a PICO.TAG system (Waters, USA) as mentioned in Ghosh *et al.* (1995) and Wang *et al.* (2007). However, tryptophan could not be quantified by this method, so we have used AIMIL Photochem colorimeter (Photochem Electric Instruments, Jodhpur, India) for this purpose as per the strategies proposed by Fischl (1960). Quantitative estimation of vitamins was also conducted by colorimetric method of Helrich (1990).

Anti-nutritional factors like tannin and oxalate were determined by the procedures proposed by Gupta *et al.* (1988). Estimation of phytin phosphorus was done according to Agbede and Aletor (2004). The value was multiplied by the factor 3.55 to obtain the phytin content (Agbede and Aletor, 2004).

Results

About 1945.85 KJ/100g of energy was recorded in the tissue samples (data not presented in tables or figures). Among minerals Ca and Mg was found in pretty good amount (8.03 and 6.01 mg/Kg respectively), followed by Fe, Zn and Cu (Figure 11.1). Fatty acids, amino acids and vitamin contents are depicted in Table 11.1. A total of six fatty acids were detected in the body tissues of *O. fuscovittata*, of which two poly unsaturated fatty acids (*i.e.* linoleic and linolenic acids) were

Table 11.1: Fatty Acid, Amino Acid and Vitamin Contents in *O. fuscovittata*

Property	*O. fuscovittata*
Fatty acid (per cent)	
Myristic acid	1.12
Palmitic acid	0.97
Oleic acid	7.24
Eicosenoic acid	15.39
Linoleic acid	27.25
Linolenic acid	2.75
Amino Acid (per cent)	
Asx	0.06
Glx	3.49
Threonine	18.48
Proline	15.61
Tyrosine	9.00
Cysteine	0.47
Valine	5.93
Leucine	4.81
Arginine	7.42
Isoleucine	1.35
Phenylalanine	3.92
Methionine	1.75
Serine	4.68
Histidine	7.86
Glycine	8.23
Alanine	3.14
Lysine	3.40
Tryptophan	2.52
Vitamin (mg/100g)	
Retinol	4.20
Ascorbic acid	7.64
Thiamin	0.20
Riboflavin	0.90
Niacin	6.25

present in significant amount. Eicosenoic acid was also found to be considerably high. However, Palmitic acid content was very low. Threonine and Proline were the dominant among amino acids in the insect of present investigation (more than 18 per cent and 15 per cent respectively). Additionally Arginine, Histidine, Glycine and Tyrosine were also found in good amounts, but Aspartic acid and Asparagine in combination (*i.e.* Asx), as well as Cysteine was noted as considerably low (less than

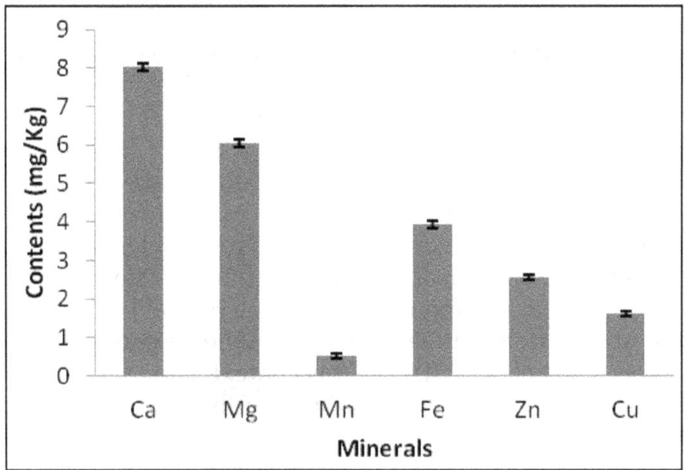

Figure 11.1: Mineral Contents (mg/kg) in the Body Tissues of *O. fuscovittata* on Dry Matter Basis. Data are presented as means ± SD.

1 per cent). Among the five estimated vitamins, Retinol, Ascorbic acid and Niacin was 4.20 mg/100g, 7.64 mg/100g, and 6.25 mg/100g respectively, but the other two B-vitamins, *i.e.* Thiamine and Riboflavin contents were meagre. Analyses of four anti-nutritional factors were carried out for the dried body tissues of *O. fuscovittata* (Figures 11.2 a,b). Results were very much encouraging because all of them were detected in very low amount. Only the content of Tannin was 1.05 per cent, rest of all had values less than 0.5 per cent.

Discussion

As reported by Ladron de Guevara (1995) monogastric animals do not have the need of "protein" as such, rather they need 9-10 amino acids that their body cannot synthesize. These are Isoleucine, Leucine, Lysine, Methionine, Phenylalanine, Threonine, Tryptophan, Valine and Histidine, and are known as indispensible

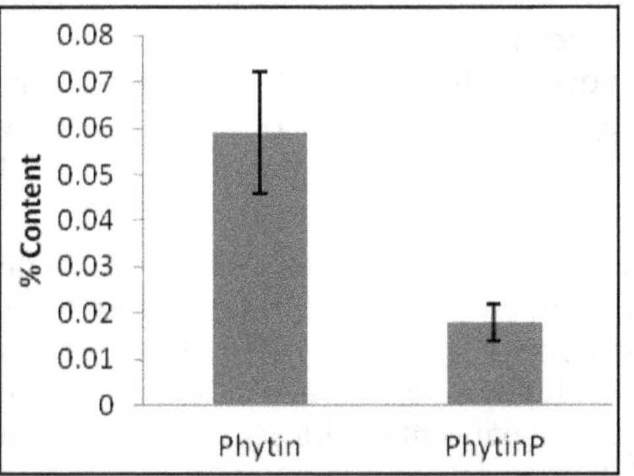

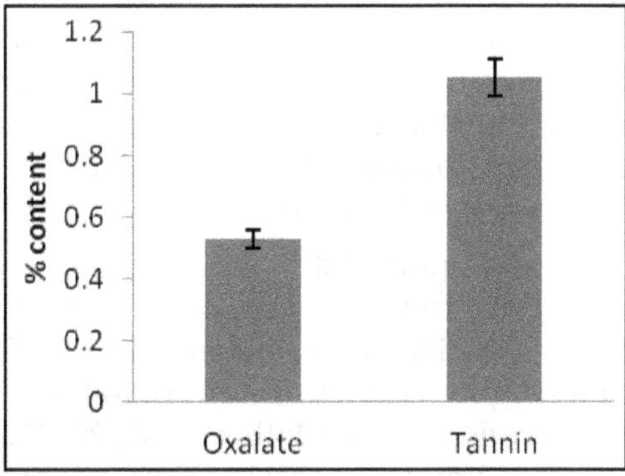

Figure 11.2: Anti-nutrient Contents (mg/kg) in the Body Tissues of *O. fuscovittata* on Dry Matter Basis. Data are presented as means ± SD.

amino acids. It was interesting to observe that all of the indispensible amino acids were extant in *O. fuscovittata*, though Methionine and Isoleucine were found to be limiting (less than 2 per cent) in this insect. Among the the non-essential amino acids Asx and Cysteine were also found in very low amount (less than 0.5 per cent). The findings were in agreement with the report of Finke (2012) that sulfur amino acids are usually sparse in edible insects. All the remaining amino acids were detected in quite good amount where Threonine was the most prevalent one (more than 18 per cent). A very high amount of Threonine was also reported in another grasshopper *Oedaleus abruptus* (Ganguly *et al.*, 2013). Despite having a good amount of protein and amino acids a critical look on its energy content is always necessary because animals cannot utilise protein unless and until there is enough calorific value in their meals (Rivas-Vega et. al, 2013). In this context a good quantity of energy in *O. fuscovittata* yet again showed the credibility of this insect as a candidate edible insect species.

We have detected six fatty acids, but their presence and magnitude was inconsistent with the previous reports. This could be explained by the fact that usually insects do not follow any pattern in their fatty acid contents because it depends on the food that they consume (St-Hilaire*et al.*, 2007). However, the two PUFAs, along with Eicosenoic acid were the dominant ones in the present case as reported in many other edible insects such as termites (Bukkens, 1997; Ekpo, 2009) and grasshoppers (Kinyuru *et al.*, 2010). Ramos-Elorduy *et al.* (1988) observed that edible insects are rich in vitamin B group such as Thiamin, Riboflavin and Niacin. But in the current study Thiamin and Riboflavin were observed to be limiting, which was rather similar to the reports on a tettigonid grasshopper *Ruspolia differens* (Kinyuru *et al.*, 2010).Various titer of ascorbic acid have been reported in edible insects till date, such as Banjo *et al.* (2006) reported more than 3mg/100g of ascorbic acid in termites, while according to Finke (2002) the same vitamin varied from less than 1.0 mg/100g in silkworm to 5.4mg/100 g in adult mealworm. In contrast, the presence of thisvitaminwasquite higher in the present study. In one of our previous investigations on *O. abruptus* also a very high level of ascorbic acid was detected (Ganguly *et al.*, 2013). This might be an indicator that acridids (*i.e.* short-horn grasshoppers) are a good source of this vitamin. Retinol was also perceived in quite good amount which was fairly higher than the reports on various other orthopterans (Alamu, 2013). Minerals were not identified in very appreciable amount in the insect of our interest that was analogous with the view of Ramos-Elorduy (2005) that orthopteroid insects are low in mineral contents. Nevertheless, Ca and Mg were the dominant minerals present in *O. fuscovittata*.

Adeduntan (2005) reported tannin percentage in grasshoppers of Ondo state of Nigeria to be 1.05 per cent. This corresponded exactly with our results of *O. fuscovittata*. However, the same author also reported 1.1 per cent of Phytate that was quite higher than our findings. Rather the value was in between 0.025 per cent of *Coptotermes gestroi* (Mathew *et al.*, 2013) and 0.18 per cent of *Oryctes monoceros* larva (Ifie and Emeruwa, 2011). Although oxalate was detected in a little higher extent than reported in other edible insects like 0.011 per cent in *Zonocerus variegates* (Sani *et al.*, 2014), 0.1 per cent in *Bombyx mori* (Omotoso, 2015), *etc.*, yet again this value is under tolerance limit because similar or higher level of oxalate is already reported in various edible plant materials such as roasted almond (Chai and Liebman, 2005), spinach (Tsai *et al.*, 2005), silver beet (Wadamori*et al.*, 2014) *etc.*

Conclusion

We can conclude the present study with the notion that *O. fuscovittata* is a nutrient rich feed supplement that not only contains a good amount of protein and energy, but also necessary amino acids. Two very important PUFAs were also detected in considerable amount. Additionally it was also found that this insect is rich in Retinol, Ascorbic acid and Niacin. The findings ultimately revealed that this insect has a great prospect to be utilised in feed industry as a descent nutrient supplement in the diets of various livestock.

References

Adeduntan, S.A., 2005. Nutritional and anti-nutritional characteristics of some insects foraging in Akure Forest Reserve, Ondo State, Nigeria. Journal of Food Technology 3:563-567.

Agbede, J.O. and Aletor, V.A., 2004. Chemical characterization and protein quality evaluation of leaf protein concentrates from *Glyricidia sepium* and *Leucaenaleuco cephala*. International Journal of Food Science and Technology 39:253-361.

Akinnawo, O. and Ketiku, A.O., 2000. Chemical composition and fatty acid profile of edible larva of *Cirinaforda* (Westwood). African Journal of Biomedical Research 3:93-97.

Alamu, O. T., Amao, A.O.I, Nwokedi, C. I., Oke, O.A. and Lawa, I. O., 2013. Diversity and nutritional status of edible insects in Nigeria: A review. International Journal of Biodiversity and Conservation 5:215-222.

Anand, H., Ganguly and A., Haldar, P., 2008b. Biomass production of acridids as possible animal feed supplement. Journal of Environment and Sociobiology 5:181-190.

Anand, H., Ganguly, A. and Haldar, P., 2008a. Potential value of Acridids as high protein supplement for poultry feed. International Journal of Poultry Science 7:722-725.

Banjo, A.D., Lawal, O.A. and Songonuga, E.A., 2006.The nutritional value of fourteen species of edible insects in Southwestern Nigeria.African Journal of Biotechnology 5: 298- 301.

Bettelheim, F.A. and Landesberg, J.M., 2001. Laboratory experiments for general, organic and biochemistry. Saunders College Publishing, USA, pp. 567.

Bukkens, S. G. F., 1997. The nutritional value of edible insects. Ecology of Food and Nutrition 36:287–319.

Chai, W. and Liebman, M., 2005.Oxalate content of legumes, nuts, and grain-based flours. Journal of Food Composition and Analysis 18: 723–729

Das, M. and Mandal, S.K., 2014. *Oxyahylahyla* (Orthoptera: Acrididae) as an Alternative Protein Source for Japanese Quail. International Scholarly Research Notices 2014: 1-14.

Das, M., Ganguly, A. and Haldar, P., 2012. Annual biomass production of two acridids (Orthoptera: Acrididae) as alternative food for poultry. Spanish Journal of Agricultural Research 10:671-680.

de Guevara, O. L.,Padilla, P.,García, L., Pino, J. M. M. and Ramos-Elorduy, J., 1995. Amino Acid Determination in Some Edible Mexican Insects.Amino Acids 9:161-173.

Ekpo, K. E., Onigbinde, A. O. and Asia, I. O., 2009.Pharmaceutical potentials of the oils of some popular insects consumed in southern Nigeria. African Journal of Pharmacy and Pharmacology 3:51–57.

Finke, M.D., 2002. Complete nutrient composition of commercially raised invertebrates used as food for insectivores. Zoo Biology 21:286–293.

Finke, N.D., 2012. Complete Nutrient Content of Four Species of Feeder Insects. Zoo Biology 00: 1–15.

Fischl, J., 1960. Quantitative colorimetric determination of tryptophan. Journal of Biological Chemistry 235:999-1001.

Ganguly, A., Chakravorty, R. and Haldar, P., 2013. Assessment of consumption, utilisation and growth of *Oedaleus abruptus* (Thunberg) and *Spathosternum prasiniferum prasiniferum* (Walker) (Orthoptera: Acrididae) fed with various food plants in laboratory conditions. Annales de la Sociétéentomologique de France (N.S.): International Journal of Entomology 49: 2160-171.

Ganguly, A., Chakravorty, R., Sarkar, A. and Haldar, P., 2010. Johnson Grass [*Sorghum halepense* (L.) Pers.]: A Potential Food Plant for Attaining Higher Grasshopper Biomass in Acridid Farms. Philippine Agricultural Scientist 93:329-336.

Ganguly, A., Chakravorty, R., Sarkar, A., Mandal, D.K., Haldar, P., Ramos-Elorduy, J. and Moreno, J.M.P., 2014. A Preliminary Study on *Oxyafuscovittata* (Marschall) as an Alternative Nutrient Supplement in the Diets of *Poecillia sphenops* (Valenciennes). PLoS ONE 9:e111848.

Ghosh, A.K., Naskar, A.K., Jana, M.L., Khowala, S. and Sengupta, S., 1995. Purification and characterization of an amyloglucosidase from *Termitomyces clypeatus* that liberates glucose from xylan. Biotechnology Progress 11:452-456.

Ghosh, S., Haldar, P. and Mandal, D.K., 2014.Suitable food plants for mass rearing of the short-horn grasshopper Oxyahylahyla (Orthoptera: Acrididae). European Journal of Entomology 111:448-452.

Gupta, P.C., Khatta, V.K. and Mandal, A.B., 1988.Analytical techniques in animal nutrition. Haryana Agricultural University, Hisar,pp.97.

Helrich, K., 1990. Official methods of analysis of the Association of Official Analytical Chemists. A.O.A.C. Philadelphia, USA, pp. 1422.

Ifie, I. and Emeruwa, C., H., 2011.Nutritional and anti-nutritional characteristics of the larva of *Oryctesmonoceros*. Agriculture and Biology Journal of North America 2:42-46.

Kinyuru, J. N., Kenji, G. M., Muhoho, S. N. and Ayieko, M. A., 2010. Nutritional potential of longhorn grasshopper (Ruspoliadifferens) consumed in Siaya District, Kenya. Journal of Agricultural Science and Technology 12:32–46.

Mathew, T.J., Ndamitso, M. M., Shaba, E.Y., Mustapha, S., Muhammed, S.S. and Adamu, A., 2013. Phytochemical, Physicochemical, Anti-Nutritional And Fatty Acids Composition Of Soldier Termites (Coptotermesgestroi) From Paikoro Local Governvernt, Niger State, Nigeria. IOSR Journal of Environmental Science, Toxicology and Food Technology 7:71-75.

Omotoso, O. T., 2015. An Evaluation of the Nutrients and Some Anti-nutrients in Silkworm, *Bombyx mori* L. (Bombycidae: Lepidoptera). Jordan Journal of Biological Sciences. 8:45 – 50.

Ramos-Elorduy J.B., Moreno, J.M.P. and Víctor-Hugo M. C., 2012. Could Grasshoppers Be a Nutritive Meal?.Food and Nutrition Sciences 3:164-175.

Ramos-Elorduy, J., 2005. Insects: A hopeful food source. In: G. M. Paoletti, (ed.), Ecological Implications of Minilive- stock (Potential of Insects, Rodents Frogs and Snails). Science Publishers, Inc., Plymouth, pp. 262-292.

Ramos-Elorduy, J., de Morales, L.J., Pino, J.M. and Nieto, Z., 1988. Contenido de tiaminariboflavina y niacina en algunos comestibles de Mexico. RevistaTecnologia deAlimentos. 23:21.

Rivas-Vega, M.E., Gil-Romero, A., Miranda-Baeza, A., Sandoval-Muy, M.A., López-Elías, J.A. and Nieves-Soto, M., 2013.Effect of Protein to Energy Ratio on Growth Performance, Body Composition and Enzymatic Digestive Activity of Juvenile Tilapia (*Oreochromis niloticus* x *O. mossambicus*) Reared in Seawater. Current Research Journal of Biological Sciences 5:30-35.

Sani, I., Haruna, M., Abdulhamid, A., Warra, A.A., Bello, F. and Fakai, I.M., 2014. Assessment of Nutritional Quality and Mineral Composition of Dried Edible *Zonocerus variegates* (Grasshopper). Research and Reviews: Journal of Food and Dairy Technology. 2:1-6.

Siriamornpun, S. and Thammapat, P., 2008.Insects as a Delicacy and a Nutritious Food in Thailand. In Robertson, G.L. and Lupien, J.R. (eds.) Using Food Science and Technology to Improve Nutrition and Promote National Development, International Union of Food Science and Technology, pp. 1- 12.

St-Hilaire, S., Cranfill, K., McGuire, M.A., Mosley, E.E., Tomberlin, J.K., Newton, L., Sealey, W., Sheppard, C., Irving, S., 2007b. Fish offal recycling by the black soldier fly produces a food stuff high in omega-3 fatty acids. Journal of World Aquaculture Society 38:309–313.

Tsai, J.Y., Huang, J.K., Wu, T.T. and Lee, Y.H., 2005. Comparison of Oxalate Content in Foods and Beverages in Taiwan.Journal of Taiwan Urological Association 16:93-98.

Wadamori,Y., Vanhanen, L. and Savage, G.P., 2014. Effect of Kimchi Fermentation on Oxalate Levels in Silver Beet (*Beta vulgaris* var. cicla). Foods 3: 269-278.

Wang, D.,Zhai, S., Zhang, C., Zhang, Q. and Chen, H., 2007.Nutritional value of the Chinese grasshopper Acridacinerea (Thunberg) for broilers. Animal Feed Science and Technology 135:66-74.

Yen, A.L., 2015. Foreword: Why a Journal of Insects as Food and Feed? Journal of Insects as Food and Feed 1: 1-2.

Zhou, J. and Han, D., 2006. Proximate, amino acid and mineral composition of pupae of the silkworm *Antheraeapernyi* in China. Journal of Food Composition and Analysis 19:850–853.

– Segment 7 –

Insect Pest Management

Chapter 12

Role of Semiochemicals in Vegetable Insect Pests Management

☆ *Jaydeep Halder and Atanu Seni*

ABSTRACT

Pheromones are the chemical signals used for communication between members of the same insect species. Amongst the several roles that have been elucidated for pheromones include attraction, aggression, aggregation, alarm signalling etc. This eco-friendly and green technology is now widely used in pest management across the crops including vegetables. In India, sex-pheromones of brinjal shoot and fruit borer (Leucinodes orbonalis), okra shoot and fruit borer (Earias vittella and E. insulana), diamond back moth (Plutella xylostella), tobacco caterpillar (Spodoptera litura), gram pod borer (Helicoverpa armigera) and recently introduced tomato pin worm (Tuta absoluta) etc. are widely used for their management by the vegetable growers of the country. The kinetics, biological role, advantages and limitations of pheromones are discussed in the chapter.

Introduction

Vegetables are important in human nutrition as they are rich sources of essential proteins, vitamins and minerals and complement the starchy staple foods well. In addition to this, they are often considered as high value crops and provide excellent income-generating opportunities to small and marginal farmers. One of the major constraints in vegetable production is the infestation of various insect pests. In India, crop losses due to different pests range from 10 to 30 per cent each year depending upon the severity of pest attack and losses to the tune of 30-40 per cent have been reported in vegetable crops (Rai *et al.*, 2014). To combat such problems growers are using many chemicals indiscriminately as well as non-judiciously. This not only increases the cost of cultivation but often causes problems like resistance to insecticides, resurgence of sucking insects, and secondary pest outbreak; in addition it leaves harmful residues that finally leads to contamination of groundwater, adverse effect on human health, and wide spread killing of non-target organisms (Halder *et al.*, 2011, 2014). Many farmers also apply different insecticides even without any prior knowledge about present pest status in the field. By this, they lose huge money for the management of insect pests. To minimize the risk of indiscriminate use of insecticides, now a days emphasis is given on integrated pest management. It includes many approaches *i.e.*, cultural, physical, mechanical, biological, chemical *etc.* Use of semiochemicals in pest management is one of the ecofriendly approaches under biorational pest management because these are non-toxic and biodegradable in nature. A detailed account of semiochemicals and pheromone is furnished below.

Semiochemicals and Pheromone

The word 'Semiochemical' has been derived from the Greek word '*Semeon*', means a signal. It can be defined as a chemical substance or mixture that carries a message. These chemicals act as messengers within or between the species. It is usually used in the field of chemical ecology to encompass pheromones and allelochemicals. Allelochemical includes allomones, kairomones, attractants and repellents.

Allelochemical

Chemicals significant to organisms of a species different from their source, for reasons other than food as such (Nordlund and Lewis, 1976). These could be of the following types:

Allomone

A substance, produced or acquired by an organism, which, when contacts an individual of another species in the natural context, evokes in the receiver a behavioral or physiological reaction adaptively favorable to the emitter but not to the receiver (Nordlund and Lewis, 1976). An allomone is a chemical or a mixture of chemicals that if released by one organism to induce a response in an individual of another species is advantageous to the releaser (Srivastava, 2001).

Kairomone

A substance, produced, acquired by, or released as a result of the activities of an organism, which, when contacts an individual of another species in the natural context, evokes in the receiver a behavioral or physiological reaction adaptively favorable to the receiver but not to the emitter (Nordlund and Lewis, 1976). A kairomone is a chemical or a mixture of chemicals produced by one species to induce response in an individual of a different species but is advantageous to the recipient (Srivastava, 2001).

Synomone

A substance produced or acquired by an organism, which, when contacts an individual of another species in the natural context, evokes in the receiver a behavioral or physiological reaction adaptively favorable to both emitter and receiver (Nordlund and Lewis, 1976)

Apneumone

A substance emitted by a non-living material that evokes a behavioral or physiological reaction adaptively favorable to a receiving organism, but detrimental to an organism, of another species, which may be found in or on the non-living material (Nordlund and Lewis, 1976).

Pheromone

These are intraspecific chemicals produced in exocrine glands and emitted. Critically these are the molecules that are evolved signals, in defined ratios in the case of multiple component pheromones, which are emitted by an individual and received by a second individual of the same species, where they cause a specific reaction, for example, a stereotyped behavior or a developmental process (Wyatt, 2009). A pheromone is a chemical or a mixture of chemicals released by an organism to the outside (in the environment) that cause a specific reaction in a receiving organism (Srivastava, 2001).

Table 12.1: Time Scale of Development of Insect Pheromones

Period	*Major Development*
1870s	French naturalist Jean- Henri Fabre notices a female peacock moth is able to attract 150 male peacock moths from miles away.
	New York entomologist Joseph A. Lintner suggests the chemical scents emitted by insects could be used to control insect pests.
1957	German biologist Dietrich Schneider develops the electroantennogram (EAG), a method for using the antenna of a moth to detect pheromones electrically.
1959	German biochemist Peter Karlson and Swiss entomologist Martin Lüscher, 1959 coin the term "Pheromone" to describe a compound an animal gives off that triggers a specific behavioral or developmental reaction in a member of the same species.
	German chemist Adolf Butenandt isolates and characterizes the first insect pheromone from *Bombyx mori*
1960s	Pheromone researchers begin to use GC, MS and NMR along with EAG to identify insect pheromones.
1961	C. G. Butler identifies the pheromone of the honey bee, the first pheromone that regulates the development of an insect.
1966	Chemist Robert Silverstein and entomologist David Wood demonstrate that all three components of the bark beetle's pheromone blend are required to attract the beetles—a phenomenon known as synergism.
1967	Entomologist Harry Shorey shows that pheromones can be used to disrupt the mating of cabbage looper moths in the field.
1970s	British biologist John Kennedy develops the wind tunnel assay.
1971	Wendell Roelofs uses EAG as an analytical tool to identify the *Cydia pomonella* pheromone.
1978	First pheromone is registered in the USA for commercial use in mating disruption— against the *Pectinophora gossypiella*.

Period	Major Development
1980	Pheromones are used in more than a million traps to capture more than four billion beetles, curbing an epidemic of bark beetles in the forests of Norway and Sweden.
1990s	Pheromones used for mating disruption effectively help curb insect damage in stone-pitted fruit orchards, tomato, rice, cotton, and grape fields.

"Insect pheromones mastering communication to control pests" In: www.beyonddiscovery.org visited on 17/10/2013.

Types of Insect Pheromones

- **Sex pheromone**: Substance produces to attract opposite sex, usually produced by female to attract male. Generally work over long distances and used by highly mobile, dispersed species. Also used by species with short lived adults.
- **Aggregation pheromone**: Substance produced by one or both sexes that brings insects together to feed or to reproduce. Attract both males, females, adults and larvae of same species. Generally acts over short distances. Used by species with long lived adults.
- **Alarm pheromone**: Alarm pheromones alert conspecifics in case of any threats. Generally, the response behavior results in dispersion of congeners. These pheromones, characteristic of social or gregarious insects, occur in some important insect pests including Aphididae and Thripidae. This class of pheromones has potential in IPM (Verheggen *et al.*, 2010).
- **Marking pheromones**: Compounds used by insects to mark the boundaries of a territory.
- **Trail pheromones:** Among social insects, compounds used by workers to mark the way to a food source.

Classification of Social Insect Pheromones

Pheromones are classified as **releaser** and/or **primer** pheromones.

- **Releaser pheromones** release a behavioral response that is mediated by the nervous system. Releaser pheromones elicit an immediate behavioral response.
- **Primer pheromones** alter endocrine, reproductive and neurological systems. Changes are not immediate but occur within hours to days.

Many social insect pheromones are releaser and primer pheromones that are blend of many components.

Releaser pheromones		
Pheromone	*Source*	*Behavioral Response*
2-heptanone	Mandibular gland	Alarm (aggression)
Isopentyl acetate	Venom	Defensive (stinging)
Queen mandibular pheromone	Mandibular gland	Retinue behaviour

Primer pheromones		
Pheromone	*Source*	*Behavioral Response*
Queen mandibular pheromone	Mandibular gland	Partial inhibition of workers' ovary development
Brood pheromone	Larvae	Stimulate hypopharyngeal gland development
Brood pheromone	Larvae	Partial inhibition of workers' ovary development

Application of Semiochemicals in IPM Strategies

- ☆ Monitoring (Indirect control)
- ☆ Mass trapping (Direct control)
- ☆ Mating disruption (Lepidopterans mainly)
- ☆ Biological control – Attraction of natural enemies using kairomones

Table 12.2: Sex Pheromone of some Major Vegetable Insect Pests

Name of the Insect	*Chemical Composition*	*Ratio*
Leucinodes orbonalis	(E)-11-hexadecenyl acetate and (E)-11-hexadecen-1-ol	100:1
Helicoverpa armigera	(Z) 11-Hexadecenal and (Z) 9-Hexadecenal	97:3
Spodoptera litura	(Z,E) 9,11-tetradecadenyl acetate and (Z,E) 9,12-dienyl acetate	19:1
Earias insulana	(E,E) 10,12-Hexadecadienal	-
Trichoplusia nii	(Z) 7-Dodecenyl acetate	-
Plutella xylostella	(Z)-11 hexadecenal and (Z)-11 hexadecenyl acetate	-
Tuta absoluta	(3*E*, 8*Z*, 11Z)-3,8,11-tetradecatrien-1-yl acetate and (3*E*, 8*Z*)-3,8-tetradecadien-l-yl acetate	9:1

For Monitoring

Insect pheromones widely can be used as surveying and early monitoring of insect pests. Mapping of pest distribution in a particular area as well as for population dynamics study this can be used. Initiation of timely control measure based on economic threshold level (ETL) can be adjudged. Dispersal and migration behaviour of the insects, pest risk analysis and occurrence of the insecticide resistance of the insects can also be worked out.

Mass Trapping

Another important usage of insect pheromone is the mass trapping of the insects. Sex pheromone baited trap can catch the male moth continuously and thereby preventing the mating and further multiplication of the pest. A trapping efficiency of more than 90 per cent is required in case of Lepidoptera males to achieve the desired results (Jutsum and Gordon, 1989). Sex pheromones for mass trapping of some important vegetable lepidopteran insect pests mainly brinjal shoot and fruit borer (*Leucinodes orbonalis*), bhendi shoot and fruit borer (*Earias vittella* and *E. insulana*), tomato fruit borers (*Spodoptera litura* and *Helicoverpa armigera*) *etc.* are available commercially. Installation of sex pheromone traps 15 days after transplanting, and lure replacement after 25-30 days are found effective against brinjal shoot and fruit borer, *L. orbonalis* (Shivalingaswamy *et al.*, 2006).

Mating Disruption

Another important usage of sex pheromone is the mating disruption. High concentrations of sex pheromones saturate the environment and suppress the normal pheromones emitted by the calling females. Thus confuse the males and prevent the mating.

The melon fruit fly, *Bactrocera cucurbitae*, one of the major pests of cucurbitaceous vegetables can also be controlled by using cue lure. It is also found that, leaf extract of *Ocimum sanctum*, which contain eugenol (53.4 per cent), beta-caryophyllene (31.7 per cent) and beta-elemene (6.2 per cent) as the major volatiles, when placed on cotton pads (0.3 mg), it attract fruit flies from a distance of 0.8 km (Roomi *et al.*, 1993; Dhillon, 2005). Melon fruit fly can also be managed through growing of *O. sanctum* as the border crop of cucurbitaceous vegetables and sprayed on them with protein bait (protein derived from corn, wheat or other sources) containing spinosad as a toxicant (Dhillon, 2005). Mitchell, (2002) found that a sex pheromone with toxicant, the blend of Z-11-hexadecenyl acetate, 27 per cent : Z-11-hexadecen-1-ol, 1 per cent : Z-11-tetradecen-1-ol, 9 per cent : Z-11-hexadecenal, 63 per cent, and the insecticide permethrin (0.16 per cent and 6 per cent w/w of total formulated material, respectively) was highly attractive to male moths *Plutella xylostella*, a major pest of cruciferous vegetables. Further, this lure can be used as to suppress the sexual communication of diamondback moth. Cocco *et al.* (2013) reported that reductions in leaf and fruit damages of tomato by *Tuta absoluta* occurred in high-containment greenhouses by disruption of the sexual communication with a dose of 60 g/ha (dispensers loaded with 60 mg of sex pheromone at a density of 1,000 dispensers/ ha). The female sex pheromone of *T. absoluta* consists of two components. The major component, which represents about 90 per cent of the volatile material found in the sex gland of calling females, is (3*E*, 8*Z*, 11Z)-3,8,11-tetradecatrien-1-yl acetate. The minor constituent (10 per cent) was identified as (3*E*, 8*Z*)-3,8- tetradecadien-l-yl acetate (Megido *et al.*, 2013). Bolckmans, (2009) suggested that for monitoring and mass trapping of *T. absoluta* in plant propagation greenhouses, 10–20 traps/ha are necessary. For lure and kill of *T. absoluta*, a matrix formulated with 0.3 per cent sex pheromone and 3 per cent cypermethrin is suitable (Megido *et al.*, 2013).

Biological Control – Attraction of Natural Enemies using Kairomones

Tricosane, a compound present in the scale of many lepidopteran insects including rice moth, *Corcyra cephalonica*, is the example of a kairomone. Rubber septa impregnated with synthetic Tricosane was found to attract the *Trichogramma* spp in the cotton and rice fields.

Slow Release Pheromones (SRS)

Most of the commercially available insect pheromones are not effective throughout the cropping season. Frequent replacements are necessary and thereby increase the cost of production. So, a slow release formulation is an alternative option for this. Slow release semiochemicals can be defined as "the dispenser or formulation which release semiochemical at a constant rate throughout the flight period of insect and is independent of weather condition". To be efficient in IPM strategies, semiochemical slow-release devices must have particular specifications: the aerial concentration after release must be sufficiently high to be detected by insects; the release of semiochemicals must be effective during all the period of insect occurrence; the production of dispenser must be reproducible. The application of dispensers must be realized early in the season when the pest density is not too high, given that their release rates, for the majority of devices, decrease with time (Witzgall, 2001).

Release Rate of Slow Release Pheromone Depends on

Two factors

i. Diffusion speed – characteristic of dispenser

ii. Evaporation speed – environmental parameters

$$C_0 = C_t e^{-kt}$$ (Mottus *et al.*, 2001)

C_0 = Amount of compound at the beginning

C_t = Amount of compound at time 't'

k = Evaporation rate constant

t = Time

Different Slow Release Pheromone (SRP) Dispensers and Sprayable Formulations

Solid Matrix Dispensers

Solid matrix dispensers are hand-applied on foliage. Here the chemicals are incorporated in a solid matrix. Based up on the constituting materials of the matrix, the release rates for a single molecule can differ significantly from another. The effective lifetime of the biggest solid matrix dispensers can range from 60 to 140 days.

Examples of Solid Matrix Dispensers:

a. **Polythelene sachets:** Used for trapping of tsetse fly kairomone (1-Octen-3-ol, 4-methylphenol and 3-n-Propyl phenol) (*Glossina* sp.). Pheromone release rate was independent of the amount present in the dispensers (zero order release kinetics), are positively related with surface area and temperature and negatively related with the surface area. Move directly to the surface (Torr *et al.*, 1997). Another example of this SRP is male aggregation pheromone of dynast beetle (*Scapanes australis*) (Rochat *et al.*, 2002)

b. **Polythelene vials and tubes:** Mainly developed against lepidopteran pests in orchards. Release rate depends on nature of dispenser, wind velocity and atmospheric temperature. Mating disruption of codling moth (*Cydia pomonella*), sex pheromones of light brown apple moth (*Epiphyas postvittana*) etc.

c. **Rubber septa:** Molecular sizes of the pheromone are one of the important criteria for determining the evaporation rate of the pheromone from the rubber septa. So, alcohol and acetone molecules were found as sex pheromones are ideal for these rubber septa. Sex pheromones of *Cydia pomonella* was used by these rubber septa (Butler *et al.*, 1981; Kehat *et al.*, 1994).

d. **Hollow fibres:** First developed for pink boll worm of cotton, *Pectinophora gossypiella* (Gelechiidae: Lepidoptera) and here the pheromones were impregnated within the hollow fibres. Total solvent extraction method was followed for the measurement for pheromone release rate (Zhang *et al.*, 2008).

e. **Plastic dispenser:** Employed for sex pheromones of *Helicoverpa zea* in cotton and maize field. Linear decrease of release rate with time was observed (Lopez *et al.*, 1991). Similarly, sex pheromones of rice yellow stem borer, *Scirpophaga incertulus* was also developed in plastic dispenser. Here the half-life of the pheromone was found to decrease with atmospheric temperature. The pheromone release rate followed the first order kinetics (Cork *et al.*, 2008).

f. **Sol-Gel matrix:** Sol-gel matrix exhibits glass characterization. Cross linking during polymerization process rendering optimal release rate. Entrapped pheromones are chemically stable and safe for dispersing. Release rate for peach twig borer, *Anarsia lineatella* constant rate @ 14-45 µg/day (Anat *et al.*, 2009).

Sprayable Slow-Release Formulations

These are generally composed of a biodegradable liquid matrix compound in which the desired semiochemical is dissolved. Other components *viz.*, UV-stabilizers, antioxidants, stickers and surfactants can be added to protect the semiochemicals. Frequently, the sprayable formulation consists in a micro-emulsion, resulting in polymeric micro-beads containing the semiochemicals (micro-encapsulated pheromones) dispersed in a liquid matrix (de Vlieger, 2001). The time of efficiency of such formulations ranges from days to weeks depending on environmental factors, microbeads size, release capacities, and the pheromones chemical properties (Welter, 2005). The major advantage of sprayable formulations compared to solid matrix dispensers is that the entire crop can be treated.

Different Types of Sprayable Formulations

a. **Paraffin emulsions:** Atterholt *et al.*, 1999 reported the Oriental fruit moth, *Grapholita molesta* (Tortricidae: Lepidoptera) mating disruptant in paraffin emulsions where blend of 3 compounds impregnated with paraffin. Release rate of pheromones depends on temperature and surface areas.

b. **Microcapsules:** Here the pheromone immobilized on a porus substrate is coated with a polymer film membrane. Sex pheromones of codling moth and gypsy moth (*Lymantria dispar*) have been developed in the microcapsules. Release rate depends on coating of the microcapsules, surface areas and micropore volume (Stipanovic *et al.*, 2004).

c. **Reservoir with silicon diffusion area:** Blend of tsetse fly allomone (Carboxylic acid, Ketone, δ-Octalactone and 2-Methoxy-4-methyl phenol) was developed in the reservoir with silicon diffusion area. Pheromone release rate depends on the atmospheric temperature and one compound interacts with each other (Shem *et al.*, 2009).

Advantages of Slow Release Pheromones

- ☆ Non-toxic and active at very low concentration
- ☆ Longevity almost throughout the season
- ☆ Can be combined with other IPM techniques
- ☆ So far there is no report of human and animal health hazard
- ☆ Species specific, so safe for natural enemies and pollinators
- ☆ No resistance and no report of any resurgence

Limitations

- ☆ Surrounding field or orchards must be part of the programme
- ☆ Not available for all the pests
- ☆ Degradation by UV-light and oxygen
- ☆ Application are laborious and some are costly
- ☆ Do not give immediate result

Conclusion

Pheromones are non-toxic, ecofriendly species-specific chemicals that affect insect behavior. They are active (*e.g.* attractive) in extremely low doses and are used to bait traps or confuse a mating population of insects. Pheromones can play an important role in integrated pest management for major vegetable insect pests under Indian conditions.

References

Anat Z, Falach L and John AB. 2009. Development of sol–gel formulations for slow release of pheromones. *Chemoecology*, 19:37–45.

Atterholt CA, Delwiche MJ, Rice RE and Krochta JM. 1999. Controlled release of insect sex pheromones from paraffin wax and emulsions. *Journal of Controlled Release*, **57**: 233-247.

Bolckmans K. 2009. Integrated pest management of the exotic invasive pest *Tuta absoluta*. *In:* International Biocontrol Manufacturers Association and Research Institute of Organic Agriculture, eds. *Proceedings of the 4th Annual Biocontrol Industry Meeting Internationals, Lucerne, Switzerland*.

Butler LI and McDonough LM. 1981. Insect sex pheromones: evaporation rates of alcohols and acetates from natural rubber septa. *Journal of Chemical Ecology*, 7(3):627-633.

Cocco A, Deliperi S and Delrio G. 2013. Control of *Tuta absoluta* (Meyrick) (Lepidoptera: Gelechiidae) in greenhouse tomato crops using the mating disruption technique. *Journal of Applied Entomology*, 137 (1-2):16-28.

Cork A, D'Souza HDR, Jones OP, Casagranade E, Krishnaih K and Syed Z. 2008. Development of a PVC-resin-controlled release formulation for pheromones and use in mating disruption of yellow rice stem borer, *Scirpophaga incertulas*. *Crop Protection*, **27**: 248-255.

De Vlieger JJ. 2001. Development of a sprayable slow-release formulation for the sex pheromone of the Mediterranean corn borer, *Sesamia nonagroides*. *IOBC WPRS Bulletin*, **24**(2): 101-106.

Dhillon MK, Singh R, Naresh JS and Sharma HC. 2005. The melon fruit fly, *Bactrocera cucurbitae*: A review of its biology and management. *Journal of Insect Science*, 5:40.

Halder J, Kodandaram MH and Rai AB. 2011. Differential responses of major vegetable aphids to newer insecticide molecules. *Vegetable Science*, **38**: 191-193.

Halder J, Rai AB and Kodandaram MH. 2014. Parasitization preference of *Diaeretiella rapae* (McIntosh) (Hymenoptera: Braconidae) among different aphids in vegetable ecosystem. *Indian Journal of Agricultural Sciences*, **84**(11):1431–1433.

Jutsum AR and Gordon RFS. 1989. Insect Pheromones in Plant Protection, John Wiley and Sons, New York, USA.

Karlson P and Lüscher M. 1959. 'Pheromones': a new term for a class of biologically active substances. *Nature*, 183:55–56.

Kehat M, Anshelevich E, Dunkelblum P, Fraishtat and Greenberg S. 1994. Sex pheromone traps for monitoring the codling moth: effect of dispenser type, field aging of dispenser, pheromone dose and type of trap on male captures. *Entomologia Experimentalis et Applicata*. 70:55–62.

Lopez JD, Leonhardt BA and Shaver TN. 1991. Performance criteria and specifications for laminated plastic sex pheromone dispenser for *Helicoverpa zea* (Lepidoptera: Noctuidae). *Journal of Chemical Ecology*, **17**(11): 2293-2305.

Megido RC, Haubruge E, Verheggen FJ. 2013. Pheromone-based management strategies to control the tomato leafminer, *Tuta absoluta* (Lepidoptera: Gelechiidae). A Review. *Biotechnology, Agronomy, Society and Environment*, 17(3), 475-482.

Mitchell ER. 2002. Promising new technology for managing diamondback moth (Lepidoptera: Plutellidae) in cabbage with pheromone. *Journal of Environmental Science and Health Part B - Pesticide Food Contamination*, 37(3):277-290.

Mõttus E, Liblikas I, Ojarand A, Kuusik S, Nikolaeyva Z, Ovsjannikova E, Karlson AKB. 2001. Calculation and using of pheromone communication channel parameters for optimization of pheromone dispensers. *In:* Metspalu L. and

Mitt S., eds. *Proceedings of the international workshop, Practice oriented results on the use of plant extracts and pheromones in pest control, 24-25th January 2001, Tartu, Estonia,* 101-125.

Nordlund DA, and Lewis WJ. 1976. Terminology of chemical releasing stimuli in intraspecific and interspecific interactions. *Journal of Chemical Ecology*. 2:211–220.

Rai A B, Halder J and Kodandaram M H. 2014. Emerging insect pest problems in vegetable crops and their management in India: An appraisal. *Pest Management in Horticultural Ecosystems*, 20(2):113-122.

Rochat D, Morin JP, Kakul T, Ollivier LB, Prior R, Relou M, Mallosse I, Stathers T, Embupa S and Laup S. 2002. Activity of male pheromone of Melanesian rhinoceros beetle *Scapanes australis*. *Journal of Chemical Ecology*, **28**(3):479-500.

Roomi MW, Abbas T, Shah AH, Robina S, Qureshi AA, Hussain SS and Nasir KA. 1993. Control of fruit flies (*Dacus* spp.) by attractants of plant origin. Anzeiger fur Schadlingskunde, Aflanzenschutz, Umwdtschutz. 66:155–157.

Shem PM, Shiundu PM, Gikonyo N K, Ali AH and Saini R K. 2009. Release kinetics of a synthetic tsetse allomone derived from waterbuck odour from a Tygon silicon dispenser under laboratory and semi field conditions. *American Eurasian Journal of Agriculture and Environmental Sciences*, 6: 625-636.

Shivalingaswamy TM, Satpathy S, Rai AB and Rai M. 2006. Insect Pests in vegetable crops: Identification and management. Technical Bulletin No.30, IIVR, Varanasi, pp. 6.

Srivastava KP. 2001. A text book of applied entomology, Vol-I, Kalyani Publishers, Ludhiana-141008, India, pp-238-255.

Stipanovic AJ, Hennessy PJ, Webster FX and Takahashi Y. 2004. Microparticle dispensers for the controlled release of insect pheromones. *Journal of Agriculture and Food Chemistry*. **52**: 2301-2308.

Torr SJ, Hall DR, Phelps RJ and Vale GA. 1997. Methods for dispensing odour attractants for tsetse flies (Diptera: Glossinidae). *Bulletin of Entomological Research*, **87**: 299-311.

Verheggen FJ, Haubruge E and Mescher MC. 2010. Alarm pheromones. *In:* Litwack G., ed. *Pheromones*. Amsterdam, The Netherlands: Elsevier.

Welter SC. 2005. Pheromone mating disruption offers selective management options for key pests. *California Agriculture*. **59**(1):16-22.

Witzgall P. 2001. Pheromones – future techniques for insect control? Pheromones for insect control in orchards and vineyards. *IOBC WPRS Bulletin*. **24**(2):114-122.

www.beyonddiscovery.org visited on 17/10/2013

Wyatt TD. 2009. Fifty years of pheromones. *Nature* 457(15th January): 262-263.

Zhang A, Kuang LF, Maisin N, Karumuru B, Hall DR, Virdiana I, Lambert S, Purung HB, Wang S and Hebbar P. 2008. Activity Evaluation of cocoa pod borer sex pheromone in cacao fields. *Environmental Entomology*, **37**(3):719-724.

Chapter 13

Semiochemicals and Insect Pest Management

☆ *Norma Robledo, Humberto Reyes Prado and René Arzuffi*

ABSTRACT

Semiochemicals, as an alternative to synthetic pesticides, have been used for the management of phytophagous insects during the last forty years. The majority of this chapter will discuss three types of semiochemicals: sexual pheromones, aggregation pheromones and plant volatiles. Distant phytophagous insects can perceive and be attracted to these chemical substances when released by an organism. The identity of these volatile compounds, the blend proportions and attraction that they provoke has been the subject of numerous studies. Pheromones have been most studied and utilized in diverse management strategies within the IPM (Integral Pest Management) framework, particularly due to the increasingly strict regulations regarding the use of insecticides on agricultural crops. Although the synthesis of some pheromones has not been possible, due to high costs, the volatile compounds emitted by host plants represent a good alternative. These compounds, although not as specific as pheromones, are characterized by their simplicity; furthermore, the recent use of blends of available compounds combined with pheromones has produced promising results. In this chapter, the most important characteristics of semiochemicals are described as well as applied strategies such as monitoring, large-scale trapping, attract-annihilate, host–searching disruption and 'push-pull' have been discussed. It is possible that the implementation of these strategies will increase with the synthesis of new compounds, field research and procedure development.

Semiochemicals

Basic Concepts

Semiochemicals are chemical compounds released by living organisms (*e.g.* insects, plants, *etc.*) that elicit a behavioural or physiological response in other individuals. These compounds can be classified into two groups: pheromones and allelochemicals. The former are intraspecific and the latter are interspecific mediators (Nordlund and Lewis, 1976; Dicke and Sabelis, 1988; Wyatt, 2003). Allelochemicals include allomones (compounds that when released, benefit the emitter), kairomones (compounds that when released, benefit the receptor) and synomones (released compounds that benefit both emitter and receptor). Host-plant volatiles fall within the kairomone category (Nordlund and Lewis, 1976; Dicke and Sabelis, 1988).

Insect pheromones, principally sexual and aggregation pheromones, are the most widely used semiochemicals in agriculture, and are commercially available for hundreds of species of insect (El-Sayed, 2016). Sexual pheromones mediate behavioural interactions between sexes and demonstrate a multitude of functions related to insect reproduction. Sexual attraction is one of the most important functions of sexual pheromones, which mediate the initial steps of activation and movement towards the "calling behaviour", specifically pheromone release (Matthews and Matthews, 2010). In contrast to sexual pheromones, which only provoke a response from one sex, aggregation pheromones are typically produced by one or both sexes and elicit a response in both (Matthews and Matthews, 2010). These species-specific pheromones are frequently multifunctional, for example they attract conspecific individuals to a host plant while also congregating individuals of different sexes for reproduction.

Although insect pheromones have been and continue to be the most used semiochemicals for insect management (Witzgall *et al.*, 2010), knowledge on the important role that plant volatiles play in host localization, has resulted in their increasing use for crop protection.

Plants generate and emit a large variety of chemical substances (Croteau *et al.*, 2000; Mello and Silva-Filho, 2002) that can play a role in the communication of insects, nearby plants and pathogenic microorganisms (Landolt and Quilici, 1996; Sepúlveda *et al.*, 2003; Das *et al.*, 2013). For some insects, these chemical substances act as food or host stimuli (Landolt and Quilici, 1996; Mello and Silva-Filho, 2002; Herrmann, 2010); however, they can also be dissuasive for phytophagous insects, thus demonstrating an ambivalent effect of both attraction and repellence (Dicke and Loon, 2000).

Below, the main strategies where semiochemicals are used for the management of insect pests are outlined.

Detection and Monitoring of Insect Populations

Pheromone traps are particularly useful for providing an early warning of the presence of potential pest species and monitoring population cycles (Witzgall *et al.*, 2010).

Currently, there is a growing use of pheromone attractant traps for the detection and interception of invasive species and the delineation of their displacement areas from their original sites of introduction and establishment. Pheromone traps also constitute a very important aspect of detection efforts carried out by regulatory agencies for invasive species (Howse *et al.*, 1998).

The use of trapping methods for pest monitoring has enabled the eradication of small localized pest populations, avoiding damage to crops, high costs incurred for the control of subsequent pest populations in addition to an unnecessary input of harmful pesticides into the environment.

Furthermore, insect pheromones and related compounds have played an important role in eradication processes, both in terms of their use for trapping and killing insects, and by providing a sensible method to determine when a specific crop can be considered pest free.

The growth in global commerce of agricultural products has resulted in a rapid increase of exotic pests around the world; therefore an increase in the use of semiochemicals as part of monitoring strategies is to be expected.

Management of Insects with Semiochemicals by Mass Trapping and Attract-annihilate

Although technically feasible, practical and economic considerations have limited its use. The costs of acquisition, implementation and maintenance of a sufficient number of traps can be excessively high. Nevertheless, under certain circumstances, pheromone based mass-trapping has the potential to both economic and effective, and therefore is used for the control of numerous species in diverse systems. The most efficient and largest pheromone based mass-trapping has been applied for the management of tropical weevils such as *Rhynchophorus palmarum, R. ferrugineus* and related species (Oehlschlager *et al.*, 2002; Faliero, 2006). One of the key factors in ensuring the success of these programs is that the reproduction rate of weevils is so slow that it cannot compensate for the removal of individuals by mass-trapping, thus leading to a reduction in population numbers. Pheromone-based mass-trapping is the principal component of IPM (integral pest management) for coconuts, dates, palm oil and banana in different parts of the world (Faliero, 2006).

A combination of biological and economic circumstances, pheromone-based mass-trapping can be highly effective and economic, in the case of the aforementioned tropical weevils, provided a solution to a specific problem that would otherwise have been untreatable. The pheromone based attract-annihilate method involves the combination of a pheromone source (synthetic compound) and an insecticide agent (toxic compound or entomopathogen). Theoretically, this method has several advantages over other strategies, for example it results in mating perturbation, given that the attracted pest insect is annihilated, poisoned or infected when in contact with the attracticide, thus effectively removed from the system. One disadvantage of the attracticide method is that it uses toxic substances; therefore, can imply more regulatory obstacles than other methods. In practice, the number of applications of the attract-annihilate method is relatively small.

Currently, attracticide technology has been predominantly implemented for the control and eradication of fruit flies (tephritids) (Millar, 1995; Shelly *et al.*, 2014).

Management of Insects by Perturbation of Mating and Host Plant Localization

As many insects depend entirely on the use of sexual pheromones so that male and female individuals can find each other and mate, the potential to control insects by manipulating this aspect of their lives has been recognized for many years (Witzgall *et al.*, 2010).

The mating perturbation technique, using synthetic compounds of sexual pheromones in large quantities, is used mainly to control populations of moths in orchards. In these insects, the females release sexual pheromones in order to attract males, which are present at relatively far-off distances (several kilometres), for mating. Females lay their eggs in orchard trees and the larvae develop inside the fruit, thus impeding their commercialization. Mating perturbation consists in affecting the male behaviour of searching for a female to mate by using large quantities of synthetic pheromones in the atmosphere. Perturbation can be attained by affecting different biological mechanisms which were originally defined by Bartell (1982). These mechanisms have been revised by Miller *et al.* (2006a, 2006b) and Stelinski (2007). When the moth population is very large, mating perturbation can be complemented by the regular and local application of pesticides.

Host plant volatiles can be used to disturb host searching behaviour in an area of crops. For example, aspersions of oil formulations resulted in a reduction in pest infestations of nut trees (Van Steenwyk and Barnett, 1987).

Dispensers

Due to their chemical structure, the majority of volatile semiochemicals (pheromones and plant released compounds) are extremely unstable; therefore they have to be formulated in such a way that they are protected from degradation caused by UV light and oxygen. Furthermore, the formulation should ensure a controlled release of semiochemicals. To be efficient management strategies the controlled liberation of semiochemicals, must have the following characteristics: i) post-release aerial concentration must be sufficiently high to be detected by the insects ii) the release of semiochemicals must be effective throughout the entire period that the insect is present, iii) the production of the dispenser must be reproducible iv) the application of the dispensers must be carried out during the initial phases, when pest density is still not high, given that in the majority of cases, its release rate decreases with time (Witzgall, 2001).

Distinct formulations and dispensers, with various release capacities, have been developed and commercialized. There are three groups of dispensers: solid matrix dispensers, liquid formulations and formulation reservoirs.

The solid matrix dispensers are manually applied in orchards and crop fields. The semiochemicals are incorporated within a solid matrix. As different materials can be used as a matrix, the release rate for an individual molecule can differ significantly from one device to another (Golub *et al.*, 1983). The most common solid matrices are: polyethylene tubes, polyethylene bags, (Torr *et al.*, 1997), polyethylene vials (Johansson *et al.*, 2001; Zhang *et al.*, 2008), membranes, spiral polymers

(Tomaszewska *et al.*, 2005), polymer films, cork septas (McDonough, 1991; Möttus *et al.*, 1997), rubber wicks, polyvinyl chloride (PVC), hollow fibres (Golub *et al.*, 1983), impregnated fabric, waxes and gel matrices (Atterholt *et al.*, 1999).

With regard to solid matrix dispensers, it is difficult to maintain a zero order release kinetics (constant release rate) over a prolonged period of time, and the concentration of the semiochemical in the air decreases with an increase in distance from the dispenser. These dispensers are only efficient at attracting and trapping insects at short distances. One solution to this problem is increasing the number of dispensers (on their own or inside a trap, depending on the management method used) in the orchard or crops; however, this involves a considerable increase in costs. An additional disadvantage is that the matrices are not biodegradable (Stipanovic *et al.*, 2004).

The slow release pulverisable formulations are composed of a biodegradable liquid matrix in which the semiochemicals are dissolved. Normally, UV stabilizers, antioxidants and surfactants are added in order to protect the semiochemicals. Pulverisable formulations frequently consist of a micro-emulsion which produces microspheres containing semiochemicals (*e.g.* micro-encapsulated pheromones) dispersed in a liquid matrix (de Vlieger, 2001). The efficiency time of these formulations ranges from days to weeks, depending on environmental factors, microsphere size, release capacity and the chemical properties of the semiochemicals (Welter *et al.*, 2005). When compared with the solid matrix dispensers, the biggest advantage of the pulverisable formulations is that the whole crop can be treated.

The reservoir dispensers generally consist of two parts, a reservoir and a diffusion area. Hofmeyr and Burger (1995) described a dispenser made from a glass tube that was impermeable to the pheromone and acted as a reservoir; this was connected to a polyethylene tube through which the pheromone could disperse.

The Aerosol emitters (*e.g.* Suttera® puffer), which consist of reservoirs that are electronically programmed to release the formulation, emit large quantities of pheromone by means of a pressurized aerosol. The emissions occur at regular intervals. This system is advantageous in that it uses only a few dispensers within the treated area. Reservoir systems are the most appropriate for attaining a zero order release dynamic for semiochemicals (Atterholt *et al.*, 1999).

The remainder of this chapter will describe the main characteristics of sexual and aggregation pheromones as well as host plant volatiles, which are commercially available. In addition, several examples of successful insect management through the use of semiochemicals will be presented.

Sexual and Aggregation Pheromones

Definition and Functions of Pheromones

Pheromones are chemical compounds emitted by an organism and which provoke a specific behavioural reaction in another organism of the same species. These are classified according to their function, such as aggregation, sexual, alarm, searching, territorial *etc.* As mentioned previously, sexual and aggregation pheromones are the most frequently used in strategies for pest insect management; the former are defined as compounds emitted by individuals of one sex that attract members of the opposite sex, leading to the localization of the emitter and subsequent mating; the latter, are compounds that attract members of the same species, with both sexes being attracted (Nordlund and Lewis, 1976; Dicke and Sabelis, 1988; Yew and Chung, 2015; Symonds and Gitau-Clarke, 2016).

Chemical Nature of Pheromones and their Synthesis and Release

The production sites of pheromones can be individual cells located in different parts of the body, or groups of cells located in the antenna, head, thorax and abdomen; for example, in moths, the pheromone gland is located between the eighth and ninth abdominal segment (Blomquist and Vogt, 2003). The majority are modified epidermis cells that belong to class I, II or III, depending on the pathway the pheromone secretions follow when crossing the cuticular barrier (Jurenka, 2004).

Pheromone synthesis has been studied in several insect orders and shares similarities with biochemical pathways related to lipid metabolism(Blomquist and Bagneres, 2010). In lepidoptera, the synthesis of the type I *de novo*, originates from fatty acids that are present in the haemolymph, and type II pheromones derived from linoleic or linolenic acid obtained from the diet (Arzuffi and Castrejón, 2012).

Pheromone synthesis is under neural (Central Nervous System) and hormonal control (juvenile hormone, 20-hydroxyecdysone and neuropeptides, activators of pheromone synthesis), and pheromone release is only under neural control (Blomquist and Vogt, 2003). This synthesis takes place through specific enzymatic reactions, including several classes of enzymes; for example, the enzyme Δ^{11} desaturase, characteristic of lepidopterans, produces a variety of precursors by chain reactions, the action of reductases and acetyltransferases produces the acetate pheromonal compound, characteristic of lepidopterans (Roelofs, 1995). Due to its potential for pest management, enzyme characterization has been subject to a great deal of research (Blomquist and Bagneres, 2010).

The composition of sexual pheromones can vary, for example, in lepidopterans, sexual pheromones form 12, 14 and 16 carbon chains and a functional group (alcohols, aldehydes or esters acetate); in dipterans, hydrocarbon chains of up to 21 carbons; in coleopterans, the pheromones are isoprenoid in origin, furthermore, these insects present high chemical diversity as the females can produce fatty acids, alkaloids, amino acid derivatives, terpene compounds, sulphurandphenol derivatives (Blomquist and Vogt, 2003). In the composition of the pheromone, the configuration of the *E* or *Z* double bonds is of importance in that it can affect the reception and response of insects, as the incorrect isomer can inhibit the response of the pheromone (Jurenka, 2004).

The sexual pheromones of insects are generally composed of more than one component, each one potentially responsible for the distinct behavioural phases involved in mate localization and subsequent mating; for example, in lepidopterans, males display a characteristic orientation behaviour that is guided by pheromones (Arzuffi and Castrejón, 2012). The proportion of pheromone blend to which males are exposed influences the following behavioural response (Evenden and Gries, 2008).

There are two important properties with respect to the sexual pheromone blend: quality (the chemical composition of the components) and the quantity (proportions of the components) (Hildebrand, 1995).

Applications of Sexual and Aggregation Pheromones

The most used semiochemicals in pest management strategies are sexual pheromones produced by lepidopterans, followed by aggregation pheromones and other attractants of coleopterans and dipterans. The application of pheromones for pest control is focused on the detection and monitoring of populations, as well as direct control methods (Witzgall *et al.*, 2010). The latter are mainly based on two modes of action: the attraction towards traps and sexual confusion (Yew and Chung, 2015).

Sexual and aggregation pheromones can be used in the following preventive and remedial strategies:

Monitoring

Currently, various species of pest insects, predominantly lepidopterans, are monitored by using pheromone septa commercialized by distinct manufacturers (Witzgall *et al.*, 2010).

The traps, together with the pheromone impregnated devices, represent the majority of the attractants used for monitoring; however, other food, plant and oviposition attractants are also used. Pheromones have been used for the monitoring of lepidopterans, dipterans and coleopterans; a successful example of this has been the monitoring of the pink bollworm *Pectinophora gossypiella* in cotton fields within the Central Valley of California since 1970. In this case, the capture of males in traps progressed to the implementation of a pink bollworm management strategy consisting of releasing sterile male moths that compete with wild males for females, with the aim of reducing the population density of *P. gossypiella* (Baker *et al.*, 1990). This strategy has benefited the economy of farmers and had a positive impact on the environment due to a marked reduction in the use of insecticides.

Other examples of successful cases include several moth species that are pests on fruit such as *Cydia pomonella* (Knight, 2010; Knight and Light, 2012), *Grapholita molesta* (Yang *et al.*, 2002), *Anarsia lineatella* (Kehat *et al.*, 1994; Ivanova *et al.*, 2010), on grapes *Lobesia botrana* (Anshelevich *et al.*, 1994), citric fruit *Ceratitis capitata* (Papadopoulos *et al.*, 2001), olives *Bactrocera oleae* (Noce *et al.*, 2009), cotton *Tuta absoluta* (Taha *et al.*, 2014), rice *Chilo suppressalis* (Kondo and Tanaka, 1994) and leguminous plants and vegetables *Heliothis armígera* (Srivastava and Srivastava, 1995; Guerrero *et al.*, 2014) and *Sitona lineatus* (Quinn *et al.*, 1999).

Mass-trapping

Mass-trapping has been successfully implemented for the control of horticultural, agricultural, forestry and grain store pests (Witzgall *et al.*, 2010), using mainly aggregation or sexual pheromones and to a lesser extent host-plant attractants (El-Sayed *et al.*, 2006)

An example of successful mass capture is the South American palm weevil *R. palmarum* which is an important pest of coconut and oil palms. This weevil is the vector of *Bursaphelenchus cocophilus,* a parasitic nematode responsible for the red ring disease that results in high losses in crop yields. The implementation of traps with sexual pheromones to capture adult weevils has resulted in a considerable reduction of this palm tree disease (Oehlschlager *et al.,* 2002).With regard to the cotton weevil, *Anthonomus grandis,* the aggregation pheromone has been used in large scale trapping when population density is low. The real impact of traps with pheromones against *A. grandis* has been difficult to assess; there was a reduction between 50 and 70 per cent in insecticide application; however, although traps with pheromones were the predominant tool, this method wasonly one component of an integrated management programme (Hardee *et al.,* 1969; Hardee *et al.,* 1970).

Mass-trapping has also been implemented for the management of several species of lepidopterans, such as the eggplant fruit and shoot borer *Leucinodes orbonalis* (Cork *et al.,* 2005). This vegetable is grown in large quantities and it is estimated that around 15 per cent of eggplant producers in India use this management technique. Other examples of pheromone mass trapping are with thetomato moth *T. absoluta,* and the corn moth *Spodoptera frugiperda;* the cotton pests belonging to the *Noctuidae* or owlet moths family *Helicoverpa armigera, Spodoptera litura* and *Earias vitella,* as well as the rice pest *Scirpophaga incertulas* (Witzgall *et al.,* 2010).

For the mass-trapping of Dipterans such as *B. oleae,* traps with ammonium sulphate, hydrolysed protein and sexual pheromones have been evaluated (Varikou *et al.,* 2014).

The mass capture of insect pests has been successfully practiced on its own or in combination with insecticides or biological control techniques. Indirectly, these methods reduce the use of insecticides and facilitate the conservation of natural enemies. Although, the use of mass-trapping has been studied for many species, the last few years has witnessed an increase in research on attracticides, particularly because of the potential to lower application costs by reducing the quantity of pheromones (Yew and Chung, 2015).

Attract-annihilate

An alternative to mass trappingis the use of attract-annihilate devices, where depending on the attractant, it is possible to capture both sexes of the insect pest. The insect pest management strategy of attract-annihilate, implicates the use of formulations that are directed at a specific pest species, thus reducing the development of pest resistance to insecticides. For example, for the management of *T. absoluta* in tomato crops, a sexual pheromone formulation (E,Z,Z)-3,8,11-tetradecatrienyl acetate and cypermethrin has been successful (Al-Zaidi, 2010; Hassan and Al-Zaidi, 2010).

In the case of the olive fly *B. oleae,* the attract-annihilate method has been implemented in olive groves where the population density of the fly is low. A combination of pheromones and ammonium bicarbonate has maintained reduced population densities of this pest fly. This method could progressively substitute the use of insecticides for the management of the olive fly (Mazomenos *et al.,* 2002).

It has been demonstrated that for *C. pomonella,* a formulation that includes pheromone, insecticide, and inert ingredients implemented in the attract-annihilate strategy, maintains infestation rates in apple crops that are lower than the economic loss threshold, with the exception of when the population density is high (Mansour, 2010).

Mating Disruption

Currently, pheromone-based mating disruption has been and continues to be the most important control strategy, both at the research and application level. The management of the codling moth *C. pomonella* is the most successful example of mating disruption for pest control (Brunner *et al.*, 2001), with over 200,000 hectares of apple orchards implementing this strategy (Witzgall *et al.*, 2010). The management of *C. pomonella* using pheromone-based mating disruption has led to a 75-80 per cent reduction in the use of insecticides for the control of this pest insect (Sumedrea *et al.*, 2015). The tea plant pest, *Adoxophyes honmai*, was managed by pheromone-based mating disruption, using the single component (Z)-11-tetradecenyl acetate, which is the major component of its sexual pheromone, however, its affectivity declined after a decade of use (Mochizuki *et al.*, 2002). The effectiveness of applying this technique for managing this particular pest insect returned when three minor components (Z)-9-tetradecenyl acetate, (E)-11- tetradecenyl acetate and 10-methyldodecyl acetate, were added to the pheromone blend (Tabata *et al.*, 2007).

The management of *L. botrana* (Witzgall *et al.*, 2010) is another example of the successful implementation of a mating disruption strategy. As in the previously mentioned case of *C. pomonella*, the management of *L. botrana* in vineyards demonstrated that this strategy was successful in reducing the population of the pest species when sustained over several growing seasons, a result of a direct decrease in the reproduction of the pest species and the stabilization of natural enemy populations that are frequently decimated by the application of insecticides (Reddy and Guerrero, 2010; Sumedrea *et al.*, 2015). It has been reported that the females of *L. botrana* are attracted by their own sexual pheromone; thus in addition to reducing the capacity of males to locate females by mating-disruption, there is a direct effect on both sexes and a consequent decrease in reproductive success (Harari *et al.*, 2015).

Another example of a successful mating disruption strategy is the management of the rice pest *C. suppressalis,* which has experienced a marked population decrease associated with a gradual reduction in the quantity of pheromone required for its control (Chen *et al.*, 2014).

Plant Host Volatiles

Definition and Functions

Volatiles plant may be allelochemicals, chemical compounds that are divided into three groups: allomones, kairomones and synomones. Allomones act as repellents or also as potential prey attractants. Herbivorous insects are attracted to kairomones as they use them to locate host plants while parasitoids use them to locate their prey. Synomones are alellochemicals that attract natural enemies of herbivorous insects, thus benefitting the plant (Rodriguez-Saona and Stelinski, 2009). In this chapter, only kairomones will be considered.

Insects are highly selective to host volatiles (Herrmann, 2010), which they use to find their host plant. At specific distances, these chemical stimulants are detected by insects which then gather around the plants before targeting a particular part (Bernays and Chapman, 1994). Some examples of this include: frugivorous dipterans from the Tephritidae family *e.g. Toxotrypana curvicauda* found in the fruit of papaya; coleopterans from the Curculionidae family *e.g. Scyphophorus acupunctatus*, localized in the bulb, and lepidopterans that consume plant tissue, such as those from the Noctuidae family *e.g. Copitarsia decolora*, which is found in the heart of cabbages.

Due to their physicochemical characteristics, the volatility of compounds released from plants enables them to be perceived by insects at great distances. These compounds present structural diversity (Herrmann, 2010), including short chain alcohols, aldehydes, ketones, esters, phenols, aromatics, lactones, monoterpenes and sesquiterpenes (Bernays and Chapman, 1994; Degenhardt, 2008). These compounds originate predominantly from three metabolic pathways: terpenoids, phenylpropanoids/benzenoids and C6-aldehydes (green-leaf volatiles) (Das *et al.*, 2013).

Chemical Nature, Production, and Release of Volatile Compounds

There are two types of volatile compounds released: constitutive and induced (Mello and Silva-Filho, 2002; Das *et al.*, 2013). Constitutive release takes place as a defence mechanism against natural enemies by repelling them directly (Bruce *et al.*, 2010; Das *et al.*, 2013); and induced release occurs when the plant tissue is damaged by herbivores (Mello and Silva-Filho, 2002). Volatile compounds such as terpenoids and phenylpropanoids are produced and stored in the glandular trichomes and are constitutive volatile compounds; in contrast, green-leaf volatiles and terpenoids are volatile compounds produced by herbivores, are induced compounds (Das *et al.*, 2013).

The basis of terpenoids is isopentenyl pyrophosphate which is produced by two pathways: the Mevalonate pathway in the cytosol and the methylerythritol phosphate pathway in the plastids (Das *et al.*, 2013). The basic units are hemiterpenes (C5) that form monoterpenes (C10), sesquiterpenes (C15), diterpenes (C20) through elongation (Croteau *et al.*, 2000; León and Guevara-García, 2007). Terpenoids that are synthesized, produced and stored in various plant parts, prevent herbivores from laying their eggs or feeding on roots and rhizomes, *e.g.* 1,8-Cineole, synthesised in the roots of Arabidopsis, a plant from the Brassicaceae family (Chen *et al.*, 2004). Monoterpenes, produced in the flowers, prevent damage to these reproductive parts of the plant (Das *et al.*, 2013), while pollinators are attracted by terpenoids and benzenoids, *e.g.* linalool (Okamoto, *et al.*, 2007), phenylacetaldehyde and benzaldehyde (Omura and Honda, 2005).

The compounds phenylpropanoids/benzenoids, which include an aromatic ring within their structure, are derived from Shikimic acid via phenylalanine (Das *et al.*, 2013). These compounds play an important role in defending plants from various threats, including pathogenic microorganisms (Sepúlveda *et al.*, 2003); in addition, phenylpropanoids participate in pollination (Das *et al.*, 2013).

Green-leaf volatiles are considered as direct defence compounds, they are produced by the lipoxygenase pathway and are composed of six carbons: Z-3-hexenal, E-3-hexenal, E-3-hexenol, Z-3-hexen-1-ol, E-2-hexenal and E-2-hexenol; their release, together with that of the terpenes (E)-β-ocimene, (E,E)-α-farnesene, (E)-β-farnesene), indole and (Z)-3- hexenyl acetate, is associated with feeding damage by herbivorous insects (Paré and Tumlinson, 1999; Das *et al.*, 2013).

Applications of Plant Volatile Compounds

For the management and control of insect pests, a variety of plant volatile compounds with different formulations and modes of action, have been studied. Knowledge on the volatile compounds that attract or elicit responses from insect pests is important (Bruce *et al.*, 2005), in addition to host-plant condition (specific part of the plant, intact plants, plants infested by conspecifics and plants with mechanical damage) and the physiological state of the herbivore (Rodríguez-Saona and Stelinski, 2009). Consequently, there are many studies on volatile identification of host plants and experiments where herbivore attraction is evaluated (Piñero and Ruiz-Montiel, 2012; Rojas, 2012; Das *et al.*, 2013). Knowledge of herbivore behaviour is indispensable in order to use the compounds that are involved in host-searching and those which insects use for finding a mate, as releasing them together could have a synergetic attraction effect (Light *et al.*, 2001; Rodriguez-Saona and Stelinski, 2009). When managing pest insects it is important to consider whether insects are specialists (those that feed on one or a few types of plant), or generalists (feed on a larger number of plants) (Bernays and Chapman, 1994), since this determines the use of plant volatile compounds to manipulate the host-searching behaviour of herbivorous insects (Rodriguez-Saona and Stelinski, 2009).

As part of the research into the use of plant volatile compounds, the following modes of action have been studied: insecticide, repellent, attractants of insect pests or of their predators (Thacker and Train, 2010). These effects are also determined by the concentration (Rodriguez-Saona and Stelinski, 2009) and blend (Bruce and Pickett 2011) of volatile compounds.

Over the last few decades, plant volatile compounds have been used for the monitoring and control of pests (mass-trapping, attract-annihilate and host-searching modification). Similar applications have been made using sexual pheromones; however, the use of plant volatiles have some advantages over pheromones, given that they can attract females or both sexes, are simpler compounds, and can be obtained commercially, and thus eliminating high costs associated with their synthesis. Some disadvantages include: their low specificity and the fact that many plants possess compounds with identical metabolic pathways, such as green-leaf volatiles and terpenoids. However, they are a good alternative to pheromones, particularly when the quantities of pheromones are very small, there are no available synthetics or the pheromone is totally absent (Rodríguez-Saona and Stelinski, 2009).

Monitoring

Volatile compounds have been used for monitoring programs primarily because they are environmentally friendly. As these compounds require preparation from specific blends and in precise proportions, considerable workload is involved. The majority of monitoring has been carried out on lepidopterans and coleopterans (Curculionidae) (Rodríguez-Saona and Stelinski, 2009; Piñero and Ruiz-Montiel, 2012; Rojas, 2012). For example, (Z)-2-pentenol and methyl eugenol has been used for the monitoring of *Otiorhynchus sulcatus* (VanTol *et al.*, 2012). In the case of tephrids, food attractants have been implemented, examples being trimedlure for *C. capitata* (Jang and Light, 1996) and cuelure for *Bactrocera cucurbitae* (Manoukis and Gayle, 2016).

Mass-Trapping

The monitoring strategy using plant volatiles can be extended to mass-trapping (Rodríguez-Saona and Stelinski, 2009), *e.g.* for *Leptinotarsa decemlineata*, a synthetic blend of host volatiles,(Z)-3-hexenyl acetate, (7)-linalool and methyl salicylate, placed in field traps yield more efficient results than that of the conventional methods (Martel *et al.*, 2005).

Attract-Annihilate

As with monitoring, mass-trapping can be extended to a strategy of attract-annihilate. Generally, pheromones and food attractants are used, although the extensive use of host plant volatiles increases trap efficiency (Rodríguez-Saona and Stelinski, 2009). This type of strategy has been used with methyl eugenol and cuelure to attract and malathion to annihilate. The traps were efficient for *Bactrocera dorsalis* and *B. cucurbitae* (Vargas *et al.*, 2000). In another study, Martel *et al.* (2007) tested the same attractant for mass-trapping of *L. decemlineata*, and a pyrethroid was additionally used for annihilation, with the aim of reducing the use of synthetic insecticides.

Host-Searching Disruption Behaviour

This strategy consists of "saturating" the crop with kairomones, so that the herbivorous pest is unable to find the host (Rodríguez-Saona and Stelinski, 2009). The majority of these studies have been focused on parasitoid hosts, in order to evaluate the impact of host search disruption on the oviposition behaviour of the parasitoid (Stelinski *et al.*, 2006).

There have been numerous studies that have identified compounds in plants which attract females from the Lepidopterans. However, these have not been commercialized for monitoring and control (Rojas, 2012). Van Steenwyk and Barnett (1987) provide another example of the implementation of host-searching disruption, this time to reduce the infestation of nuts by the lepidopteran *Amyelois transitella*. Aspersions of 5 per cent nut essential oil on the crop resulted in a decrease in infestation.

There is the potential to develop more applications with plant compound volatiles and host-searching disruption, over the next few years.

'Push-pull'

The 'push-pull' strategy is a new and effective tool for IPM, using a combination of stimuli that modify the behaviour of pest insects and their natural enemies in order to manipulate their abundance and distribution (Cook *et al.*, 2007). In this strategy, pests are repelled or dissuaded from the protected resource (push) by stimuli that disrupt the process of host plant localization and transform these plants into unattractive or inadequate hosts for pest feeding and oviposition. By using attractive stimulants, the individuals of the target pest species are simultaneously attracted (pull) to a specific source where subsequently they become concentrated, facilitating their elimination and thus protecting the resource (Khan and Pickett, 2004). In this stimuli strategy, both push and pull components are generally not toxic; therefore, they are frequently accompanied by other population reduction methods, such as insecticides, natural enemies and mass-trapping, with biological control methods used most frequently. The 'push-pull' factors, maximise the efficacy of the behaviour-modifying stimuli by means of additive and synergetic effects that integrate the use of several methods for the reduction of pest populations.

The most successful example of this strategy was developed in Africa for subsistence farmers. Although the strategy was initially aimed at poor producers, the experience can be learnt and applied to organic or agro-ecological production systems. Hundreds of farmers in Africa use "push-pull" strategies to protect their crops of maize and sorghum (Khan and Pickett, 2004). The development of these strategies has been focused mainly on pest problems in intensive agriculture systems, despite dependence on cheap insecticides and the commercial unavailability of these techniques (Eigenbrode *et al.*, 2016), they are now seriously considered as a viable solution within a sustainable development perspective, for the management of pest insects in different parts of the world.

Conclusions

Currently the main application of semiochemicals is the management of pest species as a monitoring tool, where semiochemicals are used as attractants for different types of traps. A reduction in populations of pest insects that attack economically important crops has been a difficult challenge and has only been accomplished in a few cases. Today, several formulations are employed, but applying semiochemical volatiles to large areas of crops and ensuring that they are released during the entire growing season, has yet remained problematic. In addition, the synthesis of commercially available formulations would enable the appropriate modification of previously developed methods, and one could hope that integration of several these strategies would finally lead to a more efficient and successful implementation of management strategies on agricultural crops.

Regarding 'push-pull' strategies, although these are promising in facing the problems of pests in agricultural systems, they are more complex and require monitoring research, decision-making systems and imply higher operational costs than those associated with the application of conventional insecticides. However, if the indiscriminate use of insecticides continues, 'push-pull' strategies can provide a suitable alternative, albeit one that requires a considerable increase in research if this method is to result viable and successful in many agricultural regions of the world.

The study and application of semiochemicals is an interdisciplinary approach to research is required, where various areas are involved, biological, chemical, plant and insect physiology, the synthesis of chemical compounds and fieldwork (Herrmann, 2010).

References

Anshelevich, L., Kehat, M., Dunkelblum, E., Greenberg, S., 1994. Sex pheromone traps for monitoring the european vine moth, *Lobesia botrana*: Effect of dispenser type, pheromone dose, field aging of dispenser, and type of trap on male captures. Phytoparasitica. 22, 281-290.

Al-Zaidi, S., 2010. Manejo de*Tuta absoluta* mediante feromonas. *Phytoma*.**217,** 41.

Arzuffi, B.R., Castrejón, A. F., 2012. El papel del estímulo químico durante la búsqueda de hospedero por lepidópteros herbívoros, in: Rojas, J., Malo, E. (Eds), Temas selectos de Ecología Química de Insectos. El colegio de la Frontera Sur. Chiapas, México. pp. 72-94.

Atterholt, C.A., Delwiche, M.J., Rice, R.E., Krochta, J.M., 1999. Controlled release of insect sex pheromones from paraffin wax and emulsions. J. Controlled Release, 57, 233-247.

Baker, T.C., Staten, R.T., Flint, H.M., 1990. Use of pink bollworm pheromone in the southwestern United States, in: Ridgway, R.L., Silverstein, R.M.,Inscoe, M.N. (Eds.), Behavior-Modifying Chemicals for Insect Management, New York: Marcel Dekker. pp. 417–436.

Bartell, R.J., 1982. Mechanisms of communication disruption by pheromone in the control of Lepidoptera: a review. Physiol. Entomol. 7, 353-364.

Bernays, E.A., Chapman, R.F., 1994. Host plant selection by phytophagous insects. Chapman and Hall. Nueva York. pp: 14-60.

Blomquist, G., Vogt, R., 2003. Biosynthesis and detection of pheromones and plant volatiles-introduction and overview, in:Blomquist, G.J., Vogt, R.G. (Eds.)Insect Pheromone Biochemistry and Molecular Biology. Academic Press, London. pp 3-18.

Blomquist, G. J., Bagneres, A.G, 2010. Insect Hydrocarbons: Biology, Biochemistry, and Chemical Ecology. Cambridge: Cambridge University Press 506 p.

Bruce, T.J.A., Wadhams,L.J., Woodcock, C.M., 2005. Insect host location: a volatile situation. Trends Plant Sci. 10, 269-274.

Bruce, T.J.A., Midega, C.A.O.,Birkett,M.A., Pickett,J.A.,Khan.Z.R., 2010. Is quality more important than quantity? Insect behavioural responses to changes in a volatile blend after stemborer oviposition on an African grass. Biol. Letters 6, 314-317.

Bruce, T.J.A., Pickett, J.A., 2011. Perception of plant volatile blends by herbivorous insects - Finding the right mix. 2011. Phytochemistry. pp. 1605-1611.

Brunner, J., Welter, S., Calkins, C., Hilton, R., Beers, E., Dunley, J., Unruh, T., Knight, A., Steenwyk R., Buskirk, P., 2001. Mating disruption of codling moth: a perspective from the Western United States. IOBC/WPRS Bulletin. 25, 207- 215.

Chen, F, Ro. D.K., Petri, J., Gershenzon, J., Bohlmann, J., Pichersky. E., Tholl, D., 2004. Characterization of a root-specific Arabidopsis terpene synthase responsible for the formation of the volatile monoterpene 1,8-cineole. Plant Physiol. 135, 1956-1966.

Chen, R.Z., Klein, M.G., Sheng, C.F., Li, Q.Y., Li, Y., Li, L.B., Hung, X., 2014. Mating disruption or mass trapping, compared with chemical insecticides, for suppression of *Chilo suppressalis* (Lepidoptera: Crambidae) in Northeastern China. J. Econ. Entomol.107, 1828-1838.

Cook, S.M., Khan, Z.R., Pickett, J.A., 2007. The Use of Push-Pull Strategies in Integrated Pest Management. Annu. Rev. Entomol. 52, 375–400.

Cork, A., Alam, S.N., Rouf, F.M.A., Talekar, N.S., 2005. Development of mass trapping technique for control of brinjal shoot and fruit borer, *Leucinodes orbonalis* (Lepidoptera: Pyralidae). Bull. Entomol. Res. 95, 589-596.

Croteau, R, Kutchan, T.M, Lewis. N.G., 2000. Natural Products (Secondary Metabolites), in: Buchanan B., Gruissem W., Jones R. (Eds). Biochemistry and Molecular Biology of Plants American Society of Plant Physiologists. Rockville, MD, pp. 1250-1268.

Das, A., Lee S. H., Hyun T. K., Kim S. W., Kim, J. Y., 2013. Plant volatiles as method of communication. Plant Biotechnol. Rep. 7, 9-26.

Degenhardt, J., 2008. Ecological Roles of Vegetative Terpene Volatiles, in: Schaller A. (Ed.), Induced Plant Resistance to Herbivory, Springer. Dordrecht, pp. 433-442.

De Vlieger, J.J., 2001. Development of a sprayable slow- release formulation for the sex pheromone of the Mediterranean corn borer, *Sesamia nonagroides*. IOBC wprs Bull. 24, 101-106.

Dicke, M., Sabelis M., 1988. Infochemicals terminology: based on cost-benefit analysis rather than origin of compounds?.Funct. Ecol. 2, 131-139.

Dicke, M., Loon J.J., 2000. Multitrophic effects of herbivoreinduced plant volatiles in an evolutionary context. Entomol. Exp. Appl. 97, 237-249.

Eigenbrode, S.D., Birch, A.N.E., Lindzey, S., Meadow, R., Snyder, W.E., 2016. A mechanistic framework to improve understanding and applications of push-pull systems in pest management. J. Appl. Ecol. 53, 202-212.

El-Sayed, A.M., Suckling, D.M., Wearing, C.H., Byers, J.A., 2006. Potential of mass trapping for long-term pest management and eradication of invasive species. J. Econ. Entomol. 99, 1550-1564.

El-Sayed, A.M., 2016. The Pherobase: Database of Pheromones and Semiochemicals. http://www.pherobase.com. Last access: June 2016.

Evenden, M.L., Gries, R., 2008. Plasticity of male response to sex pheromone depends on physiological state in a long-lived moth. Anim. Behav. 75, 663-672.

Faliero, J.R., 2006. A review of the issues and management of the red palm weevil *Rhynchophorus ferrugineus* (Coleoptera: Rhynchophoridae) in coconut and date palm during the last one hundred years. Int. J. Tropical Insect Sci. 26, 135–154.

Guerrero, S., Brambila, J., Meagher, R.L., 2014. Efficacies of four pheromone-baited traps in capturing male *Helicoverpa* (Lepidoptera: Noctuidae) moths in northern Florida. Fla. Entomol.97, 1671-1678.

Golub, M., Weatherston J., Benn, M.H., 1983. Measurement of release rates of gossyplure from controlled release formulations by mini-airflow method. J. Chem. Ecol. 9, 323-333.

Hardee, D.D., Cross, W.H., Mitchell, E.B., Huddleston, P.M., Mitchell, H.C., Merkl, M.E., Davich, T.B., 1969. Biological factors influencing responses of the female boll weevil to the male sex pheromone in field and large-cage tests. J. Econ. Entomol. 62, 161-165.

Hardee, D.D., Cross, W.H., Huddleston, P.M., Davich, T.B., 1970. Survey and control of the boll weevil in West Texas with traps baited with males. J. Econ. Entomol. 63, 1041-1048.

Harari, A.R., Zahavi, T., Steinitz, H., 2015. Female detection of the synthetic sex pheromone contributes to the efficacy of mating disruption of the European grapevine moth, *Lobesia botrana*. Pest Manag. Sci. 71, 316-22.

Hassan, N., Al-Zaidi S., 2010.*Tuta absoluta*-pheromone mediated management strategy.*Int. PestControl*.**52**, 158-160.

Herrmann, A., 2010. Volatile – An Interdisciplinary Approach, in: Herrmann A. (Ed). The Chemistry and Biology of volatiles. Wiley and Sons Ltd. West Sussex. pp. 1-10.

Hildebrand, J., 1995. Analysis of chemical signals by nervous systems, in: Eisner, T., Meinwald, J. (Eds), Chemical Ecology. The Chemistry of Biotic Interaction. National Academy Press, Washington, D. C. pp. 161-181.

Hofmeyr H., Burger, B.V., 1995. Controlled-release pheromone dispenser for use in traps to monitor flight activity of false codling moth. J. Chem. Ecol., 21, 355-363.

Howse P.E., Stevens I.D., Jones, O.T., 1998. Insect pheromones and their use in pest management. Chapman and Hall. London. 369 pp.

Ivanova, L., Kutinkova, H., Dzhuvinov, V., 2010. Flight monitoring of oriental fruit moth, *Cydia molesta*, and peach twig borer, *Anarsia lineatella*, by pheromone traps in apricot orchard of north-east bulgaria. ActaHortic. 862, 465-470.

Jang, E. B., Light, D.M., 1996. Olfactory semiochemicals of tephritids, in: McPheron BA, Steck GJ. (Eds.), Fruit fly pests: a world assessment of their biology and management. St. Lucie Press, Delray Beach, FL, pp. 73–90.

Johansson, B.G., Anderbrant, O., Simandl, J., Avtzis, N.D., Salvadori, C., Hedenström, E., Edlund, H., Högberg, H. E., 2001. Release rates for pine sawly pheromones from two types of dispensers and phenology of *Neodiprion sertifer*. J. Chem. Ecol. 27, 733-745.

Jurenka, R., 2004. Insect pheromone biosynthesis. Top. Curr. Chem. 239, 97-132.

Kehat, M., Anshelevich, L., Dunkelblum, E., Greenberg, S., 1994. Sex pheromone traps for monitoring the peach twig borer, *Anarsia lineatella* Zeller: effect of pheromone dose, field aging of dispenser, and type of trap on male captures. Phytoparasitica. 22, 291-298.

Khan, Z.R., Pickett J.A. 2004. The 'push-pull' strategy for stemborer management: a case study in exploiting biodiversity and chemical ecology, in:Gurr, G.M., Wratten, S.D., Altieri, M.A. (Eds), Ecological Engineering for Pest Management: Advances in Habitat Manipulation for Arthropods, CABI, Wallington, Oxon, UK. pp. 155-164.

Knight, A., 2010. Improved monitoring of female codling moth (Lepidoptera: Tortricidae) with pear ester plus acetic acid in sex pheromone-treated orchards. Environ Entomol. 39, 1283-1290.

Knight, A.L., Light, D.M., 2012. Monitoring codling moth (Lepidoptera: Tortricidae) in sex pheromone-treated orchards with (E)-4,8-dimethyl-1,3,7-nonatriene or pear ester in combination with codlemone and acetic acid. Environ Entomol. 41, 407-414.

Kondo, A., Tanaka, F., 1994. Suitable location of sex-pheromone traps for monitoring rice stem borer moth, *Chilo suppressalis* (Walker) (Lepidoptera: Pyralidae). Jpn. J. Appl. Entomol. Z. 38,283-287.

Landolt, P. J., Quilici, S., 1996. Overview of research on behaviour of fruit flies, in: Mc Pheron, B.A., Steck, G.J. (Eds). Fruit fly pest: A world assessment of their biology and management. St. Lucie Press. Delray Beach. pp: 19-26.

León P., Guevara-García, A., 2007. La síntesis de isoprenoides a través de la vía MEP; un nuevo blanco de manipulación para la salud y el beneficio humano, in: Oria Hernández, J., Rendón Huerta, E., Reyes Vivas, H., Romero Álvarez, I., Velázquez López, I. (Eds.). Mensaje Bioquímico, Vol. XXXI. Depto. Bioquímica, Fac. Medicina, Universidad Nacional Autónoma de México. Cd. Universitaria, México, D.F., pp. 77-91.

Light, D.M., Knight, A.L., Henrick, C.A., Rajapaska, D., 2001. A pear-derived kairomone with pheromonal potency that attracts male and female codling moth, *Cydia pomonella* (L.) Naturwissenschaften. 88, 333–338.

Mansour, M., 2010. Attract and kill for codling moth *Cydia pomonella* (Linnaeus) (Lepidoptera: Tortricidae) control in Syria. J. Appl. Entomol.134, 234-242.

Manoukis, N.C., Gayle, S.M., 2016. Attraction of wildlike and colonyreared *Bactrocera cucurbitae* (Diptera: Tephritidae) to cuelure in the field. J. Appl. Entomol. 140, 241-249.

Martel, J.W., Alford A.R., Dickens J.C., 2005. Synthetic host volatiles increase efficacy of trap cropping for management of Colorado potato beetle, *Leptinotarsa decemlineata* (Say). Agr. Forest Entomol. 7, 79-86.

Martel, J.W., Alford, A.R., Dickens, J.C., 2007. Evaluation of a novel host plant volatile-based attracticide for management of Colorado potato beetle, *Leptinotarsa decemlineata* (Say). Crop Prot. 26, 822-827

Matthews, R.W., Matthews, J.R., 2010. Insect behavior: Springer. New York. 514 pp.

Mazomenos, B., Pantazi-Mazomenou, A., Stefanou, D., 2002. Attract and kill of the olive fruit fly *Bactrocera oleae* in Greece as a part of an integrated control system. IOBC wprs Bulletin 25, 1-11.

McDonough, L.M., 1991. Controlled release of insect sex pheromones from a natural rubber substrate, in: Hedin, P. A. (Ed.) Naturally occurring pest bioregulators. ACS Symposium serie 449. Washington, DC, USA. American Chemical Society, pp: 106-124.

Mello, M.O., Silva-Filho, M.C., 2002. Plant Insect interactions and evolutionary arms race between two distinct defense mechanisms. Braz. J. Plant Physiol. 14, 71-81.

Millar, J.G., 1995. An overview of attractants for Mediterranean fruit fly. Proceedings of a Workshop on the Medfly Situation in California: Defining Critical Research, Riverside, CA, 11–13. November 1994, pp. 123–144. College of Natural and Agricultural Sciences, Riverside CA.

Miller, J.R., Gut, L.J., de Lame, F.M., Stelinski, L.L., 2006a. Differentiation of competitive vs non-competitive mechanisms mediating disruption of moth sexual communication by point sources of sex pheromone (Part I): theory. J. Chem. Ecol. 32, 2089-2114.

Miller, J.R., Gut, L.J., de Lame, F.M., Stelinski, L.L., 2006b. Differentiation of competitive vs non-competitive mechanisms mediating disruption of moth sexual communication by point sources of sex pheromone (Part II): case studies. J. Chem. Ecol., 32, 2115- 2143.

Möttus, E., Nômm, V., Williams, I.H., Liblikas, I., 1997. Optimization of pheromone dispensers for diamondback moth *Plutella xylostella*. J. Chem. Ecol. 23, 2145-2159.

Mochizuki, F., Fukumoto, T., Noguchi, H., Sugie, H., Morimoto, T., Ohtani, K., 2002. Resistance to a mating disruptant composed of (Z)-11-tetradecenyl acetate in the smaller tea tortrix, *Adoxophyes honmai* (Yasuda) (Lepidoptera: Tortricidae). Appl. Entomol. Zool. 37, 299-304.

Noce, M. E., Belfiore, T., Scalercio, S., Vizzarri, V., Iannotta, N., 2009. Efficacy of new mass-trapping devices against *Bactrocera oleae* (Dipteratephritidae) for minimizing pesticide input in agroecosystems. J. Environ. Sci. Health B44, 442-448.

Nordlund, D.A., Lewis, W.J., 1976. Terminology of chemical releasing stimuli in intraspecific and interspecific interactions. J. Chem. Ecol. 2, 211-220.

Oehlschlager, A.C., Chinchilla, C., Castillo, G., Gonzalez, L., 2002. Control of red ring disease by mass trapping of *Rhynchophorus palmarum* (Coleoptera: Curculionidae). Fla. Entomol. 85, 507- 513.

Okamoto, T., Kawakita, A., Kato, M., 2007. Interpecific variation of floral scent composition in Glochidion and its association with host-specific pollinating seed parasite (Epicephala). J. Chem Ecol. 33,1065-1081.

Omura, H., Honda, K., 2005. Priority of color over scent during flower visitation by adult *Vanessa indica* butterflies. Oecologia. 142, 588-596.

Papadopoulos, N.T., Katsoyannos, B.I., Kouloussis, N.A., Hendrichs, J., Carey, J. R., Heath, R.R., 2001. Early detection and population monitoring of *Ceratitis capitata* (Diptera: Tephritidae) in a mixed-fruit orchard in northern Greece. J. Econ. Entomol.94, 971-978.

Paré, W., Tumlinson, J.H., 1996. Plant volatile signals in response to herbivore feeding. Fla. Entomol. 79, 93-103.

Piñero, J.C., M., Ruiz-Montiel C., 2012. Ecología Química y Manejo de Picudos (Coléoptera:Curculionidae) de importancia económica, in: Rojas, J., Malo, E. (Eds), Temas selectos de Ecología Química de Insectos. El colegio de la Frontera Sur. Chiapas, México, pp. 361-400.

Quinn, M., Bezdicek, D., Smart, L., Martin, J., 1999. An aggregation pheromone system for monitoring pea leaf weevil (Coleoptera: Curculionidae) in the Pacific Northwest.J. Kans. Entomol. Soc.72, 315-321.

Reddy, G.V.P., Guerrero, A., 2010. New pheromones and insect control strategies. Vitam. Horm. 83, 493-519.

Roelofs, W., 1995. The chemistry of sex attraction, in: Eisner, T., Meinwald, J. (Eds), Chemical Ecology. The Chemistry of Biotic Interaction. National Academy Press, Washington, D. C. pp. 103-117.

Rodríguez-Saona, C.R., Stelinski L. L., 2009. Behavior-modifying strategies in IPM: Theory and practice. Integrated pest management: Innovation-development process. Springer. pp. 263-315.

Rojas, J.C., 2012. El papel del estímulo químico durante la búsqueda de hospedero por lepidópteros herbívoros, in: Rojas, J., Malo, E. (Eds), Temas selectos de Ecología Química de Insectos. El colegio de la Frontera Sur. Chiapas, México. pp. 287-314.

Sepúlveda G, Porta H., Rocha, M., 2003. La participación de los metabolitos secundarios en la defensa de las plantas. Rev. Mex. Fitopatol. 21, 355-363.

Shelly, T., Epsky N., Jang E.B., Reyes-Flores J., Vargas R., 2014. Trapping and the detection, control, and regulation of tephritid fruit flies. Springer. New York. 638 pp.

Sumedrea, M., Florin-Cristian, M., Calinescu, M., Sumedrea, D., Iorgu, A., 2015. Researches regarding the use of mating disruption pheromones in control of apple codling moth - *Cydia pomonella* L. Agric. Agric. Sci. Procedia. 6, 171-178.

Srivastava, C.P., Srivastava, R.P., 1995. Monitoring of *Helicoverpa armigera* (Hbn.) by pheromone trapping in chickpea (*Cicer arietinum* L.). J. Appl. Entomol. 119, 607-609.

Stelinski, L.L., Pelz-Stelinski K.S., Liburd, O.E., Gut L.J., 2006. Control strategies for *Rhagoletis mendax* disrupt host-Wnding and ovipositional capability of its parasitic wasp, *Diachasma alloeum*. Biol. Control. 36, 91-99.

Stelinski, L.L., 2007. On the physiological and behavioral mechanisms of pheromone-based mating disruption. Pestycydy. 3, 27-32.

Stipanovic, A.J., Hennessy, P.J., Webster, F.X., Takahashi, Y., 2004. Microparticle dispensers for the controlled release of insect pheromones. J. Agric. Food Chem. 52, 2301-2308.

Symonds, M.R.E., Gitau-Clarke, C.W., 2016. The Evolution of aggregation pheromone diversity in bark beetles, in: Claus, T., Gary, J. B. (Eds.), Advances in Insect Physiology Academic Press. 50, 195-234.

Tabata, J., Noguchi, H., Kainoh, Y., Mochizuki, F., Sugie, H., 2007. Sex pheromone production and perception in the mating disruption-resistant strain of the smaller tea leafroller moth, *Adoxophyes honmai*. Entomol. Exp. Appl.122, 145-153.

Taha, A.M., Emara, T., Hanafy, A.R.I., Hassan, G., 2014. Evaluation of pheromone lures for trapping the tomato borer moths, *Tuta absoluta* in tomato fields in Egypt. Int. J. Environ. Sci. Eng. 5, 99- 109.

Thacker, J.R.M., Train M.R., 2010. Use of Volatiles in Pest Control, in: Herrmann, A. (Ed.), The chemistry and biology of volatiles. Wiley and Sons Ltd. West Sussex, pp.151-168.

Tomaszewska, E., Hebert, V.R., Brunner, J.F., Jones, V.P., Doerr, M., Hilton, R., 2005. Evaluation of pheromone release from commercial mating disruption dispensers. J. Agric. Food Chem. 53: 2399-2405.

Torr, S.J., Hall, D.R., Phelps, R.J., Vale, G.A., 1997. Methods for dispensing odour attractants for tsetse flies (Diptera: Glossinidae). Bull. Entomol. Res. 87: 299-311.

Van Tol, R.W., Bruck, D.J., Griepink, F.C., Kogel, W.J., 2012 Field attraction of the vine weevil *Otiorhynchus sulcatus* to kairomones. J. Econ. Entomol. 105,169-175.

Van Steenwyk, R.A.,Barnett, W.W., 1987. Disruption of navel orangeworm (Lepidoptera: Pyralidae) oviposition by almonds by-products. J. Econ. Entomol. 80, 1291-1296.

Vargas, R. I., Stark, J.D., Kido, M.H., Ketter, H.M., Whitehand, L.C., 2000. Methyl Eugenol and Cue-Lure Traps for Suppression of Male Oriental Fruit Flies and Melon Flies (Diptera: Tephritidae) in Hawaii: Effects of Lure Mixtures and Weathering. J. Economic. Entomol. 93, 81-87.

Varikou, K, Garantonakis, N., Birouraki, A., 2014. Comparative field studies of *Bactroceraoleae* in olive orchards in Crete. Crop Prot. 65, 238-243.

Welter, S.C.,Pickel, C., Millar, J.G., Cave, F., Van Steenwyk, R.A.Dunley, J., 2005. Pheromone mating disruption offers selective management options for key pests. California Agric., 59, 16-22.

Witzgall, P., 2001. Pheromones future techniques for insect control? Pheromones for insect control in orchards and vineyards. IOBC wprs Bull. 24, 114-122.

Witzgall, P., Kirsch, P., Cork A., 2010. Sex pheromones and their impact on pest management.*J.Chem. Ecol.* **36**, 80-100.

Wyatt, T.D., 2003. Pheromones and animal behaviour: communication by smell and taste. Cambridge University Press. Cambridge. 389 pp.

Zhang, A., Kuang, L.F., Maisin, N., Karumuru, B., Hall, D.R., Virdiana, I., Lambert, S., Purung, H.B., Wang, S., Hebbar, P., 2008. Activity evaluation of cocoa pod borer sex pheromone in cacao fields. Environ. Entomol. 37, 719-724.

Yang, C.Y., Jung, J.K., Han, K.S., Boo, K.S., Yiem, M.S., 2002. Sex pheromone composition and monitoring of the oriental fruit moth, *Grapholita molesta* (Lepidoptera: Tortricidae) in naju pear orchards. J. Asia-Pac. Entomol. 5, 201-207.

Yew, J.Y., Chung, H., 2015. Insect pheromones: An overview of function, form, and discovery. Prog. Lipid Res.59, 88-105.

Chapter 14

Desert Locust Management: Present and Future Prospects

☆ *Sory Cisse*

ABSTRACT

Desert locust has an ability to swarm widely causing considerable damage to cultivated lands and pastures over 31 million km² extending from West Africa to India. During upsurge, curative chemical control is the most effective way to stop the damage and to limit socio-economic effects. The financial cost of chemical control is too expensive. Also, it is known that chemical insecticides used for locust control have harmful impacts on humans and environment. The slow kill time and the high cost of current biological control agents justify the continued use of chemical insecticides. Consequently, the strategy recognized and adopted by the international community as the only sustainable strategy against Desert Locust threat is the preventive control management. This strategy is focused on early warning and early reaction to stop an outbreak to develop to upsurge by controlling first grouping populations as early as possible, before large hopper bands and swarms are created. It was proposed as early as the 1930s (Uvarov, 1938) with the understanding of phase polyphenism. It aims essentially to maintain constant field monitoring and regular control activities with timely and well-targeted chemical control in order to lower the level of locust populations, to maintain populations below a critical value of density threshold of gregarization. Small scales chemical treatments and the use of biological control agents in environmentally sensitive areas have the advantage of limiting environmental damage, reduce health concerns, and reduce costs.

Desert locust information management system is wide-ranging. It does not suffer from interference of administrative or political leaders since this ignores the boundaries between the affected countries. The information is collected and disseminated instantly and management is mainly ensured by affected-countries and FAO through donor assistance. In this chapter we describe the unique experience of the management system of Desert Locust, we also present an overview of current progress in understanding of the factors that may influence the population dynamics. The impact of climate change on this historic pest is also widely discussed.

Introduction

The desert locust, *Schistocerca gregaria* (Forksål, 1775) (Orthoptera: Acrididae), is the most dangerous of all migratory pests and can threat food security of more than 60 countries (Figure 14.1). The largest single swarm of Desert Locust was recorded in Kenya in 1954 and covered more than 200 km^2, contained 50 million insects per Km^2 and weighed 80000 tons (Rainey, 1954). Each locust consumes per day about 2 grams of fresh vegetation, equivalent to its own weight. One ton of locusts eat as much food in one day as about 2500 people. It's perhaps the best-known member of the Acrididae, and certainly is the most feared one for centuries (Lecoq 2001, Sword *et al.,* 2010). Farmers and nomads who have experienced a desert locust upsurge, will be frustrated and will remember forever the spectacular mass flights of swarms and the hopper bands movement.

Under appropriate environmental conditions, a few solitarious form of desert locust can multiply rapidly and change behavioral, morphological, and physiological characteristics to gregarious form in response to population density raise (Uvarov, 1966). This density-dependent phenotypic change, known as phase polyphenism, emerges gradually and may be reversible at any time depending on the density variations (Pener, 1991; Pener and Yerushalmi, 1998; Simpson *et al.,* 2011).

The gregarious form of Desert Locust is characterized by swarm formations for winged adults and marching hopper bands (Figure 14.2) threatening agriculture and pasture lands (Roffey and Magor, 2003). Gregarization is a form of adaptation of locust populations for migration facing heterogeneous environmental conditions (Sword, 2000). Migration occurs when locust numbers increased to a level where the small-scale environment cannot satisfy the food needs of locust population. Consequently, desert locust expands its range of distribution from desert areas to

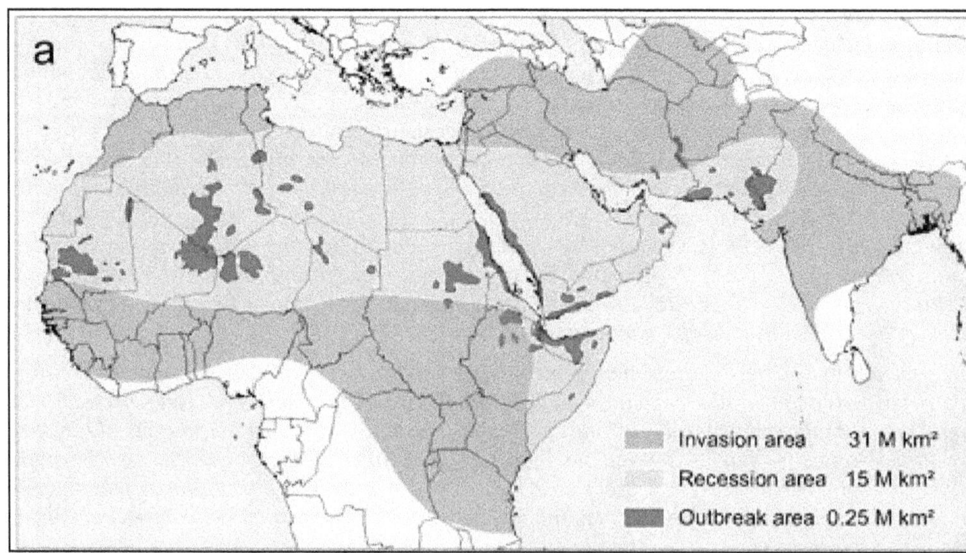

Figure 14.1: The Limits of Desert Locust Distribution Areas (Sword *et al.,* 2010).

Figure 14.2: Desert Locust *Gregarious* Hopper Band (a) and *Solitarious* (b) in Mauritania (Cisse Sory).

cultivated land and pasture. The net profits of gregarization are various: reducing predation by predators confusion (Gillet and Gonta 1978; Olson *et al.*, 2013.) eventually through the expression of aposematism (Sword, 1999), the resistance facing food shortage (Popov *et al.*, 1991; Maeno *et al.*, 2013), the possibility of mass migration over long distances, and an increase of the chance to discover suitable food supply and appropriate laying place (Roffey and Magor, 2003). Frequently, small scales group formations are recorded in the Desert Locust habitats. However, production of numerous hopper bands and swarms are less frequent, because the latter needs at least two successive generations of mass breeding (Duranton and Lecoq, 1990). After all, understanding the described process is crucial to better realize the desert locust issue and the complexity of its management. Accordingly, FAO and affected countries are continuously improving capacity of the staff members of national locust control office on various aspects such as Desert Locust biology and ecology, field monitoring, development of strategic plans, data collection and transmission, and control operations.

Desert Locust Phase Polyphenism

Desert locust phase change happen in the recession area that is equivalent to about 16 million square km located in the remote deserts extending over 25 countries (Lecoq, 2005). Locust densities, ecological conditions such as green vegetation, soil moisture, warm temperature and landscape structure influence considerably the phase change process (Ellis and Ashall, 1957; Collet *et al.*, 1998; Cisse *et al.*, 2013). Also, under optimal conditions, Desert Locust lives about three months and can multiply in number from one generation to another by about 16 times compared to a starting population (Duranton and Lecoq, 1990). The main factor triggering the transformation from solitary to gregarious is contact (Ellis, 1959; Despland 2000; Hassanali *et al.*, 2005; Maeno *et al.*, 2011), mainly by stimulating the external surface of the femur (Simpson *et al.*, 2001). The process involves several discrete events that may occur unnoticed (Despland, 2000; Rogers *et al.*, 2014). It has been noticed that the behavioral phase can be transferred across generations through epigenetic transmission (Simpson *et al.*, 1999; Rogers *et al.*, 2014). Also, physiological studies of *Schistocerca gregaria* showed that in the presence of high population densities, a chemical change occurs in the brain, which is responsible for changing behavior (Anstey *et al.*, 2009; Ott and Rogers 2010; Ott *et al.*, 2011).

The gregarious forms are different from the solitary forms in various traits such as coloration, wing length in adults, hind femur length, a bulge in the cheeks, a depression in the crest of the pronotum, a shortening of the prozona, and a constriction in the thorax (Duranton and Lecoq, 1990). In the intermediate forms, called transient forms (congregans or dissocians, depending on phase change direction), phase characteristics are independently and continuously induced or shifted to either direction in response to changes in the density of the population (Rao, 1942, Pener and Yerushalmi 1998, Pener and Simpson 2009). In most studies, prominence has been given to behavior rather than appearance in characterizing the phase status of individuals because morphological changes could be induced later in the phase change process (Rogers *et al.*, 2014). At low densities, the locusts are harmless, discreet, and cryptic and are dispersed in their traditional habitat in the distant deserts (Figure 14.3). If locust numbers are high and suitable ecological conditions are very low, individuals might concentrate on the limited green vegetation that is available, as long as these conditions remain favorable for feeding and breeding. This period may be long enough to cause gregarization and subsequently outbreak begins (Roffey and Popov, 1968; Duranton and Lecoq, 1990).

Influence of External and Internal Factors on Desert Locust Population Dynamics

It is well known that the climate is the key factor influencing the increase in locust densities, population dynamics and their spatial distribution in habitat (Waloff 1966; van Huis *et al.*, 2007). Rainfall is essential for locust because it creates moist soil conditions that females need to lay eggs, while allowing the development of green vegetation that could feed and house the locust hatchlings. Also, warm temperature is important for egg development, sexual maturation, adult flight and migration (Cressman and Hodson, 2009). Larval development speed, such as

Figure 14.3: Desert Locust Adult *Solitarious* Form in Mating in Mauritania (Cisse Sory).

embryonic development, is a function of temperature. High temperatures (from 25 to 35°C) accelerate the development while low temperatures delay the development (Symmons and Cressman 2001). The convergence winds can cause the accumulation of individuals in areas where there is a high probability of significant rainfall occurring (Waloff, 1966; Homberg, 2015).

Field information on locust situation, ecological conditions, historical migration events and weather are used to forecast the scale, timing, and location of locust breeding and making trajectory of short term expectations. Subsequently, the preventive control strategy exploits knowledge on locust biological processes (multiplication, concentration, gregarization) and ecology, population dynamics, advances in remote sensing and control tools.

Preventive Management Strategy

Preventive control strategy is intended to minimize occurrence and frequency of plagues through using field surveys to early detect favorable areas (vegetation, soil humidity, *solitarious* density) and to make control if necessary on *transient* groups, hoppers bands and swarms leading to stop the development of major outbreak to upsurge. Understanding of the habitat conditions associated with the enormous breeding potential that could dilute the natural predation pressures, allowed considering and implementing a preventive control strategy. It is focused on field monitoring by survey officers in remote areas to detect the favorable ecological conditions for Desert Locust breeding and concentration and to stop the densities increase by conducting chemical and biological treatments. The strategy is a national and transversal management system that is well coordinated by the Food and Agriculture Organization (FAO) of the United Nations (UN). FAO's commitment stems from the fact that Desert Locust plague management costs are unbearable for most of the affected countries and the invasion effects are catastrophic for farming communities which are known as among the poorest in the world (Brader *et al.*, 2006). It's an old collaboration of over 70 years, between FAO and the affected countries, that is constantly improving, adapting to changes in the forecasting ability, remote sensing and use of new technology tools.

Indeed, since the mid 1950s FAO is mandated by Member states to coordinate the management and efforts to minimize the risk of development of upsurge by encouraging the implementation of preventive control strategy, which prevent Desert locust outbreak. Then, FAO establish a Desert Locust control committee (DLCC) to ensure the overall supervision of locust control and three regional commissions for controlling the Desert Locust have been established in the Desert Locust Habitat (The commission for controlling the Desert Locust in the Southwest Asia (1964), the commission for controlling the Desert Locust in the Central region (1966) and the Commission for controlling the Desert Locust in the North Africa (1976) the last one has been renamed 2002 to become the Commission for controlling the Desert Locust in Western region after the adhesion of 5 Western Africa countries. In addition and on behalf of States member, FAO maintains a global overview of the locust situation, issues regular forecasts, and coordinates desert locust survey and control operations during emergencies. These activities are performed at FAO Headquarters by the Locust Group, including the Desert Locust Information Service (DLIS). All affected countries are involved in the collection and transmission of field information to the FAO headquarters.

The operability of these Commissions is supported by the concept of solidarity between affected countries (annual contributions of States for establishment of trust fund) and the good governance through high commitment of Member states. So, it was first established in 1964 to fight against the Desert Locust in the South-West Asia Commission (SWAC), which is the smallest of the three regional commissions against the Desert Locust. It has four Member States: Afghanistan, India, the Islamic Republic of Iran and Pakistan. Two other Commissions will then be created, covering the other two homogeneous geographical areas from the perspective of bio-ecology of the Desert Locust. This is the Commission of the Central Region (CRC), which includes Bahrain, Djibouti, Egypt, Eritrea, Ethiopia, Iraq, Jordan, Kuwait, Lebanon, Oman, Qatar, Saudi Arabia, Sudan, Syria, UAE and Yemen; and the Commission for the Western Region (CLCPRO) which States member currently are: Algeria, Burkina Faso, Chad, Libya, Mali, Mauritania, Morocco, Niger, Senegal and Tunisia. Political instability in some states, contribution arrears and bad security condition has often disrupted the activities in these three regions. Nevertheless, all affected countries have, at least, a plant protection service. The front line countries (that have gregarization habitat) periodically deploy monitoring teams to conduct field survey ensure early warning and rapid control if necessary. For over 10 years, almost all developed upsurges were successfully controlled through the efforts of the affected countries, in coordination with the regional commissions and FAO's Headquarters (FAO, 2012).

Desert Locust Information Service in a Constant Quest to Master the Issue

Despite the political will of the leaders of the concerned countries, transboundary pest management suffers more frequently of country's intrinsic constraints particularly related to economic and political governance mode. The success of desert locust management is based on organization, training and promotion of international cooperation among affected countries and regions.

The Desert Locust Information Service (DLIS) at FAO headquarters and affected countries are actively cooperating together for early warning and early control of the desert locust (FAO, 2015). DLIS strengthens the capacity of national locust survey and control officers to record, transmit, manage and analyze field data. DLIS has developed two primary tools for this purpose: eLocust3 used by field officers to record survey and control observations and send them in real time by satellite to their national locust centre, and RAMSES, a custom GIS for the management and analysis of field data and satellite imagery. DLIS acts as a centralized clearing house of all field data and satellite imagery. Since the mid 1980s, several remote sensing products have been used to estimate rainfall and green vegetation in the desert locust habitat and breeding areas. The latest products in use are rainfall estimates produced by Columbia University's International Research Institute (IRI) and 250-m resolution MODIS-derived 10-day dynamic greenness maps. A custom SWARMS GIS is used to manage and analyze all of the field and remote sensing data in order to assess the current situation in all countries and forecast its developments in order to provide precise early warning on time. SWARMS's extensive database contains gridded data from 1930-1985 and thereafter, individual locations of surveys and control. The forecasts are sub-national level and focus on the scale, timing and location of breeding and migration. Finally, DLIS produces regular monthly bulletins containing country situation assessments and forecasts on the likely evolution of the locust situation. The bulletins are supplemented by updates, warnings and alerts. All information products produced by DLIS are distributed on a wide basis to countries, donors, scientists and the general public via email, the Internet (www.fao.org/ag/locusts), Facebook (www.facebook.com/faolocust), Twitter (www.twitter.com/faolocust) and other social media.

Effects of Climate Change on Desert Locust and Long-term Forecasting

Nowadays, the effects of climate change on the locust dynamics arouse great interest in the scientific and economic environments (Despland *et al.*, 2004; Trémolières, 2010; Latchininsky, 2013). This interest is related to the fact that the assumptions, in most cases, proclaim pessimistic development where the Desert Locust upsurges frequencies become important with rainfall instability (Grist and Nicholson, 2001). This risk could be particularly significant for locust affected countries in terms of their mostly vulnerable characteristic. In these countries, desertification is not an empty word but a daily reality. They are mostly facing a poor spatiotemporal distribution of rainfall since a long time. Forecasts of how climate change affects rainfall in the desert are variable and lack of consensus. So far, all research projects have focused on the cross analysis of historical data of locust outbreak frequencies, annual rainfall recorded and distribution, vegetation patterns (Despland *et al.*, 2004; Tratalos *et al.*, 2010; Piou *et al.*, 2013; Cressman, 2013). The contribution of historical data is critical to make simulations of climate change effects. Geo-referencing data and efficient archiving by DLIS and affected countries are very recent, compared to the needs of the assessment scale of climate change. On one hand, some studies suggest that we should see an increase in rainfall in the desert areas, while others argue the opposite. According to Claussen *et al.* (2003), there are some indications that precipitation could increase, due to greater than before of carbon dioxide levels, leading to vegetation and moisture augmentation. These scientists are seeing signs that the desert and surrounding areas become green because of increased precipitation. If this is the case, an increase in frequency reproduction of locusts would be very likely (Roffey and Magor, 2003; Waloff 1966). An increase in the annual number of desert locust generations would raise the risk of gregarization and expose agriculture to threat (Duranton and Lecoq, 1990). On the contrary, it is well known that desert locust numbers decrease during drought periods.

Despite the lack of consensus concerning effects of climate change on rainfall in the deserts of northern Africa, the Near East and Southwest Asia, it can be expected that there will be an effect on desert locust habitats, breeding, migration and population dynamics (Cressman, 2013). The greatest impact of warmer temperatures associated with climate change will be potentially faster development of desert locust eggs and hoppers during the winter and spring breeding periods that could lead to an extra generation of breeding during the winter under warmer temperature conditions. However, sufficient rainfall will also be required so if there predicted decline in rainfall amounts during these periods, they will offset the effect of warmer temperatures on desert locust development. Consequently, it is the combined effects of rainfall and temperature changes induced by climate change that will impact desert locust. For example, a greater frequency of unusual rainfall that is short-lived, high intensity rainfall, combined with warmer temperatures could result in the development of more desert locust outbreaks.

In addition, the long-term forecasting models have largely been developed from an analysis of the dynamic differences between recession and outbreak periods. First, it was reported that regional changes in precipitation could be the main trigger of outbreaks (Pedgley 1981). Other theoretical studies by Blackith and Albrecht (1979), Cheke and Holt (1993) and Holt and Cheke (1996) indicated that changes in desert Locust densities may be related to endogenous dynamics. However, Farrow and Longstaff (1986) argued that migration tends to prevent overcrowding and endogenous models are therefore not effective in explaining the dynamics of the Desert Locust populations. More recently, Vallebona *et al.* (2008) demonstrated a difference in climate parameters between the years of recession and outbreak. The upsurge in the Sahel and around the Red Sea can be consecutive to climate disturbances reaching the dynamics of desert locust reproduction. The longer optimal breeding conditions extend over the risk of outbreak. This is particularly true in the Arabian Peninsula where, quite often, heavy rains associated with cyclones lead to upsurges. Moreover, Magor *et al.* (2008) suggested that a general reduction in the frequency and duration of locust activity since 1965 are due to a change in rainfall patterns caused by a north-south oscillation that became narrower in the intertropical convergence zone. Other prediction models were developed from data of occurrences related to the use of remote sensing (Despland *et al.*, 2004; Piou *et al.*, 2013). These observations have examined the gregarization probabilities based on the landscape structure, wind conditions, the rainfall variation from different sizes of satellite images. Also, new products are increasingly explored and integrated in this system to improve the monitoring of population dynamics. These include Earth observation products for the detection, in desert areas, green vegetation as a potential indicator of habitat favorable to the development of Desert Locust. Convincing results were provided by SPOT-VEGETATION and MODIS in monitoring the production of dynamic greenness map (Waldner *et al.*, 2015). The project SMELLS (Soil Moisture for Desert Locust Early Survey) develop algorithms to estimate soil moisture to help identify suitable areas for egg-laying up to about 15 centimeters below the surface. This is currently being explored in some countries (Algeria, Mauritania, Mali and Morocco) with the help of the European Space Agency, IsardSAT and CIRAD (see http://smells.isardsat.com).

Otherwise, natural regulation of locust populations cannot be excluded with climate change in the case of a possible synchronization of increased density and increased pressure from natural predators (such as ants, beetles, lizards, birds) and other disease vectors (Ghaout 1990; Greathead *et al.*, 1994; Symmons and Cressman 2001). However, human impacts could also influence this hypothesis. Indeed, some birds like *Cursorius cursor* (Figure 14.4) and *Chlamydotis undulata*, are great predators of the locust larvae. They can destroy an early outbreak by predation effect. Now, these two bird species are on the list of endangered animals (IUCN, 2015; Bourass

Figure 14.4: *Cursorius cursor*, Enormous Predator of Desert Locust's Hoppers in Mauritania (Cisse Sory).

et al., 2012). The growing human presence following the armed conflicts in the desert area, the movements of vehicles and hunting by falconers contribute actively to the habitat disturbance. Nomadic people, who are indigenous in these semi-desert regions, had a tacit code of conduct that prohibited them to take more to the environment than they need for their survival. Currently this natural process of events and the ancient harmony between man and his environment are abused in the heart of the desert, corresponding to the traditional habitat of the Desert Locust.

Moreover, any prediction of the effect of climate change on Desert Locust must take into account its potential adaptability and the risk of increase/decrease in the density of natural enemies' pressure. The relationship prey/predator is important in natural control of locust populations (Greathead *et al.*, 1994). However, we believe that the locust risk remain relevant because long term forecasting models available are not robust enough and the climate remains favorable of Desert Locust outbreak. The potential effect of climate change on the evolution of natural enemies is a prospective track that could be included in the long-term forecasting models.

Conclusion

The recent research of alternatives to chemical treatment is focused on hyper-virulent and specific fungal pathogen, *Metharizium acridum* (Fang *et al.*, 2014). Some interesting results were obtained and countries now have an alternative to chemical control that can be used in sensitive areas such as near water bodies, crops and inhabitations and in national parks. There is a general trend of increasing insecurity and instability in several countries that prevents field survey operations for early warning and early reaction. This could result in an increasing number of outbreaks that may or may not be detected and hence are irrepressible. International cooperation and the role of FAO can help to alleviate some of these, but in general, this remains as a significant concern. The introduction of new tools and improvements into the system must be rigorously maintained and continuously adopted for locust-affected countries. The use of drones in field survey, currently investigating by DLIS, could facilitate access to complex areas and subsequently being helpful for early warning system. Risk management plans and implementation of initiatives for sustainable financing mechanisms must be prepared, updated and accompanied by strong political commitments of the affected countries. The maintenance of a collaborative, reliable and sustainable strategy for managing desert locust in a cost-effective and safe manner remains the best hope in safeguarding our food supplies from this ancient devastating migratory pest.

References

Anstey ML, Rogers SM, Ott SR, Burrows M and Simpson SJ., 2009. Serotonin mediates behavioral gregarization underlying swarm formation in desert locusts. *Science* 323: 627–63

Blackith RE and Albrecht FO., 1979. Locust plagues, the interplay of endogenous and exogenous control. *Acrida* 8:83–94.

Bourass K, Zaime A, Qninba A, Benhoussa A, Rguibi Idrissi H and Yves Hingrat (2012) Evolution saisonnière du régime alimentaire de l'Outarde houbara nord-africaine, Chlamydotis undulata undulata ; *Bulletin de l'Institut Scientifique*, Rabat, section Sciences de la Vie, 34 (1), 29-43.

Brader L, Djibo H, Faye FG, Ghaout S, Lazar M, Luzietoso PN and Babah MAO., 2006. Towards a more effective response to desert locust and their impacts on food security,

Ceccato P, Cressman K, Giannini A and Trazaska S., 2007. The desert locust upsurge in West Africa (2003-2005): information on desert locust early warning system and the prospects for seasonal climate forecasting. *International Journal of Pest Management* 53, 7-13.

Cheke RA and Holt J,. 1993. Complex dynamics of Desert Locust plagues. *Ecological Entomology*, 18, 109-115

Cheke RA and Tratalos JA., 2007. Migration, patchiness and population processes illustrated by two migrant pests. *Bioscience* 57:145–154

Claussen M, Brovkin V, Ganopolski A, Kubatzki C and Petoukhov V., 2003. Climate change in northern Africa: the past is not the future. *Clim Change*, 57:99–118

Cisse S, Ghaout S, Mazih A, Babah M, Benahi A and Piou C., 2013. Effet of vegetation on density thresholds of adult Desert Locust gregarization from survey data in Mauritania. *Entomologia Experimentalis et Applicata*, 149, 159-165.

Cisse S, Ghaout S, Mazih A, Babah MAO and Piou C., 2015. Estimation of density threshold of gregarization of desert locust hoppers from field sampling in Mauritania. *Entomologia Ex-perimentalis et Applicata*, 156:136–148

CollettM, DesplandE, Simpson SJ and Krakauer DC., 1998. Spatial scales of desert locust gregarization. Proceedings of the National Academy of Sciences of the USA 95: 13052–13055

Cressman K., 2001. Desert Locust Guidelines 2: Survey. FAO, Rome, Italy

Cressman K and Hodson D., 2009. Surveillance, information sharing and early warning systems for transboundary plant pests diseases: the FAO experience. *Arab Journal of Plant Protection*, 27, 226-232.

Cressman, K. ; 2013. Climate change and locusts in the WANA Region. In M.V.K Sivakumar *et al.* (eds.), Climate Change and Food Security in West Asia and North Africa. (pp. 131-143). Netherlands: Springer. DOI 10.1007/978-94-007-6751-5_7

.Despland E, Rosenberg J and Simpson SJ., 2004. Landscape structure and locust swarming: A satellite's eye view. *Ecography*, 27, 381–391

Despland E. 2000. Role of olfactory and visual cues in the attraction/repulsion responses to conspecifics by gregarious and so-litarious desert locusts. *Journal of Insect Behavior* 14: 35–46.

Duranton JF and Lecoq M.,1990. Le Criquet pèlerin au Sahel. Montpellier, *CIRAD-PRIFAS*.

Ellis PE,.1959. Learning and social aggregation in locust hoppers. Animal Behaviour 7: 91–106

Ellis PE and Ashall C.,1957. Field studies on diurnal behaviour, movement and aggregation in the desert locust (*Schistocerca gregaria* Forskal). Anti-Locust Bulletin 25: 1–94.

FAO.,2012., Rapport de la quatrième session du comité de lute contre le Criquet pèlerin. FAO, Rome

FAO.;2015. FAO Desert Locust Information Service (DLIS) helps countries to control Desert Locust. Rome (Italie) I4353, 2p.

Fang W, Lu HL, King GF and St. and Leger RJ.,2014. Construction of a Hypervirulent and Specific Mycoinsecticide for Locust Control. *Scientific Reports*, 4, 7345.

Farrow RA and Longstaff BC.,1986., Comparison of the annual rates of increase of locusts in relation to the incidence of plagues. *Oikos*, 46:207–222.

Ghaout S.,1990. Contribution à l'étude des ressources trophiques de *Schistocerca gregaria* (Forsk.) (Orthoptera, Acrididae) solitaire en Mauritanie occidentale et télédétection de ses biotopes par satellite. *Thèse Doct. Universit6 Paris-Sud*, 203p.

Gillet SD and Gonta E.,1978. Locusts as prey: factors affecting their vulnerability to predation. Animal Behaviour, 26:282-289.

Greathead DJ, Kooyman C, Launois-Luong MH and Popov GB.,1994. Les Ennemis Naturels des Criquets du Sahel. *CIRADGERDAT-PRIFAS, Montpellier, France.*

Grist JP and Nicholson SE., 2001. A study of the dynamic factors influencing the rainfall variability in the West African Sahel. *J Clim* 14:1337–1359

Hassanali A, Bashir MO, Njagi PGN and Ould Ely S., 2005. Desert locust gregarization: a conceptual kinetic model. *Journal of Orthoptera Research* 14: 223–226.

Holt J and Cheke RA., 1996. Réussites et échecs d'une règle simple de prévision des infestations du criquet pèlerin au Sahel. *Sécheresse*, 7:151–154.

Homberg U., 2015 Sky Compass Orientation in Desert Locusts—Evidence from Field and Laboratory Studies. *Frontiers in Behavioral Neuroscience*, 9(346)

van Huis A, Cressman K and Magor JI., 2007. Preventing desert locust plagues: optimizing management interventions. Entomologia Experimentalis et Applicata 122, 191-214.

IUCN, 2015. The IUCN Red List of threatened species. http://www.iucnredlist.org/

Latchininsky V., 2013. Locusts and remote sensing: a review. *Journal of Appl. Remote Sens.* 7(1).

Lecoq M., 2001. Recent progress in desert and migratory locust management inAfrica. Are preventive actions possible? *Journal of Orthoptera Research* 10: 277–291

Lecoq M., 2004. Vers une solution durable au problème du criquet pèlerin ? *Science et changements planétaires / Sécheresse*, 15, 217-224.

Lecoq M., 2005. Desert locust management: from ecology to anthropology. *Journal of Orthoptera Research*, 14, 179-186.

Maeno K, Tanaka S and Harano K., 2011. Tactile stimuli perceived by the antennae cause the isolated females to produce gregarious offspring in the desert locust, *Schistocerca gregaria. Journal of Insect Physiology* 57: 74–82.

Maeno K, Piou C, Ould Babah MA and Nakamoura S., 2013. Eggs and hatchlings variations in desert locusts: phase related characteristics and starvation tolerance. *Frontiers in Physiology* 4:345.

Magor JI, Lecoq M and Hunter DM., 2008. Preventive control and Desert Locust plagues. *Crop Protection*, 27, 1527-1533.

Olson SR, Hintze A, Dyer FC, Knoester DB and Adami C., 2013. Predator confusion is sufficient to evolve swarming behavior. Journal of Royal Society Interface, 10 20130305.

Ott SR and Rogers SM., 2010. Gregarious desert locusts have substantially larger brains with altered proportions compared with the solitarious phase. *Proceedings of the Royal Society of London B* 277: 3087–3096.

Ott SR, Verlinden H, Rogers SM, Brighton CH, Quah PS *et al.*, 2011. Critical role for protein kinase A in the acqui-sition of gregarious behavior in the desert locust. *Proceed-ings of the National Academy of Sciences of the USA* 109: E381–E387

Pedgley D.,1981. Desert locust forecasting manual. London, *COPR*.

Pener MP.,1991. Locust phase polymorphism and its endocrine relations. *Advances in Insect Physiology*, 23, 1-79.

Pener MP and Yerushalmi Y., 1998. The physiology of locust phase polymorphism: an update. *Journal of Insect Physiology*, 44, 365-377.

Pener MP and Simpson SJ. 2009. Locust phase polyphenism: an update. Advances in Insect Physiology 36: 1–272.

Piou C, Lebourgeois V, Benahi AS, Bonnal V, Jaavar MH, Lecoq M and Vassal JM., 2013 Coupling historical prospection data and a remotely-sensed vegetation index for the preventative control of Desert Locusts. *Basic and Applied Ecology*, 50728, 12.

Popov GB, Duranton JF and Gigault J., 1991. Etude Ecologique des Biotopes du Criquet Pèlerin *Schistocerca gregaria* (Foskal,1775) en Afrique Nord Occidentale: Mise en Evidence des Unités Territoriales Ecologiquement Homogènes. CIRAD-PRI-FAS, Montpellier, France.

Rainey, R.C., 1954. Recent developments in the use of insecticides from aircraft against locusts. Rep. 6th Commonw. Entomol. Conf. 6: 48-51.

Rao YR.,1942. Some results of studies on the desert locust (*Schistocerca gregaria*, Forsk.) in India. *Bulletin of Entomological Research* 33: 241–265

Roffey J and Magor JI.,2003. Desert Locust Population Parameters. AGP/DL/TS/30-29, FAO, Rome, Italy.

Roffey J and Popov G., 1968. Environmental and behavioural processes in a desert locust outbreak. *Nature* 219: 446–450.

Rogers SM, CullenDA, AnsteyML, BurrowsM, Despland E *et al.*, 2014. Rapid behavioural gregarization in the desert locust, *Schistocerca gregaria*, entails synchronous changes in both activity and attraction to conspecifics. *Journal of Insect Physiology* 65: 9–26.

Roy J., 2001. Un siècle de lutte antiacridienne en Afrique : Contribution de la France, Edition l'Harmattan, 278p.

Simpson JS, Sword G and Lo N.2011. Polyphenism in insects. *Current Biology*, 21, R738–R749.

Sword GA, Lecoq M, and Simpson SJ., 2010. Phase polyphenism and preventative locust management. *Journal of Insect Physiology*, 56, 949-957.

Sword GA., 1999. Density-dependent warning coloration. Nature 397: 217

Symmons PM and Cressman K., 2001. Directives sur le Criquet pèlerin 1 : Biologie et comportement. *Rome, FAO.*

Trémolières M.2010. Sécurité et variables environnementales : débat et liens. Incidences sécuritaires du changement climatique au Sahel. *Club du Sahel et de l'Afrique de l'Ouest* (*CSAO/OCDE*).

Uvarov BP.1938. Locust as a world problem. Première conférence internationale pour la protection contre les calamités naturelles. Commission française d'études des calamites with the support of Union Internationale de Secours. Paris, France.

Uvarov BP.,1966. Grasshoppers and Locusts. A Handbook of General Acridology. I: Anatomy, Physiology, Development, Phase-Polymorphism, Introduction to Taxonomy. Cambridge University Press, Cambridge, UK.

Vallebona C, Genesio L, Crisci A, Pasqui M, Di Vecchia A and Maracchi G.,2008. Large-scale climatic patterns forcing desert locust upsurges in West Africa. Clim Res, 37, 35–41.

Waldner F, Babah Ebbe MA and Cressman K.2015. Operational Monitoring of the Desert Locust Habitat with Earth Observation: An Assessment. *ISPRS Int. J. Geo-Inf.*, 4, 2379-2400; doi:10.3390/ijgi4042379.

Waloff Z.,1966. The upsurges and recessions of the Desert Locust plague: an historical survey. *Anti-Locust Memoir, 8, 1-111.*

Index

D

E

F

G

H

www.ingramcontent.com/pod-product-compliance
Lightning Source LLC
Chambersburg PA
CBHW050523190326
41458CB00005B/1644
9789387057593